Uncertainty Modeling and Analysis in Engineering and the Sciences

Bilal M. Ayyub
George J. Klir

Chapman & Hall/CRC
Taylor & Francis Group
Boca Raton London New York

Chapman & Hall/CRC is an imprint of the
Taylor & Francis Group, an informa business

Published in 2006 by
Chapman & Hall/CRC
Taylor & Francis Group
6000 Broken Sound Parkway NW, Suite 300
Boca Raton, FL 33487-2742

Printed in the United States of America on acid-free paper
10 9 8 7 6 5 4 3 2 1

International Standard Book Number-10: 1-58488-644-7 (Hardcover)
International Standard Book Number-13: 978-0978-1-58488-644-0 (Hardcover)

informa
Taylor & Francis Group
is the Academic Division of Informa plc.

Visit the Taylor & Francis Web site at
http://www.taylorandfrancis.com

and the CRC Press Web site at
http://www.crcpress.com

To my wife, Deena, and our children,
Omar, Rami, Samar, and Ziad.

Bilal M. Ayyub

To my wife, Milena, and our daughter, Jane.

George J. Klir

Preface

The treatment of uncertainty in analysis, design, and decision making is going through a paradigm shift from a probabilistic framework to a generalized framework that includes both probabilistic and nonprobabilistic methods. Presently, analysts, including engineers and scientists, recognize the presence of uncertainty and treat it formally. For example, engineers analyze and model uncertainty in many of their specialty fields, such as the development of building codes, analysis of natural hazards (e.g., floods, wind, and earthquakes), decision making in infrastructure maintenance expenditure, homeland security and protection of assets, and environmental risks. Similarly, scientists analyze and model uncertainty in many of their specialty fields, such as the diagnostics of diseases, health effects of food additives and toxins, pharmaceutical research for developing new drugs, understanding of physical phenomena, prediction and forecasting in economy and weather, and sociopolitical changes, trends, and evolutions. The interest in uncertainty will continue to increase as we continue to design complex systems and deal with new technologies, systems, and materials, and are increasingly required to make critical decisions with potentially high adverse consequences. Also, political, societal, and financial requirements are increasing, thereby adding new dimensions of complexity in meeting the societal demands. The expectations of society are becoming larger than ever, and its tolerance to errors is diminishing. The aggregate of these factors produces an environment that requires the formal consideration of uncertainty in decision making at all levels in a systems framework.

Problems that are commonly encountered by engineers and scientists require decision making under conditions of uncertainty, lack of knowledge, and ignorance. The lack of knowledge and ignorance can be related to the definition of a problem, the alternative solution methodologies and their results, and the nature of the solution outcomes. Based on present trends, analysts will need to solve complex problems with decisions made under conditions of limited resources, thus necessitating increased reliance on the proper treatment of uncertainty and the use of expert opinions. This book is therefore intended to better prepare future analysts, as well as assist practitioners in understanding the fundamentals of knowledge and ignorance, how to model and analyze uncertainty, and how to select appropriate analytical tools for a particular problem.

Traditionally, intelligence is defined as the ability to understand and adapt to the environment by using a combination of inherited abilities and learning experiences. This ability certainly includes the analysis of uncertainty and making decisions under conditions of uncertainty. This is true of many organisms — from ants to aardvarks to humans. Any organism that survives the remorseless rigors of evolution is sufficiently intelligent for its role in life. Likewise, machines need to be sufficiently intelligent to make decisions suitable for their functions and adapt to and deal with the presence of uncertainty. Any collectives of human decision makers and their decision-aiding machines must make, in the aggregate, good decisions. But these decisions are almost always made under conditions of uncertainty.

The term *uncertainty* can be viewed as a component of ignorance. A taxonomic breakdown of ignorance can reveal many components having a strong association with human cognition of information and knowledge construction philosophies and practices. Uncertainty and information as a pair, and ignorance and knowledge as another pair, are studied in this book since they are tightly interconnected, as the former component of each pair describes a deficiency in the respective latter component, while the latter component of a pair can be viewed as the respective capacity available to reduce the respective former component. The identification and treatment of this duality of the respective components of a pair offer opportunities to enhance understanding of underlying problems or issues and our ability to make decisions. This book covers primary components of ignorance and their impact on our practice and our ability to make decisions. This book gives an overview of the current state of uncertainty modeling and analysis, and covers emerging theories with emphasis on practical applications in engineering and the sciences.

The complexity of a particular decision situation could increase substantially by the inclusion of uncertainties, thus requiring, in many cases, the reliance on experts to shed light on the situation. The complexity of our society and its knowledge base requires its members to specialize and become experts to attain recognition and reap rewards to the society and themselves. We commonly deal with or listen to experts on a regular basis, such as weather forecasts by weather experts, stock and financial reports by seasoned analysts, suggested medication or procedures by medical professionals, policies by politicians, and analyses by world affairs experts. We know from our own experiences that experts are valuable sources of information and knowledge, and can also be wrong in their views rendered to us. Expert opinions, therefore, can be considered to include or constitute nonfactual information. The fallacy of these opinions might disappoint us, but does not surprise us since issues that require experts tend to be difficult or complex, with a lot of uncertainty, and sometimes with divergent views. The nature of some of these complex issues could only yield views that have subjective truth levels; therefore, they allow for contradictory views that might all be somewhat credible. In political and economic world affairs and international conflicts, such issues are of common occurrence. For example, we have recently witnessed the debates that surrounded the membership of the People's Republic of China to the World Trade Organization in 1999, or experts airing their views on the insoluble Arab–Israeli affairs for the last century, or analysts' views on the war in Iraq in 2003, or future oil prices in 2005. These issues have a common feature of the presence of complexity and uncertainties requiring the use of expert opinions. Such issues and situations are also encountered in engineering, the sciences, medical fields, social research, stock and financial markets, and the legal practice.

Experts, with all their importance and value, can be viewed as double-edged swords. Not only do they bring in a deep knowledge base and thoughts, but also they could infuse biases and pet theories. The selection of experts, elicitation, and aggregation of their opinions should be performed and handled carefully by recognizing uncertainties associated with this type of information, and sometimes with skepticism. A primary reason for using expert opinions is to deal with uncertainty in selected technical issues related to a system of interest. Issues with significant uncertainty, issues that are controversial or contentious, issues that are complex, issues with limited objective information, or issues that can have a significant effect on risk are most suited for expert opinion elicitation. The value of the expert opinion elicitation comes from its initial intended uses as a heuristic tool, not a scientific tool, for exploring vague and unknowable issues that are otherwise inaccessible. It is not a substitute to scientific, rigorous research.

Current techniques for visualizing information commonly do not include degrees of certainty (or the degrees and types of ignorance) associated with individual or aggregated information. For example, for a commander in a battlefield to command, she or he needs to choose. To choose is to decide — almost always on the basis of imperfect information — and momentous decisions require knowledge of threats with a degree of certainty that might not be a requisite for decisions less momentous than waging war. Battlespace visualization techniques should allow both information and uncertainty to be portrayed effectively and grouped intuitively. Intelligent agents are promising technologies that may facilitate visualization of data and information uncertainty. Civilian applications can also be constructed to meet societal needs, such as Internet information metatagging for uncertainty and uncertainty visualization of search results.

In preparing this book, the authors strove to achieve the following objectives:

1. To develop a philosophical foundation for the meaning, nature, and hierarchy of knowledge and ignorance
2. To provide background information and historical developments related to knowledge, ignorance, and the elicitation of expert opinions
3. To provide a systems framework for the analysis and modeling of uncertainty
4. To summarize and illustrate methods for encoding data and expressing information
5. To provide and illustrate methods for uncertainty and information synthesis
6. To develop and illustrate methods for uncertainty measures and related criteria for knowledge construction
7. To examine and illustrate methods for uncertainty propagation in input–output systems
8. To guide the readers of the book on how to effectively elicit opinions from experts in such a way that would increase the truthfulness of the outcomes
9. To provide methods for visualizing uncertainty
10. To provide practical applications in these areas based on recent studies

The book introduces fundamental concepts of classical sets, fuzzy sets, rough sets, probability, Bayesian methods, interval analysis, fuzzy arithmetic, interval probabilities, evidence theory, open-world models, sequences, and possibility theory. These methods are presented in a style tailored to meet the needs of practitioners in many specialty fields, such as engineering, physical and social sciences, economics, law, and medicine. The book emphasizes the practical use of these methods, and establishes their limitations, advantages, and disadvantages. Although the applications at the end of the book were developed with emphasis on engineering, technological, and economics problems, the methods can also be used to solve problems in other fields, such as social sciences, law, insurance, business, and management.

STRUCTURE, FORMAT, AND MAIN FEATURES

This book was written with a dual use in mind, as both a self-learning guidebook and a required textbook for a course. In either case, the text has been designed to achieve important educational objectives of introducing theoretical bases, guidance and applications of the analysis, and modeling of uncertainty.

The eight chapters of the book lead the readers from the definition of needs, to the foundations of the concepts covered in the book, to theory and guidance and applications. The first chapter provides an introduction that discusses systems, knowledge (its sources and acquisition), and ignorance (its categories as bases for modeling and analyzing uncertainty). The practical use of concepts and tools presented in the book requires a framework and a frame of thinking that deals holistically with problems and issues as systems. Background information on system modeling is provided also in Chapter 1. Appendix A is called out in Chapter 1 to offer a historical perspective on knowledge.

Chapter 2 presents the fundamentals of encoding data and expressing information using classical set theory, fuzzy sets, and rough sets. Basic operations for these sets are defined and demonstrated. Fuzzy relations and fuzzy arithmetic can be used to express and combine collected information. The fundamentals of probability theory, possibility theory, interval probabilities, and monotone measures are summarized as they relate to uncertainty analysis. Examples are used in this chapter to demonstrate the various methods and concepts.

Chapter 3 covers uncertainty and information synthesis based on a mission-based system definition. The chapter starts by introducing measure theory and monotone measures and includes possibility theory and Dempster–Shafer theory of evidence, and then compares and contrasts them with probability theory with some of its variations and special applications, including linguistic probabilities, Bayesian probabilities, imprecise probabilities (including interval probabilities), interval cumulative distribution functions, and probability bounds. This chapter also discusses various multivariate dependence types and their models and describes fuzzy measures and fuzzy integrals.

Chapter 4 provides definitions and classification of uncertainty measures, including nonspecificity measures, such as the Hartley, evidence, possibility, and fuzzy sets' nonspecificity measures; entropy-like measures, such as Shannon entropy, discrepancy measure, and entropy measures for evidence theory of dissonance and confusion; and fuzziness measure. The chapter also includes applications relating to combining expert opinions.

Chapter 5 introduces uncertainty-based criteria for the construction of knowledge that include a minimum uncertainty criterion, maximum uncertainty criterion, and uncertainty invariance criterion, with demonstrative examples of aggregating expert opinions. The chapter also introduces methods for open-world analysis, including statistical estimators for sequences and patterns, such as the Laplace model, add-c model, and Witten–Bell model, and an analytical estimator based on the theory of evidence, i.e., the transferable belief model for evidential reasoning and belief revision. Applications to diagnostics are discussed.

Chapter 6 focuses on a class of models in engineering and the sciences of relating input variables to output variables for a system, building on knowing the underlying physical laws, such as material mechanics, and utilizing constraints, such as boundary conditions. The numerical computations might be based on finite element methods that are used to model the entire system. The model complexity can be increased by considering nonlinearity in behavior and other special considerations, such as bifurcation, instability, logic rules, and across-discipline or across-physics interactions. This chapter also presents methods for propagating uncertainty in input–output systems. The methods presented in this chapter are illustrated using simple linear systems. These methods form the basis for potential extensions to complex cases.

Chapter 7 provides guidance on using expert opinion elicitation processes. These processes can be viewed as variations of the Delphi technique, with scenario analysis based on uncertainty models, ignorance, knowledge, information and uncertainty modeling related to experts and opinions, and nuclear industry experiences and recommendations. This chapter also demonstrates the applications of expert opinion elicitation by summarizing results from practical examples.

Chapter 8 provides techniques for visualizing uncertainty in information. Visualization techniques are needed to allow both information and uncertainty to be portrayed effectively and grouped intuitively. This need is demonstrated, and icons are introduced for uncertainty and ignorance that are called uncerticons and ignoricons, respectively.

In each chapter of the book, computational examples are given in the individual sections of the chapter, with more detailed engineering applications provided in some of the key chapters. Also, each chapter includes a set of exercise problems that cover topics discussed in the chapter. The problems were carefully designed to meet the needs of instructors in assigning homework and the readers in practicing the fundamental concepts.

For the purposes of teaching, the book can be covered in one semester. The chapter sequence can be followed as a recommended sequence. However, if needed, instructors can choose a subset of the chapters for courses that do not permit a complete coverage of all chapters, or a coverage that cannot follow the order presented. In addition, selected chapters can be used to supplement courses that do not deal directly with uncertainty modeling and analysis, such as risk analysis, reliability assessment, expert opinion elicitation, economic analysis, systems analysis, litigation analysis, and social research courses. Chapters 1 and 2 can be covered concurrently, or preferably, Chapter 2 covered after Chapter 1. Appendix A is called out in Chapter 1 to offer historical perspective on knowledge. Chapter 3 builds on some of the materials covered in Chapter 2. Chapter 4 builds on some of the materials covered in Chapter 3. Chapter 5 builds on Chapters 3 and 4 and should be covered after completing Chapter 4. Chapter 6 requires knowledge of materials

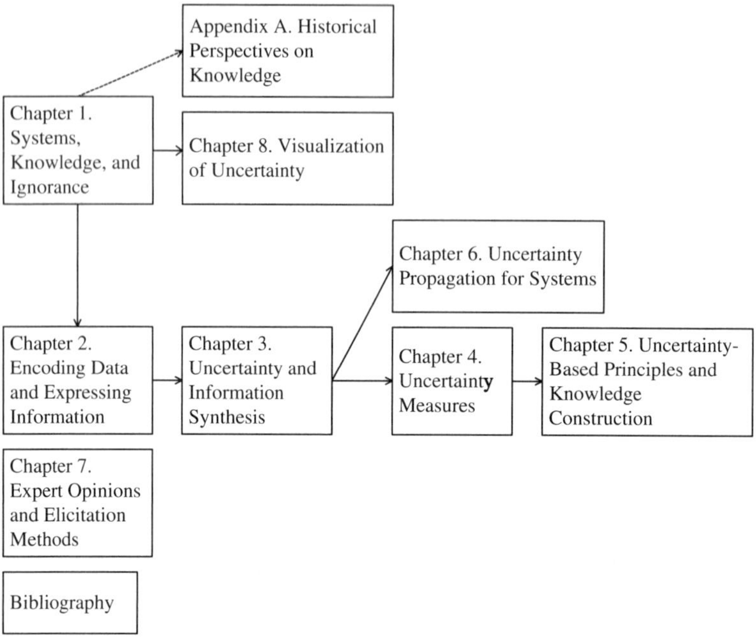

FIGURE 1 Sequence of chapters.

covered in Chapters 2 and 3. Chapter 7 provides guidance on using expert opinion elicitation and can be introduced independently. Chapter 8 also can be introduced after Chapter 1. The book also contains an extensive bibliography at its end. The accompanying schematic diagram (Figure 1) illustrates possible sequences of these chapters in terms of their interdependencies.

The authors invite users of the book to send any comments on its structure or content to the e-mail address ba@umd.edu. These comments will be used in developing future editions of the book. Also, users of the book are invited to visit the website of the Center for Technology and System Management at the University of Maryland, College Park, to find information posted on various projects and publications that can be related to uncertainty and risk analysis. The URL address is http://www.ctsm.umd.edu.

Bilal M. Ayyub and George J. Klir
2006

Acknowledgments

This book was developed over several years and draws on the experiences of the authors in teaching courses related to risk analysis, uncertainty modeling and analysis, probability and statistics, systems science, numerical methods and mathematics, reliability assessment, and decision analysis. Drafts of most sections of the book were tested in several courses at the University of Maryland, College Park, for about 3 years before its publication. This testing period has proved to be a very valuable tool in establishing its contents and the final format and structure.

The direct and indirect help received from many individuals over the years has greatly affected this book. Students who took courses and used portions of this book at the University of Maryland provided great insight on how to effectively communicate various theoretical concepts. Also, advising and interacting with students on their research projects stimulated the generation of some examples used in the book. The students who took courses on uncertainty modeling and analysis, structural reliability, risk analysis, and mathematical methods in civil engineering at the University of Maryland from 1999 to 2005 contributed to this endeavor. Their feedback was very helpful and significantly contributed to the final product. The help provided by Dr. Mark Kaminskiy and Mr. William L. McGill in proofreading chapters is greatly appreciated. The assistance of the following students in developing and solving examples and end-of-chapter exercise problems is acknowledged: K. Avrithi, R. Carvalho, A. Faghi, E. Intarakosit, D. Kim, F. Luo, K.S. Nejaim, J. Sirivansant, Z. Tang, S. Tiku, and K. Zheng.

The book manuscript reviewers' comments provided by the publisher were used to improve the book to meet the needs of readers and enhance the educational process and are greatly appreciated.

The financial support that Dr. Ayyub has received from the U.S. Air Force Office of Scientific Research, Navy, Coast Guard, Army Corps of Engineers, National Science Foundation, Department of Homeland Security, Homeland Security Institute, Analytic Services, Inc., Office of Naval Research, Ford Motor Company, Maryland Emergency Management Agency, and American Society of Mechanical Engineers over more than 22 years contributed immensely to this book by providing him with a wealth of information and ideas for formalizing the theory and developing applications — especially the support received from the National Science Foundation in 1990 of the first and second Symposia on Uncertainty Modeling and Analysis in 1990 and 1993. In particular, Dr. Ayyub acknowledges the opportunity and support provided by P. Alman, L. Almodovar, A. Ang, R. Art, N.O. Attoh-Okine, K. Balkey, J. Beach, M. Blizzard, P. Bowen, P. Capple, K. Chong, J. Crisp, S. Davis, D. Dressler, A. Engle, G. Feigel, M. Firebaugh, J. Foster, M. Frankel, D. Green, A. Haldar, P. Hess III, Z. Karaszewski, T.S. Koko, V. Krivtsov, W. Melton, R. Meisinger, D. Moser, N. Nappi, Jr., G. Remmers, J. Scouras, D. Segalman, T. Shugar, J.R. Sims, R. Taylor, M. Wade, S. Wehr, and G. White.

About the Authors

Bilal M. Ayyub is a professor of civil and environmental engineering and the director of the Center for Technology and Systems Management at the University of Maryland, College Park. He has extensive expertise in developing and using uncertainty and risk analysis methods, with over 450 professional papers and publications. Dr. Ayyub is the author and co-author of six books, with one of them as a second edition, and the editor and co-editor of seven books. His research received support from the Air Force, Navy, Coast Guard, Army Corps of Engineers, National Science Foundation, Department of Homeland Security, Homeland Security Institute, Ford Motor Company, Maryland Emergency Management Agency, Maryland State Highway Administration, American Society of Mechanical Engineers, and several engineering companies. Dr. Ayyub has served the engineering community in various capacities through societies that include ASNE (life member), ASCE (fellow), ASME (fellow), SNAME (fellow), IEEE (senior member), NAFIPS, Society for Risk Analysis, and World Future Society, among others. He chaired the ASCE Committee on the Reliability of Offshore Structures, the design philosophy panel of the SNAME Ship Structures Committee, and the *Naval Engineers Journal* Committee of ASNE. Presently, he is chairman of the Safety Engineering and Risk Analysis (SERAD) Division of ASME. He also was the general chairman of the first, second, third, and fourth International Symposia on Uncertainty Modeling and Analysis that were held in 1990, 1993, 1995, and 2003. Currently, he is member of the *Journal of Ship Research* Committee of SNAME. Dr. Ayyub is the recipient of the ASNE "Jimmie" Hamilton Award for the best papers in the *Naval Engineers Journal* in 1985, 1992, 2000, 2002, and 2004. Also, he received the ASCE "Outstanding Research Oriented Paper" in the *Journal of Water Resources Planning and Management* for 1987, the NAFIPS K.S. Fu Award for Professional Service in 1995, the ASCE Edmund Friedman Award in 1989, and the ASCE Walter L. Huber Civil Engineering Research Prize in 1997, among other awards and certificates of appreciation. Also, he was awarded the State of Maryland Governor's Citation for "positive contributions, leadership and distinguished service … in honor and appreciation of your selfless efforts on behalf of the community." His biography is included in many biographical sources, such as *Who's Who in America* and *Who's Who in the World.*

George J. Klir is currently a distinguished professor of systems science at State University of New York (SUNY) at Binghamton. He has been with SUNY since 1969. His earlier work was in the areas of systems modeling and simulation, logic design, computer architecture, and discrete mathematics. His current research interests include the areas of intelligent systems, generalized information theory, fuzzy set theory and fuzzy logic, theory of generalized measures, and soft computing. He is the author of over 300 articles and 16 books. He has also edited 10 books and has been editor of the *International Journal of General Systems* since 1974 and the *International Book Series on Systems Science and Systems Engineering* since 1985. He was president of SGSR (1981 to 1982), IFSR (1980 to 1984), NAFIPS (1988 to 1991), and IFSA (1993 to 1995). He is a fellow of IEEE and IFSA and has received numerous awards and honors, including five honorary doctoral degrees, the Gold Medal of Bernard Bolzano, the Lotfi A. Zadeh Best Paper Award, the Kaufmann's Gold Medal, SUNY Chancellor's Award for Excellence in Research, and the IFSA Award for Outstanding Achievement. His biography is included in many biographical sources, including *Who's Who in America, Who's Who in the World, American Men and Women of Science, Outstanding Educators of America, Contemporary Authors*, etc. His research has been supported for more than 20 years by grants from NSF, ONR, the U.S. Air Force, NASA, NATO, Sandia National Laboratories, and a number of industries.

Books by the Authors

BILAL M. AYYUB

Analysis and Management of Uncertainty: Theory and Applications, North-Holland-Elsevier Scientific Publishers, Amsterdam, 1992, B.M. Ayyub and M.M. Gupta (editors).

Applied Research in Uncertainty Modeling and Analysis, International Series on Intelligent Technologies, Springer Science, Heidelberg, 2005, B.M. Ayyub and N.O. Attoh-Okine (editors).

Elicitation of Expert Opinion for Uncertainty and Risks, CRC Press, Boca Raton, FL, 2001, B.M. Ayyub.

Numerical Methods for Engineers, Prentice Hall, New York, 1996, B.M. Ayyub and R. McCuen.

Probability, Statistics, and Reliability for Engineers, CRC Press, Boca Raton, FL, 1997, B.M. Ayyub and R. McCuen.

Probability, Statistics, and Reliability for Engineers and Scientists, 2nd ed., Chapman & Hall/CRC Press, Boca Raton, FL, 2003, B.M. Ayyub and R. McCuen.

Risk Analysis in Engineering and Economics, Chapman & Hall/CRC Press, Boca Raton, FL, 2003, B.M. Ayyub.

Uncertainty Analysis in Engineering and the Sciences: Fuzzy Logic, Statistics, and Neural Network Approach, Kluwer Academic Publisher, Dordrecht, Netherlands, 1997, B.M. Ayyub and M.M. Gupta (editors).

Uncertainty Modeling and Analysis in Civil Engineering, CRC Press, Boca Raton, FL, 1998, B.M. Ayyub (editor).

Uncertainty Modeling and Analysis: Theory and Applications, North-Holland-Elsevier Scientific Publishers, Amsterdam, 1994, B.M. Ayyub and M.M. Gupta (editors).

Uncertainty Modeling in Finite Element, Fatigue, and Stability of Systems, World Scientific, Singapore, 1997, A. Haldar, A. Guran, and B.M. Ayyub (editors).

Uncertainty Modeling in Vibration, Control, and Fuzzy Analysis of Structural Systems, World Scientific, Singapore, 1997, B.M. Ayyub, A. Guran, and A. Haldar (editors).

GEORGE J. KLIR

Analog and Digital Computers, Orbis, Prague, in Czech, 1964, G.J. Klir, J. Gecsei, and P. Pelikan.

Applied General Systems Research, Plenum Press, New York, 1978, G.J. Klir (editor).

An Approach to General Systems Theory, Van Nostrand Reinhold, New York, 1980, G.J. Klir. Spanish translation, Ediciones Ice, Madrid.

Architecture of Systems Problem Solving, Plenum Press, New York, 1985, G.J. Klir and D. Elias. Russian Translation, *Radio i Sviaz*, 1990. Second edition, Kluwer/Plenum, Dordrecht, Netherlands, 2003.

Basic and Applied General Systems Research: A Bibliography, SUNY-Binghamton, New York, 1977, G.J. Klir and G. Rogers (editors).

Basic and Applied General Systems Research: A Bibliography, 1977–1984, Hemisphere, Washington, D.C., 1985, G.J. Klir, R. Trappl, and W. Horn (editors).

Computer-Aided Theory and Technology, Springer-Verlag, Berlin, 1996, G.J. Klir and T. Oren (editors).

Cybernetic Modelling, SNTL, Prague, 1965, in Czech, G.J. Klir and M. Valach. British edition, Iliffe, London, 1967. American edition, Van Nostrand, New York, 1967.

Facets of Systems Science, Plenum Press, New York, 1991, G.J. Klir. Second edition, Kluwer/Plenum, Dordrecht, Netherlands, 2001.

Fuzzy Logic in Geology, Academic Press/Elsevier, Amsterdam, 2004, G.J. Klir and R.V. Demicco (editors).

Fuzzy Measure Theory, Plenum Press, New York, 1992, G.J. Klir and Z. Wang.

Fuzzy Set Theory: Foundations and Applications, Prentice Hall, Englewood Cliffs, NJ, 1997, G.J. Klir, U. St. Clair, and B. Yuan.

Fuzzy Sets, Fuzzy Logic, and Fuzzy Systems, selected papers by Lotfi A. Zadeh, World Scientific, Singapore, 1996, G.J. Klir and B. Yuan (editors).

Fuzzy Sets and Fuzzy Logic: Theory and Applications, Prentice Hall, Englewood Cliffs, NJ, 1995, G.J. Klir and B. Yuan.

Fuzzy Sets: An Overview of Fundamentals and Personal Views, Normal University Press, Beijing, 2000, G.J. Klir.

Fuzzy Sets, Uncertainty and Information, Prentice Hall, Englewood Cliffs, NJ, 1998, G.J. Klir and T. Folger. Japanese Translation, UNI, Tokyo, 1993.

Introduction to the Methodology of Switching Circuits, Van Nostrand Reinhold, New York, 1972, G.J. Klir.

Mathematical Machines, Prace, Prague, in Czech, 1961, G.J. Klir.

Methodology in Systems Modelling and Simulation, North-Holland, Amsterdam, 1979, G.J. Klir, B.P. Zeigler, M.S. Elzas, and T.I. Oren (editors).

Progress in Cybernetics and Systems Research, Vol. III, Hemisphere, Washington, D.C., 1978, G.J. Klir, R. Trappl, and L. Ricciardi (editors).

Progress in Cybernetics and Systems Research, Vol. VIII, Hemisphere, Washington, D.C., 1982, G.J. Klir, R. Trappl, and F. Pichler (editors).

Synthesis of Switching Circuits, SNTL, Prague, 1966, in Czech, G.J. Klir and L. Seidl. British edition, Iliffe, London, 1968. American edition, Gordon and Breach, New York, 1968.

Theory of Switching Circuits, SNTL, Prague, in Czech, 1965, G.J. Klir.

Trends in General Systems Theory, John Wiley, New York, 1972, G.J. Klir (editor). Polish translation, *Wydawictwa naukovo-techniczne*, Warsaw, 1976. Spanish translation, Alianza Editorial, Madrid, 1978.

Uncertainty-Based Information: Elements of Generalized Information Theory, Physica Verlag/Springer-Verlag, Heidelberg, 1998, G.J. Klir and M. Wierman. Second edition, 1999.

Uncertainty and Information: Foundations and Applications of Generalized Information Theory, John Wiley, New York, 2006, G.J. Klir.

Contents

1 Systems, Knowledge, and Ignorance

The greatest enemy of knowledge is not IGNORANCE, it is the ILLUSION of knowledge.

— **Stephen Hawking**

1.1 DATA ABUNDANCE AND UNCERTAINTY

Intelligence is defined as the ability to understand and adapt to the environment by using a combination of inherited abilities and learning experiences. This ability certainly includes the analysis of uncertainty and making decisions under conditions of uncertainty. The definition of intelligence is applicable to living systems — from ants to aardvarks to humans — as well as machines. Any organism that survives the remorseless rigors of evolution is sufficiently intelligent for its role in life. Likewise, machines need to be sufficiently intelligent to make decisions suitable for their functions and adapt to and deal with the presence of uncertainty. Any collectives of human decision makers and their decision-aiding machines must make, in the aggregate, good decisions.

The ability of a living system or machine to make appropriate decisions can be taken as a measure of intelligence. This decision-making ability requires the processing of data and information, construction of knowledge, and assessment of associated uncertainties and risks. The analysis and modeling of uncertainty enhances this ability of making appropriate decisions, thereby increasing intelligence. This need to model and analyze uncertainties also stems from the awareness that data abundance does not necessarily give us certainty, and sometimes can lead to error in decision making, with undesirable outcomes due to either overwhelming, confusing situations or a sense of overconfidence leading to an improper information use. The former situations can be an outcome of the limited capacity of a human mind in some situations to deal with complexity and data abundance, whereas the latter can be attributed to a higher order of ignorance, called the ignorance of self-ignorance.

As our society advances in many scientific dimensions and invents new technologies, human knowledge is being expanded through observation, discovery, information gathering, and propositional logic. Also, the access to newly generated information is becoming easier than ever as a result of computers and the Internet. We have entered an exciting era where electronic libraries, online databases, and information on every aspect of our civilization, such as patents, engineering products, literature, mathematics, physics, medicine, philosophy, and public opinions, are becoming a mouse-click or a few clicks away. In this era, computers can generate even more information from abundantly available online data. Society can act or react based on this information at the speed of its generation, creating sometimes nondesirable situations, for example, price or political volatilities. There is a great need to assess uncertainties associated with information and quantify our state of knowledge or ignorance. The accuracy, quality, and incorrectness of such information, and knowledge incoherence are coming under focus by philosophers, scientists, engineers, technologists, decision and policy makers, regulators and lawmakers, and our society as a whole. As a result, uncertainty and ignorance analyses are receiving a lot of attention by our society. We are moving from emphasizing the state of knowledge expansion and creation of information to a state that includes knowledge and information assessment by critically evaluating them in terms of relevance, completeness, nondistortion, coherence, and other key measures.

Our society is becoming less forgiving and more demanding from our knowledge base. The use of noncredible information, leading to questionable decisions, could place decision makers on the defensive. On the other hand, untimely processing and use of any available information, even if it might be inconclusive, can be treated worse than a lack of knowledge and ignorance. In the January 2003 State of the Union address, U.S. President George W. Bush stated, "The British government has learned that Saddam Hussein recently sought significant quantities of uranium from Africa." A few months later, after the conclusion of the war on Iraq in May 2003, senior White House officials conceded the information that former Iraqi president Hussein tried to buy uranium from Niger was inaccurate, but they said Bush's State of the Union speech was based on a broader range of intelligence. The argument that Iraq was trying to reconstitute its nuclear weapons program was a key point in the administration's rationale for war. These statements and decisions were made despite the March 2003 International Atomic Energy Agency dismissal as forgeries documents that alleged Iraq may have tried to buy 500 tons of uranium from Niger. The news elevated the problem to scandalous levels for this action on uncertain information, although inaction on uncertain intelligence, such as the "intelligence failure" in the case of the 2001 World Trade Center attacks, was treated as scandalous and was investigated due to its unacceptability. Any inaction due to noncredible information can be easily taken by our demanding society to be as erroneous as an action based on noncredible information — hence the need for uncertainty assessment, modeling, and analysis.

Making appropriate decisions commonly entails risks requiring risk control and management. Although people have control over the levels of some technology-caused risks to which they are exposed, reduction of risk needs to be pursued by governments and corporations in response to increasing demands by our society. Risk reduction generally entails a reduction of benefits to people, thus posing a serious dilemma. Moreover, the public and policy makers are required, with increasing frequency, to subjectively weigh benefits against risks and assess associated uncertainties when making decisions. Not using a systems or holistic approach, vulnerability exists for overpaying to reduce one set of risks that may introduce offsetting or larger risks of another kind. Such risk-based decisions require uncertainty modeling and analysis.

The objective of this chapter is to present a systems framework for uncertainty modeling and analysis, and to discuss knowledge, its sources and acquisition, and ignorance and its categories. The practical use of concepts and tools presented in the book requires a framework and a frame of thinking that deals holistically with problems and issues as systems.

1.2 SYSTEMS FRAMEWORK

1.2.1 Systems Definitions and Modeling

The definition and articulation of problems in engineering and the sciences is a critical task in the processes of analysis and design, and can be systematically performed within a systems framework. "The mere formulation of a problem is often far more essential than its solution," Albert Einstein said. "What we observe is not nature itself, but nature exposed to our method of questioning," Werner Karl Heisenberg said. Commonly, an engineering project can be modeled to include a segment of its environment that interacts significantly with it to define an underlying system. The boundaries of the system are drawn based on the mission, goals, and objectives of the analysis, and the class of performances (including failures) under consideration.

A generalized systems formulation allows scientists and engineers to develop a complete and comprehensive understanding of the nature of a problem, and underlying physical phenomena, processes, and activities. In a system formulation, an image or a model of an object that emphasizes some important and critical properties is defined. System definition is usually the first step in an overall methodology formulated for achieving a set of objectives. This definition can be based on observations at different system levels that are established based on these objectives. The observations can be about the different elements (or components) of the system, interactions among these elements, and the expected behavior of the system. Each level of knowledge that is obtained about an engineering problem defines a system to represent the project or the problem. As additional levels of knowledge are added to previous ones, higher epistemological levels of system definition and description are attained that, taken together, form a hierarchy of the system descriptions.

Informally, what is a system? The term *system* originates from the Greek word *systma*, which means an organized whole. According to *Webster's Dictionary*, a *system* is defined as "a regularly interacting or interdependent group of items forming a unified whole," such as a solar system, school system, or system of highways. For scientists and engineers, the definition can be stated as "a regularly interacting or interdependent group of items forming a unified whole that has some attributes of interest." Alternatively, a system can be defined as a group of interacting, interrelated, or interdependent elements that together form a complex whole that can be a complex physical structure, process, or procedure of some attributes of interest. All parts of a system are related to the same overall process, procedure, or structure, yet they are different from one another and often perform completely different functions. It follows from these definitions that the term *system* stands, in general, for a *set of things* and a *relation among the things*. It can be formally stated as

$$S = (T, R) \tag{1.1}$$

where S, T, and R denote, respectively, a *system*, a *set of things*, and a *relation* (or possibly a set of relations) defined on T. This *commonsense expression* by the pair (T, R) seems overly simple. Its simplicity, however, is only on the surface. While the definition is very simple in its form, it contains symbols, T and R, that are extremely rich in content. T stands not only for a single set with arbitrary elements, finite or infinite, but also, for example, for a power set, a power set of a power set, etc., or any arbitrary set of sets. Furthermore, things in T may have special properties by which systems are distinguished from one another. These properties can be referred to as *thinghood properties*. The content of symbol R is even richer. For each set T, with its special characteristics, the symbol stands for any conceivable relation defined on T. Formally, a relation is a subset of some Cartesian product of given sets. Even if T is only a single set, R stands for a relation from a family of distinct types of relations: $R \subset T \times T$ (*binary relations*), $R \subset T \times T \times T$ (*ternary relations*), etc. When T is a set of sets, the variety of distinct types of relations virtually explodes. For example, when T consists of just two sets, say $T = \{X, Y\}$, the number of types of relations grows quite rapidly, including, for example, the following types:

$$R \subset X \times Y \tag{1.2a}$$

$$R \subset (X \times X) \times Y \tag{1.2b}$$

$$R \subset (X \times X) \times (Y \times Y) \tag{1.2c}$$

$$R \subset (X \times X) \times (X \times Y) \tag{1.2d}$$

$$R \subset (X \times Y) \times (X \times Y) \tag{1.2e}$$

$$R \subset (X \times X \times X) \times (Y \times Y \times Y) \tag{1.2f}$$

$$R \subset (X \times Y) \times (X \times Y) \times (X \times Y) \tag{1.2g}$$

Although these few examples illustrate the great variety of possibilities represented by the single symbol R, they still do not capture the full richness of this symbol. The form of the Cartesian product on which a relation is defined is only one property of the relation. Other properties depend on the nature of elements of the relevant Cartesian product that are included in the relation. All these properties of relations can be subsumed under the suggestive name *systemhood properties*.

The simplicity of the commonsense expression of a system is, paradoxically, its weakness as well as its strength. The definition is weak because it is too general and, consequently, of little pragmatic value. It is strong because it encompasses all other, more specific definitions of systems. Due to its full generality, the commonsense expression qualifies for a criterion by which we can determine whether any given object is a system or not: an object is a system if and only if it can be described in the form that conforms to Equation 1.1.

Once we have the capability of distinguishing objects that are systems from those that are not, it is natural to define systems science as a *science whose objects of study are systems*. It is significant that this definition refers to systems, but not to any particular types of systems, such as physical systems, biological systems, social systems, or economic systems. This implies that these distinctions of systems, which are expressed solely in terms of the things involved, are not significant in systems science. This means, in turn, that systems science is concerned with systemhood properties of systems rather than their thinghood properties.

Classical science, which is predominately oriented to thinghood properties, and systems science, which is predominately oriented to systemhood properties, are two distinct perspectives from which scientific inquiry can be approached. These perspectives are complementary. Although classical scientific inquiries are almost never devoid of issues involving systemhood properties, these issues are not of primary interest in classical science and have been handled in an opportunistic, *ad hoc* fashion. There is no place in classical science for a comprehensive and thorough study of the various properties of systemhood. The systems perspective thus cannot be fully developed within the confines of classical science. It was liberated only through the emergence of systems science. While the systems perspective was not essential when science dealt with simple systems, its significance increases with the growing complexity of systems of our current interest. From the standpoint of the disciplinary classification of classical science, systems science is clearly cross-disciplinary.

Systems are traditionally grouped in various overlapping categories, such as:

1. Natural systems, e.g., river systems and energy systems
2. Human-made systems that can be embedded in the natural systems, e.g., hydroelectric power systems and navigation systems
3. Physical systems that are made of real components occupying space, e.g., automobiles and computers
4. Conceptual systems that could lead to physical systems
5. Static systems that are without any activity, e.g., bridges subjected to dead loads
6. Dynamic systems, e.g., transportation systems
7. Closed- or open-loop systems, e.g., a chemical equilibrium process and logistic systems, respectively.

Blanchard (1998) provides additional information on these categories.

1.2.2 Realism and Constructivism in Systems Thinking

The emergence of systems science is from two different views about the nature of knowledge: *realism* and *constructivism*. According to realism, a system that is obtained by applying correctly the principles and methods of science *represents* some aspect of the real world. This representation is only approximate, due to limited resolution of our sensors and measuring instruments, but the relation comprising the system is a *homomorphic image* of its counterpart in the real world. Using more refined instruments, the homomorphic mapping between entities of the system of concern and those of its real-world counterpart (the corresponding real system) becomes also more refined, and the system becomes a better representation of its real-world counterpart. Realism thus assumes the existence of systems in the real world, which are usually referred to as *real systems*. It claims that any system obtained by sound scientific inquiry is an approximate (simplified) representation of a real system via an appropriate homomorphic mapping.

According to constructivism, all systems are artificial abstractions. They are not made by nature and presented to us to be discovered, but we construct them by our perceptual and mental capabilities within the domain of our experiences. The concept of a system that requires correspondence to the real world is illusory because there is no way of checking such correspondence. We have no access to the real world except through experience. It seems that the constructivist view has become predominant, at least in systems science, particularly in the way formulated by von Glasersfeld (1995). According to this formulation, constructivism does not deal with ontological questions regarding the real world. It is intended as a theory of knowing, not a theory of being. It does not require the denial of ontological reality. Moreover, the constructed systems are not arbitrary: they must not collide with the constraints of the experiential domain. The aim of constructing systems is to organize our experiences in useful ways. A system is useful if it helps us to achieve some aims, for example, to predict, retrodict, control, make proper decisions, etc.

1.2.3 Taxonomy of Systems

Since systems science is oriented to the study of systemhood properties, its aim is to understand these properties as completely as possible. The following are key steps in pursuing this aim:

1. Dividing the spectrum of conceivable systems into significant categories defined in terms of systemhood properties
2. Studying individual categories of systems and their relationship
3. Organizing these categories into a coherent whole
4. Studying systems problems that emerge from the underlying set of organized systems categories
5. Studying methodological issues regarding the various types of systems problems
6. Studying metamethodological issues emerging from systems methodology

A prerequisite for dividing systems by their systemhood properties into significant categories is developing a conceptual framework within which these properties can properly be codified. Each framework determines the scope of systems conceived. It captures some basic categories of systems, each of which characterizes a certain type of knowledge representation, and provides a basis for further classification of systems within each category. To establish firm foundations of systems science, a comprehensive framework is needed to capture the full scope of systemhood properties.

1.2.3.1 Epistemological Categories of Systems

Several conceptual frameworks that attempt to capture the full scope of systems currently conceived have been proposed by Klir (1985), Mesarovic and Takahara (1975, 1988), Wymore (1976), and Zeigler (1976). In spite of differences in terminology and in the way in which these frameworks evolved, they have essentially the same expressive power. As an example, a particular framework developed by Klir (1985) is described here, which is known in the literature as the *general systems problem solver* (GSPS). The kernel of the GSPS is a hierarchy of *epistemological categories of systems*, which represents the most fundamental taxonomy of systems. The following is a brief outline of the basic levels in this hierarchy.

At the lowest level of the epistemological hierarchy, an *experimental frame* is defined in terms of appropriate variables and their state sets (value sets). In addition, some supporting medium (such as time, space, or population) within which the variables change their states is also specified. Furthermore, variables may be classified as input and output variables.

An experimental frame (also called a *source system*) may be viewed as a *data description language*. When actual data described in this language become available, we move to the next level in the hierarchy. Systems on this level are called *data systems*.

When variables of an experimental frame are characterized by a relationship among them, we move to a level that is still higher in the hierarchy. It is assumed on this level that the relationship among the variables is invariant with respect to the supporting medium involved. That is, it is time invariant, space invariant, space-time invariant, population invariant, etc. The relationship may involve not only variables contained in the experimental frame, but also additional variables defined in terms of the former by specific translation rules in the supporting medium. When the supporting medium is time, for example, we obtain lagged variables. Systems on this level are called *behavior systems*. Some of these systems can also be characterized conveniently as *state transition systems*.

A data system is represented by a behavior system if, under appropriate initial or boundary conditions, the support-invariant relation of the latter can be utilized for generating the data of the former. The generative capability of a behavior system extends, of course, beyond any given data. That is, a behavior system is capable to generate, for example, predictions or retrodictions of the variables involved. Moreover, it provides us with an explanation of the behavior of the variables within the given supporting medium.

Climbing further up the hierarchy involves two principles of integrating systems as components in larger systems. According to the first principle, several behavior systems (or sometimes lower-level systems) that may share some variables or interact in some other way are viewed as subsystems integrated into one overall system. Overall systems of this sort are called *structure systems*. The subsystems forming a structure system are often called its *elements*.

When elements of a structure system are themselves structure systems, this overall system is called a *second-order structure system*. *Higher-order structure systems* are defined recursively in the same way.

According to the second integrating principle, an overall system is viewed as varying (in time, space, etc.) within a class of systems of any of the other types. The change from one system to another in the delimited class is described by a replacement procedure that is invariant with respect to the supporting medium involved (time, space, etc.). Overall systems of this type are called *metasystems*.

In principle, the replacement procedure of a metasystem may also change. Then, an invariant (changeless) higher-level procedure is needed to describe the change. Systems of this sort, with two levels of replacement procedures, are called *metasystems of second order. Higher-order metasystems* are then defined recursively in the same way. Structure systems whose elements are metasystems are also allowed by the framework, similarly as metasystems defined in terms of structure systems.

The key feature of the epistemological hierarchy is that every system defined on some level in the hierarchy entails knowledge associated with all corresponding systems on lower levels and, at the same time, contains some knowledge that is not available in any of these lower-level systems.

The number of levels in the epistemological hierarchy is potentially infinite. In practice, however, only a small number of levels is considered. For each particular number of levels, the hierarchy is a semilattice. For five levels, for example, a part of the semilattice is expressed by the Hasse diagram in Figure 1.1. The circles represent the various epistemological categories of systems, and the arrows indicate the ordering from lower to higher categories. Symbols E, D, and B denote experimental frames (source systems), data systems, and behavior systems, respectively. Symbol S, used as a prefix, stands for structure systems. For example, SD denotes structure systems whose elements are data systems. Symbol S^2 denotes structure systems of second order. For example, S^2B denotes structure systems of structure systems whose elements are behavior systems. Symbols M and M^2 denote metasystems and metametasystems, respectively. The combination SM and MS denotes structure systems whose elements are metasystems and metasystems whose elements are structure systems, respectively. The diagram in Figure 1.1 describes only a part of the first five levels in the epistemological hierarchy; it can be extended in an obvious way to combinations such as S^3B, S^2MB, SMSB, M^2SB, S^2MB, etc.

Categories of systems captured by the epistemological hierarchy are actually categories in the strong sense of mathematical category theory. It is useful to further classify systems subsumed under each epistemological category by relevant methodological distinctions. The aim of this classification is to capture the relationship between classes of systems and methods applicable to problems associated with the systems. Examples of methodological distinctions are those between systems based on discrete variables and systems based on continuous variables, between deterministic and nondeterministic systems, and between dynamic and spatial systems.

In the subsequent sections, the source, data, generative, structure, and metasystems are described and illustrated in Examples 1.1 and 1.2.

1.2.3.2 Source (or Experimental Frame) Systems

At the first level of knowledge, which is usually referred to as level 0, the system is known as a *source system*. Source systems comprise three different components,

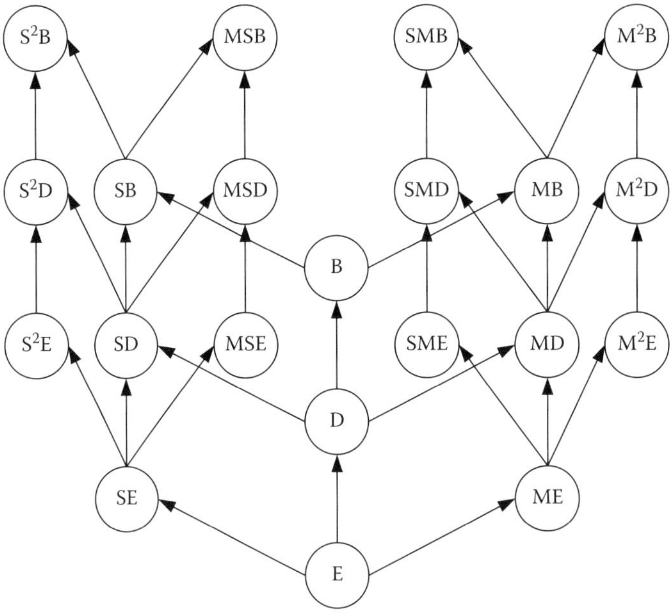

FIGURE 1.1 Epistemological hierarchy of systems categories. E = experimental frame or source system; D = data system; B = behavior system; SE, SD, SB = structure systems based on source, data, and behavior systems, respectively; S^2E, S^2D, S^2B = second-order structure systems of the three types; ME, MD, MB = metasystems based on source, data, and behavior systems, respectively; M^2E, M^2D, M^2B = second-order metasystems of the three types; SME, SMD, SMB = structure systems based on metasystems of the three types; MSE, MSD, MSB = metasystems based on structure systems of the three types.

namely, *object systems*, *specific image systems*, and *general image systems*, as shown in Figure 1.2. The object system constitutes a model of the original object. It is composed of an object, attributes, and a backdrop. The object represents the specific problem under consideration. The attributes are the important and critical properties or variables selected for measurement or observation as a model of the original object. The backdrop is the domain or space within which the attributes are observed. The specific image system is developed based on the object. This image is built through observation channels that measure the attribute variation within the backdrop. The attributes when measured by these channels correspond to the variables in the specific image system. The attributes are measured within a support set that corresponds to the backdrop. The support can be either time or space, or can be population. Combinations of two or more of these supports are also possible. Before upgrading the system to a higher knowledge level, the specific image system can be abstracted into a general format. A mapping function is utilized for this purpose among the different states of the variables to a set of generals that is used for all the variables.

There are some methodological distinctions that could be defined in this level. Ordering is one of these distinctions that is realized within state or support sets. Any

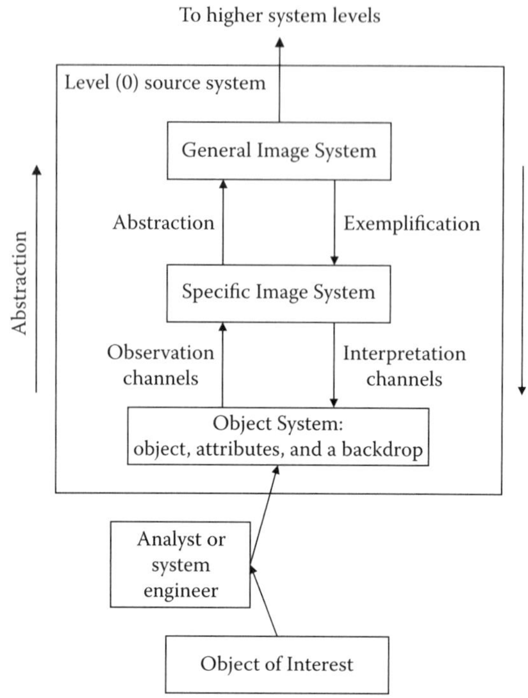

FIGURE 1.2 Source system components.

set can be either ordered or not ordered, and those that are ordered may be partially ordered or linearly ordered. An ordered set has elements that can take real values, or values on an interval or ratio scale. A partially ordered set has elements that take values on an ordinal scale; for example, military ranks are partially ordered. A nonordered set has components that take values on a nominal scale, such as gender classification of people or political party affiliations of people. Distance is another form of distinction, where the distance is a measure between pairs of elements of an underlying set. It is obvious that if the set is not ordered, the concept of distance is not valid. Continuity is another form of distinction, where variables or support sets could be discrete or continuous. The classification of the variables as input or output variables forms another distinction. Those systems that have classified input–output variables are referred to as directed systems; otherwise, they are referred to as neutral systems. The last distinctions that could be realized in this level are related to the observation channels, which could be classified as crisp or fuzzy, corresponding to nonvague and vague information channels, respectively. For example, the number of hurricanes in a year hitting a region is uncertain, but takes on discrete crisp counts, whereas the fit or comfort level associated with wearing a piece of garment can only be measured in vague terms using linguistic terms such as *comfortable*, *not comfortable*, or *partly comfortable*. Figure 1.3 summarizes methodological distinctions realized in the first level of knowledge.

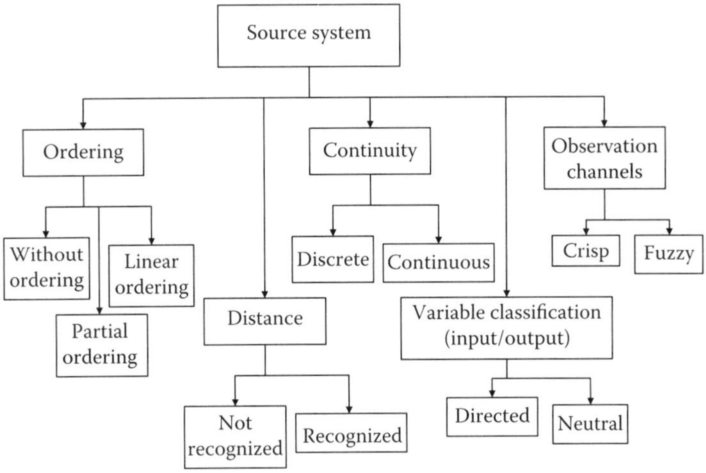

FIGURE 1.3 Methodological distinctions of source systems.

1.2.3.3 Data Systems

The second level of a hierarchical system classification is the data system. The data system includes a source system together with actual data used for the states of variables for each attribute. The actual states of the variables at the different support instances yield the overall states of the attributes. Special functions and techniques are used to infer information regarding an attribute, based on the states of the variables representing it. A formal definition of a data system could be expressed as follows:

$$D = \{S, a\} \tag{1.3}$$

where D = data system, S = the corresponding source system, and a = observed data that specify the actual states of the variables at different support instances.

1.2.3.4 Generative Systems

At the generative knowledge level, support-independent relations are defined to describe the constraints among the variables. These relations could be utilized in generating states of the basic variables for a prescribed initial or boundary condition. The set of basic variables includes those defined by the source system and possibly some additional variables that are defined in terms of the basic variables. There are two main approaches for expressing these constraints. The first approach consists of a support-independent function that describes the *behavior* of the system. A function defined as such is known as a *behavior function*. The second approach consists of relating successive *states* of the different variables. In other words, this function describes a relationship between the current overall state of the basic variables and the next overall state of the same variables. A function defined as such is known as a *state transition function*. For example, a state transition function can

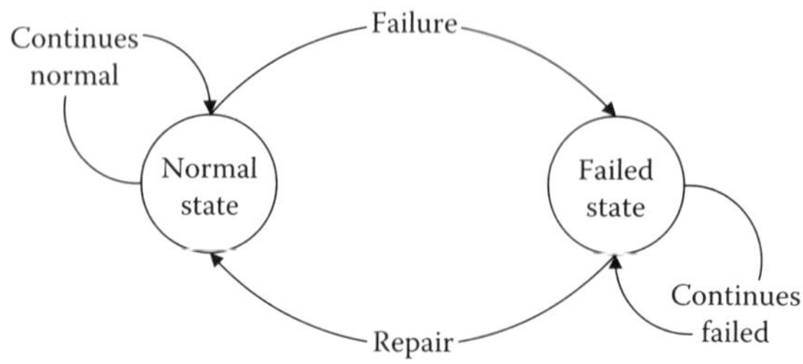

FIGURE 1.4 A Markov transition diagram for repairable systems.

be used to model repairable systems. Such systems can be assumed for the purpose of demonstration to exit in either a normal, i.e., operating, state or failed state, as shown in Figure 1.4. A system in a normal state makes transitions to either normal states that are governed by its reliability level (i.e., it continues to be normal) or failed states through failure. Once it is in a failed state, the system makes transitions to either failed states that are governed by its repairable-ease level (i.e., it continues to be failed) or normal states through repair. Therefore, four transition probabilities are needed for the following cases:

- Normal-to-normal state transition
- Normal-to-failed state transition
- Failed-to-failed state transition
- Failed-to-normal state transition

The sum of probabilities for transitions originating from the same state must add up to 1. These probabilities can be determined by testing the system or based on analytical modeling of the physics of failure and repair logistics as provided by Kumamoto and Henley (1996).

A generative system defined by a behavior function is referred to as a *behavior system*, whereas if it is defined by a state transition function, it is known as a *state transition system*. State transition systems can always be converted into equivalent behavior systems, which makes the behavior systems more general.

The constraints among the variables at this level can be represented using many possible views or perspectives that are known as *masks*. A mask represents the pattern in the support set that defines sampling variables that should be considered. The sampling variables are related to the basic variables through translation rules that depend on the ordering of the support set. A formal definition of a behavior system could be expressed as

$$E_B = (I, K, f_B) \tag{1.4a}$$

where E_B = the behavior system defined as the triplet of three items, I = the corresponding general image system or the source system as a whole, K = the chosen mask, and f_B = the behavior function. If the behavior function is used to generate data or states of the different variables, the sampling variables should be partitioned into generating and generated variables. The generating variables represent initial conditions for a specific generating scheme. The system in this form is referred to as a *generative behavior system*. The formal definition for such a system could be expressed as

$$E_{GB} = (I, K_G, f_{GB}) \tag{1.4b}$$

where E_{GB} = the generative behavior system defined as the triplet of three items; I = the corresponding general image system or the source system as a whole; K_G = the chosen mask partitioned into submasks, namely, a generating submask that defines the generating sampling variables and a generated submask that defines the generated variables; and f_{GB} = the generative behavior function, which should relate the occurrence of the general variables to that of the generating variables in a conditional format.

Most engineering and scientific models, such as the basic Newton's law of force computed as the product of mass of an object and its acceleration, or computing the stress in a rod under axial loading as the applied force divided by the cross-sectional area of the rod, can be considered generative systems that relate basic variables such as mass and acceleration to force, or axial force and area to stress, respectively. In these examples, these models can be considered behavior systems.

Several methodological distinctions can be identified in this level. One of these distinctions is the type of behavior function used. For nondeterministic systems where variables have more than one potential state for the same support instant, a degree of belief or a likelihood measure to each potential state in the overall state set of the sampling variables should be assigned. They can be used to quantify uncertainty using uncertainty measures, discussed in detail in Chapter 4. Each one of these measures is considered to form a certain distinction within the generative system. Probability distribution functions and possibility distribution functions are widely used to construct behavior functions as introduced in Chapters 3 and 4. The determination of a suitable behavior function for a given source system, mask, and data is not an easy task. Potential behavior functions should meet a set of conditions to be satisfactorily accepted. These conditions should be based on the actual constraints among the variables. They also relate to the degree of generative uncertainty and complexity of the behavior system. Another distinction at this level could be identified in relation to the mask used. If the support set is ordered, the mask is known as *memory dependent*; otherwise, the mask is referred to as *memoryless*. Figure 1.5 summarizes the different distinctions identified in this knowledge level.

1.2.3.5 Structure Systems

Structure systems are sets of smaller systems or subsystems, as previously discussed. The subsystems could be source, data, or generative systems. These subsystems may be coupled due to having common variables or due to interaction in some other form. A formal definition of a structure system could be expressed as follows:

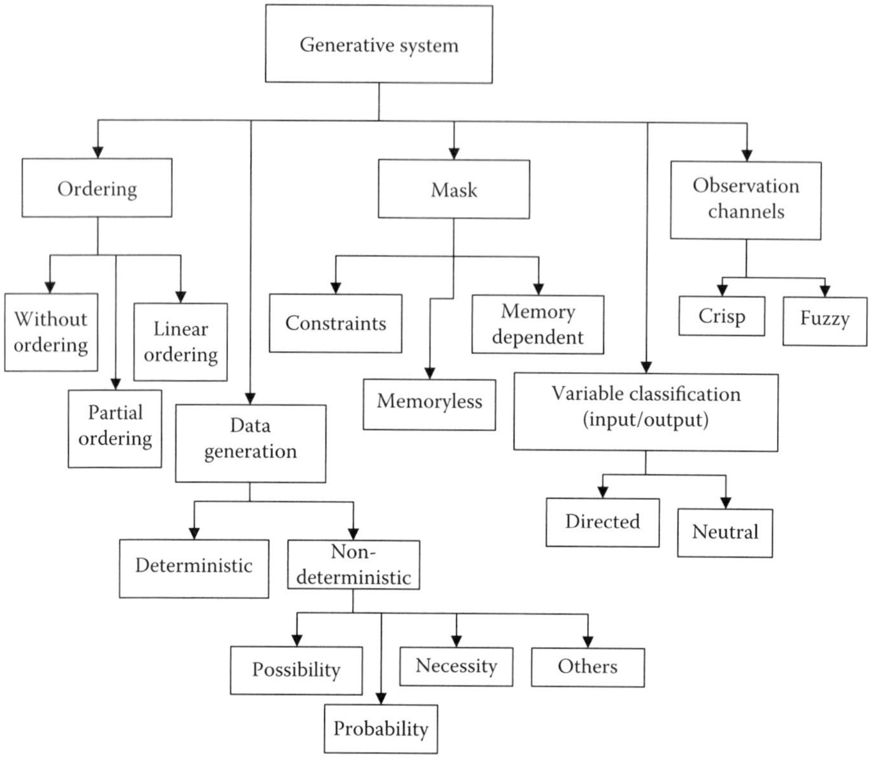

FIGURE 1.5 Example methodological distinctions for generative systems.

$$SE_B = \{(V_i, E_B^i), \text{for all } i \in e\} \tag{1.5}$$

where SE_B = a structure system whose elements are behavior systems, V_i = the set of sampling variables for the element of the behavior system, E_B^i = i^{th} behavior system, and e = the total number of elements or subsystems in the structure system with all i that belong to e, i.e., for all $i \in e$.

1.2.3.6 Metasystems

Metasystems are introduced for the purpose of describing changes within a given support set. The metasystem consists of a set of systems defined at some lower knowledge level and some support-independent relation. Referred to as a replacement procedure, this relation defines the changes in the lower-level systems. All the lower-level systems should share the same source system. There are two different approaches whereby a metasystem could be viewed in relation to the structure system. The first approach is introduced by defining the system as a structure metasystem. The second approach consists of defining a metasystem of a structure system whose elements are behavior systems.

EXAMPLE 1.1 SYSTEM DEFINITION OF CONSTRUCTION OPERATIONS

Construction management concerns itself, among other things, with the real-time control of construction or production activities. However, in order to develop a control system for a construction activity, this activity has to be suitably defined depending on its nature and methods of control using a *hierarchical control system* (Abraham et al., 1989; Ayyub and Hassan, 1992a, 1992b, 1992c). The hierarchical system classification enables the decomposition of the overall construction activity into subsystems that represent the different processes involved in each activity. Then each process could be decomposed into tasks that are involved in performing the process. For construction activities, a set theory framework is suitable for representing the variables of the problem. The ability to infer information about the overall system, knowing the behavior of its components, can be dealt with using special system prediction techniques (Chestnut, 1965; Hall, 1962, 1989; Klir, 1969, 1985; Wilson, 1984). In this example, levels of an epistemological hierarchy are defined for the purpose of real-time control.

Source Systems

For the purpose of illustration, the construction activities of concrete placement are considered and their knowledge level upgraded throughout the course of this example. The first step in defining the system for these construction activities is to identify a goal, in this case construction control by safely placing high-quality concrete efficiently and precisely. This goal can be defined through some properties or attributes of interest that can include safety, quality, productivity, and precision. Considering only two attributes of construction activities, i.e., safety and quality, the variables or factors that affect those attributes should be identified. Only two variables are assumed to affect the safety attribute. These variables could be quantitatively or qualitatively defined depending on their nature. For qualitative variables, linguistic terms are used and can be modeled using fuzzy set theory (which is formally introduced in Chapter 2) to define the potential states, together with a suitable observation channel that yields a quantitative equivalent for each state (Klir, 1985; Klir and Folger, 1988; Zimmerman, 1985). An example of this variable type is labor experience (v_1), which is used herein. This variable is assumed to have four potential states: fair, good, moderate, and excellent. These linguistic measures can be defined using fuzzy sets. Using a scale of 0 to 10 for the level of experience, these measures can be defined as shown in Figure 1.6. The vertical axis in the figure represents the degree of belief that the corresponding experience value belongs to the fuzzy sets of fair, good, moderate, or excellent experience, where experience is on a scale of 0 to 10 (0 = absolutely no experience and 10 = the absolute highest experience). A mathematical operator can then be defined in order to get a quantitative equivalent for each state. A one-to-one mapping function is used in order to define the corresponding general states of the variable (v_1). The second variable (v_2) is the method of construction. This variable could have three potential states, e.g., a traditional method, slip form method, and precast element method. This is a crisp variable, and its observation channel is represented by an engineer who decides which method should be used. A similar one-to-one mapping function is used to relate the different construction methods to the corresponding general states of the variable (v_2).

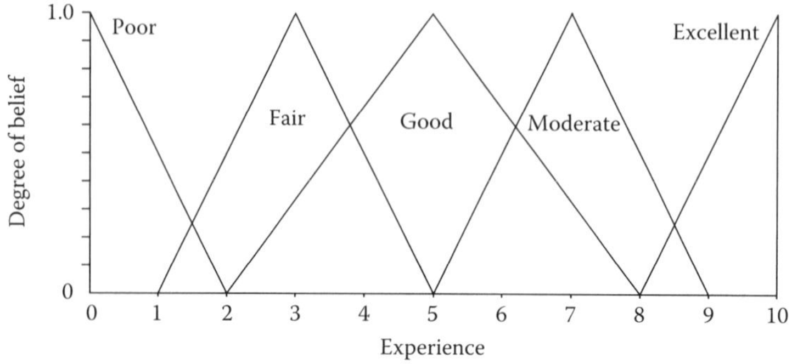

FIGURE 1.6 Fuzzy definitions of experience.

The next step in the definition of this system is the identification of the different supports, i.e., backdrops. In this example, the supports include time, space, and population. The time support is needed in measuring the progress of the different variables during the construction period. Assuming a construction period of 2 months with weekly observations, the time support set has eight elements that correspond to the weeks during the construction period. In other words, the elements are week 1, week 2, …, week 8. The space support is used in relating the current state of each variable at a specific time support instant to a specific location in space within the system. As an example, a space support set with elements that represent the type of structural element under construction is considered. These elements are columns, beams, slabs, and footings. Such a classification constitutes a space support set with four potential elements. The population support is used to represent the performance of units having the same structure with respect to the same variables. The population support set in this example can represent the set of different crews involved in the construction activity. This support set could have four potential elements: a falsework crew, a rebar crew, a concreting crew, and a finishing crew. The overall support set, which represents the domain within which any of the defined variables can change, is defined by the Cartesian product of the three support sets. In other words, each variable is measured at a specific time instant in a specific location for a specific working crew. Therefore, the overall state of the attribute at a specific time instant is related to the performance and location of the working crew at that time. This fine classification allows for a complete identification of the reasons and factors that are responsible for a measured state of an attribute. This process enables construction control, and results in much more precise and accurate corrective actions. Table 1.1 summarizes different potential states for each variable together with observation channels (o_i), a specific variable (v_i), and corresponding general variables (v_i'). This example is based on the assumption that personnel with poor experience are not used in the construction activities. The observation channel is taken as a maximum operator to obtain the specific variable (v_i). For example, using the maximum operator on *poor* produces 2 from Figure 1.6. The mapping from v_i to v_i' is a one-to-one mapping that can be made for abstraction purposes to some generalized states. The tabulated values under v_i' in Table 1.1 were selected arbitrarily for demonstration purposes of such a mapping. Table 1.2 summarizes the different elements for each support set. Table 1.3 shows the overall support set for a combination of two of the supports considered in this example of time and space. For

TABLE 1.1
States of Variables

Variable	States	Observation Channel o_i	Specific Variable v_i	Mapping Type	General Variable v_i'
v_1	Poor	Maximum	2	One-to-one	0
	Fair	Maximum	5	One-to-one	1
	Good	Maximum	8	One-to-one	2
	Moderate	Maximum	9	One-to-one	3
	Excellent	Maximum	10	One-to-one	4
v_2	Traditional method	One-to-one	Method 1	One-to-one	10
	Slip form method	One-to-one	Method 2	One-to-one	20
	Precast method	One-to-one	Method 3	One-to-one	30

TABLE 1.2
Elements of the Different Support Sets

Support	Specific Element	Mapping Type	General Element
Time	Week 1	One-to-one	11
	Week 2	One-to-one	21
	Week 3	One-to-one	31
	Week 4	One-to-one	41
	Week 5	One-to-one	51
	Week 6	One-to-one	61
	Week 7	One-to-one	71
	Week 8	One-to-one	81
Space	Columns	One-to-one	12
	Beams	One-to-one	22
	Slabs	One-to-one	32
	Footings	One-to-one	42
Population	Falsework crew	One-to-one	13
	Rebar crew	One-to-one	23
	Concreting crew	One-to-one	33
	Finishing crew	One-to-one	43

example, the pair [12, 11] in Table 1.3 indicates columns (i.e., general element 12 according to Table 1.2) and week 1 (i.e., general element 11 according to Table 1.2).

The source system defined as such is classified as neutral since an input–output identification was not considered. The variables used herein are discrete. The time support set is linearly ordered, while the space and population support sets are not ordered. Observation channels for variable v_1 are linearly ordered, while those for variable v_2 are not ordered. Observation channels for variable v_1 are fuzzy, while those for variable v_2 are crisp. Figure 1.7 shows a procedure diagram of the source system for this example.

TABLE 1.6
The Data Matrix of Method of Construction (v_2)

	Time (Week)							
Space	11	21	31	41	51	61	71	81
12	10	10	10	20	20	20	20	20
22	20	20	20	10	10	10	20	20
32	10	10	20	20	20	10	10	10
42	10	10	10	10	10	10	10	10

TABLE 1.7
A Behavior Function Evaluation for Variables v_1 and v_2

	Variable v_1		Variable v_2		Degree of Belief of Overall State (C)	Likelihood of Occurrence (N_c)	Behavior Function (f_B)
Overall State (C_i)	State	Degree of Belief	State	Degree of Belief			
C_1 (1, 10)	1	0.8	10	1	0.8		
	1	0.7	10	1	0.7		
	1	0.5	10	1	0.5	2.3	0.354
	1	0.3	10	1	0.3		
C_2 (3, 10)	3	0.4	10	1	0.4		
	3	0.7	10	1	0.6	1.7	0.262
	3	0.6	10	1	0.7		
C_3 (2, 10)	2	0.5	10	1	0.5		
	2	0.8	10	1	0.8	2.2	0.338
	2	0.9	10	1	0.9		
C_4 (0, 10)	0	0.2	10	1	0.2		
	0	0.1	10	1	0.1	0.3	0.046

was accompanied by state 10 for the construction method variable (v_2), as shown in the table. Accordingly, the overall state C_1 = (1, 10) has support-variant degrees of belief. Using a minimum operator, for example, as an aggregation function, the degree of belief of state (C_1) can be calculated at the different support instants as shown in Table 1.7. In other words, the degree of belief of the combination of states 1 and 10 is the minimum of the two degrees of belief of the separate states. It should be noted that since variable v_2 is a crisp variable. Its degree of belief was taken to be one at any support instant. The likelihood of occurrence of each overall state C_1 was then calculated as follows:

$$N_c = \sum_{all\, t} d_{s,t} \tag{1.6}$$

where N_c = likelihood of occurrence, $d_{s,t}$ = aggregated degree of belief of state(s) at support instant t, and the summation was performed over the support instances. The corresponding probability of overall state (C_1) is then calculated using the following formula (Klir, 1969, 1985):

$$f_B(C_1) = \frac{N_c}{\sum_{all\,c} N_c} \tag{1.7}$$

where $f_B(C_1)$ = probability of having state C_1, which corresponds to the value of the behavior function for that state; N_c = likelihood of occurrence of state C_1; and the summation was performed over all the overall states. The expressions provided by Equations 1.6 and 1.7 were chosen for illustration purposes. The resulting probabilities (f_B) for selected potential overall states are shown in Table 1.7.

A state transition system can be expressed as

$$E_n = (I, K, f_n) \tag{1.8}$$

where E_n = a state transition system, I = the corresponding general image system, K = the chosen mask, and f_n = the state transition function. An important interpretation of the state transition concept in construction is the state table approach as used by Abraham et al. (1989). The state table format could be viewed as a *state transition function* in a feedback control framework. Table 1.8 shows an example of such a table that describes the process of giving some command, the current state of a certain variable, the next state for the same variable, and feedback information for control purposes. The main concept in this framework is the relationship developed through the table, between the consecutive states of the different variables. Figure 1.8 shows a procedure diagram for constructing a generative system. It should be emphasized here that although variables 1 and 2 for attribute 1 have the same names as variables 3 and 4 for attribute 2, this does not mean that they would take the same values. In other words, the same variables have different impacts on different attributes according to the nature of each attribute.

TABLE 1.8
A State Table Format

Command	State	Feedback	Next State	Output	Report
Place concrete for a foundation	Forms without concrete	Concrete overflow	Structural member	Concrete member	Concrete quantities
Place concrete for a column	Forms without concrete	Concrete overflow	Structural member	Concrete member	Concrete quantities
Place concrete for a beam	Forms without concrete	Concrete overflow	Structural member	Concrete member	Concrete quantities
Place concrete for a slab	Forms without concrete	Concrete overflow	Structural member	Concrete member	Concrete quantities

Source: Adapted from Abraham, D.M. et al., *J. Comput. Civ. Eng.*, 3, 320–332, 1989.

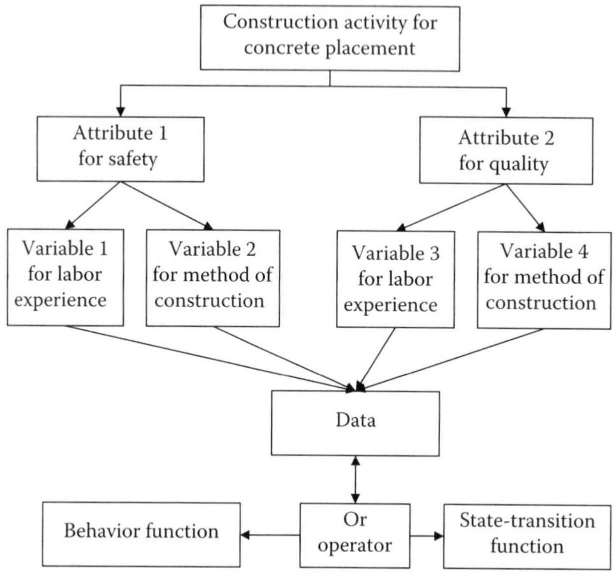

FIGURE 1.8 A generative system for a construction activity.

Structure Systems

Structure systems are sets of smaller systems or subsystems, as illustrated in Figure 1.9. The subsystems could be source, data, or generative systems. In this example, the construction activity consists of a number of processes that should be accomplished. These processes depend on each other in some manner. Considering concrete placement as a construction activity, the different processes involved include falsework construction, rebar placement, concrete pouring, and concrete finishing. These processes represent interrelated subsystems within the structure system. Each process is defined as a generative system. The interrelation among the subsystems represents the dependence of each process on the preceding one. Another form of the interrelationship is the input–output relation between the successive processes. A nested structure system could also be defined on the same example by defining each of the subsystems as another structure system whose elements are generative systems. It should be noted that each of the described subsystems and their corresponding elements should be defined on the same source system.

Metasystems

The construction activity is defined as a structure system whose elements are metasystems. Each metasystem represents the change in its behavior system. For a concrete placement activity, the processes include falsework construction, rebar placement, and concrete pouring. However, in order to represent the actual behavior of this system within the overall support set required, the behavior system in this case can only be defined using more than one behavior function. Each one of these functions is valid for only a subset of the overall support set. Stating the problem in this format, a metasystem should be defined to describe the change in each one of these subsystems.

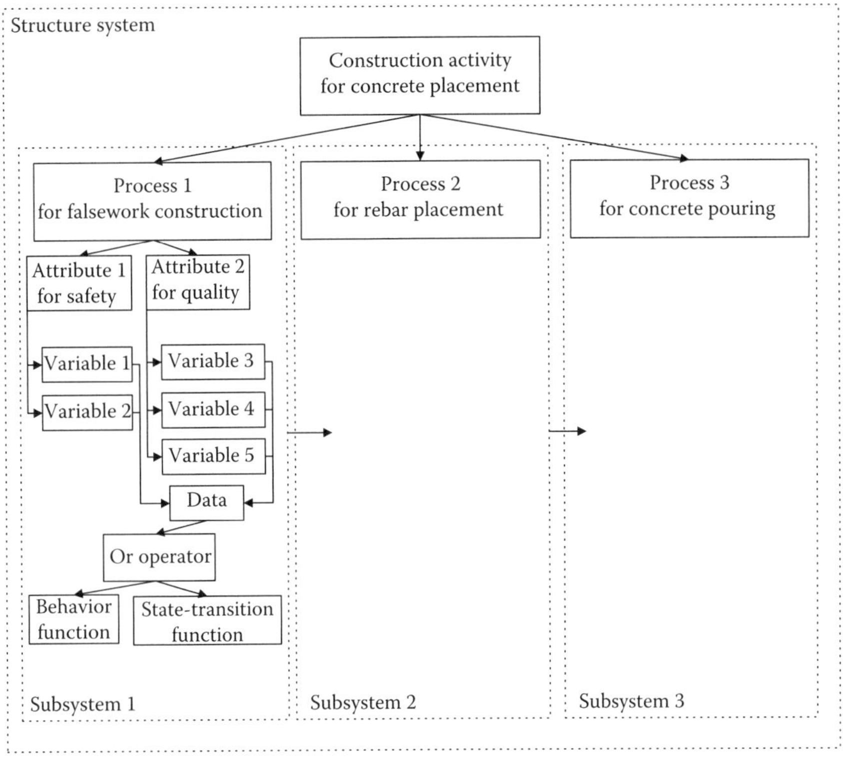

FIGURE 1.9 A structure system for a construction activity.

The replacement procedure is required in order to decide which behavior function should be used. This decision should be taken based on the states of some basic variables specified for this purpose. Referring to the behavior functions previously defined, i.e., the probability/possibility distributions, more than one distribution might be necessary to fit the available data. Each one of these distributions is valid within a subset of the overall support set. For example, some distribution might fit variable v_1 for the first month, i.e., 4 weeks, while a different distribution might more appropriately represent the same variable during the next four weeks. Thus, a replacement procedure is required in order to calculate the current time and choose the appropriate distribution that represents the data during this period. Figure 1.10 shows a graphical representation of a metasystem.

The metasystem can be defined as a structure system whose elements are behavior systems. Applying this concept to the example under consideration, a metasystem is defined on the construction activity that represents a structure system. In other words, the construction activity is the structure system with the different processes defined as subsystems. This structure system changes with time, where time is an example support. At some instant of time, falsework construction and rebar replacement as two subsystems might be in progress, whereas the other processes, i.e., concrete pouring and concrete finishing, might not be started yet. Therefore, the components of the structure system at this instant of time do not include the ones that have not been started. This

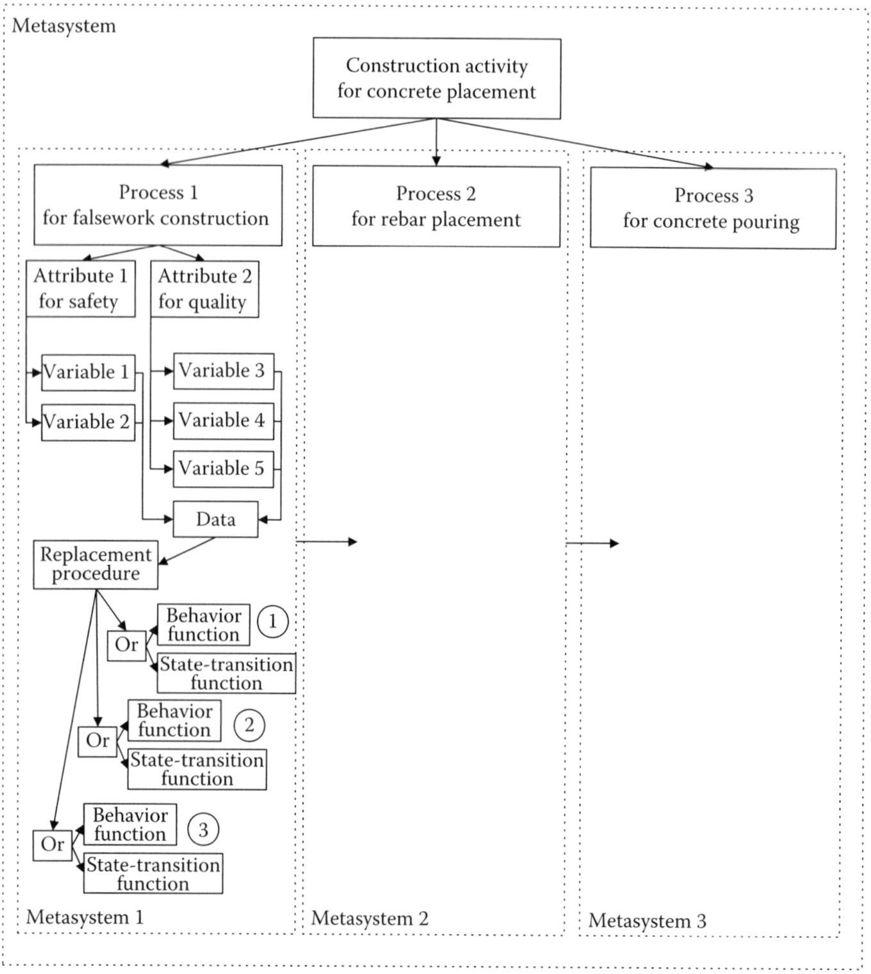

FIGURE 1.10 A metasystem for a construction activity.

composition would probably be different after some time, where all the processes might be in progress at the same time. The replacement procedure in this case should observe the change in each process such that the composition of the structure system could be defined at any support instant.

Example 1.2 System Definition of an Office Building

An information-based system can be defined for an office building in a hierarchical manner consisting of five knowledge levels, in which the higher level contains more details than the lower levels. For an office building, each level is briefly defined as follows:

- *The source system* defines the physical objects comprising an office building for the purpose of structural analysis. The objects herein include the column, beams, slabs, connections, occupancy loads, and the ground assumed as a fixed support. The system boundaries herein exclude the soil properties and the soil–structure interactions.

- *The data system* contains the actual data relating to the defined source system. For example, the data include member sizes, material properties, and structural details.
- *The generative systems* define and relate the constraints among the variables of the system and their use to predict behavior attributes of interest. For example, the stiffness matrix and its use to compute member forces and stresses are included in this system definition level.
- *The structure systems* are sets of subsystems. For an office building, these subsystems are the stiffness matrices of all the components assembled in the global stiffness matrix, loads and their characteristics, and other components, such as window and cladding.
- *The metasystem* describes the changes in the lower-knowledge-level subsystems due to some time-dependent characteristics or subsystems included in the definitions, for example, changes in climate or environmental conditions, properties of construction materials, and degradation of materials.

1.2.4 Disciplinary Roots of Systems Science

Systems science as a phenomenon emerged from what is usually referred to as *systems movement*. In general, systems movement may be characterized as a loose association of people from different disciplines of science, engineering, philosophy, and other areas, who share a common interest in ideas (concepts, principles, methods, etc.) that are applicable to all systems and that, consequently, transcend the boundaries between traditional disciplines.

Systems movement emerged from three principal roots: *mathematics*, *computer technology*, and a host of ideas that are well captured by the general term *systems thinking*. Mathematics is clearly a source of the various systemhood properties. It also provides us with methodological tools pertaining to these properties. While the classical (analytic) mathematical methods are applicable only to problems that involve a small number of variables related to each other in a predictable way, the applicability of statistical methods has exactly opposite characteristics: they require a very large number of variables and a high degree of randomness. These two types of methods are thus highly complementary. When one of them excels, the other totally fails. Unfortunately, despite their complementarity, these types of methods cover only problems that are clustered around the two extremes of the complexity and randomness scales. Most problems are somewhere in between these two extremes. They are typical in life, behavioral, social, and environmental sciences, as well as in applied fields such as modern technology or medicine.

To deal with the broader range of problems mathematically required more expressive mathematical theories, and that required, in turn, to generalize existing theories. The following are some of these generalizations:

- From quantitative theories to *qualitative theories*
- From functions to *relations*

- From classical (additive) measures to *nonadditive measures*
- From classical set theory to *fuzzy set theory*
- From graphs to *hypergraphs*
- From ordinary geometry (Euclidean as well as non-Euclidean) to *fractal geometry*
- From ordinary automata to *dynamic cellular automata*
- From linear theories to *nonlinear theories*
- From regularities to *singularities* (*catastrophe theory*)
- From precise analysis to *interval analysis* and to *fuzzy analysis*
- From regular languages to *developmental languages*
- From special algebras to *universal algebra* and to *category theory*
- From single-objective criteria optimization *to multiple-objective criteria optimization*

Each generalization of a mathematical theory usually results in a conceptually simpler theory. This is a consequence of the fact that some properties of the former theory are not required in the latter. At the same time, the more general theory always has a greater expressive power, which, however, is achieved only at the cost of greater computational demands. This explains why these generalizations are closely connected with the emergence of computer technology and steady increases in computing power.

In addition to mathematics and computer technology, systems science has also been influenced by a host of ideas for which we use the general term *systems thinking*. Perhaps the most important of them are ideas associated with *holism*, which emerged at the beginning of this century as an antithesis of *reductionism*, a methodological view predominant in science since about the 16th century. The latter claims that properties of a whole are explicable in terms of properties of constituent elements. Holism rejects this claim and maintains that a whole cannot be analyzed in terms of its parts without some residuum.

There were other developments besides holism that paved the way for systems science. Among them was the increasing awareness that there were many phenomena and problems that could not be studied within the boundaries of individual disciplines of science. This eventually leads to the emergence of *interdisciplinary areas* such as biophysics, biochemistry, physiological psychology, or social psychology. The existence of these interdisciplinary areas was probably the first step leading to the recognition that systems may be defined across disciplinary boundaries. We may say that it was the first step in recognizing the notion of systemhood. Another step was the recognition of analogies (isomorphies) between systems describing different phenomena (e.g., mechanical, electrical, acoustic, thermal), which made it possible to construct and use analog computers and resulted eventually in the formulation of the *theory of similarity*. Further progress was made by actually identifying some fundamental systemhood properties, such as information and control in Wiener's *Cybernetics* (1948).

The ideas of holism, the emergence of interdisciplinary areas in science, and the increasing recognition of the existence and utility of isomorphies between disciplines of science created a growing awareness among some scholars that certain concepts, ideas, principles, and methods were applicable to systems in general, regardless of their disciplinary categorization. This eventually led to the notions of general systems, general theory of systems, general systems research, and the like. These notions, originated and promoted primarily by Ludwig Von Bertalanffy (1968), formed an intellectual basis from which the organized systems movement emerged.

1.2.5 Systems Knowledge, Methodology, and Metamethodology

As explained previously, systems science is not a new science in the traditional sense, but rather a new dimension in science. Yet, the two dimensions of science have significant parallels: systems science, as any of the classical sciences, contains a body of knowledge regarding its domain, a methodology for acquisition of new knowledge and for dealing with relevant problems within the domain, and a metamethodology, by which methods and their relationship to problems are characterized and critically examined.

In every traditional discipline of science, we develop systems models of various phenomena of the real world. Each of these models, when properly validated, represents some specific knowledge regarding the relevant domain of inquiry. In systems science, the domain of inquiry consists of knowledge structures themselves — the various categories of systems that emerge from the conceptual framework employed. That is, the objects of investigation in systems science are not objects of reality, but systems of certain specific types.

Knowledge pertaining to systems science, or *systems knowledge*, is thus different from knowledge in traditional science. It is not knowledge regarding various aspects of reality, as in traditional science, but rather knowledge regarding the various types of systems in terms of which knowledge in traditional science is organized. That is, it is *knowledge concerning knowledge structures*. This knowledge is, of course, applicable to the processes of acquisition, management, and utilization of knowledge in every discipline of traditional science.

Systems knowledge can be obtained either mathematically or experimentally. Mathematically derived systems knowledge is the subject of the various mathematical systems theories, each applicable to some class of systems. It consists of theorems regarding issues such as controllability, stability, state equivalence, information transmission, decomposition, homomorphism, self-organization, self-reproduction, and many others.

Systems knowledge can also be obtained experimentally. Although systems (knowledge structures) are not objects of reality, they can be simulated on computers and in this sense made real. We can then experiment with the simulated systems for the purpose of discovering or validating various hypotheses in the same way as other scientists do with objects of their interests in their laboratories. In this sense, computers may be viewed as laboratories of systems science. Experimentation with systems on computers is not merely possible, but it may give us knowledge that is otherwise unobtainable.

The computer has, in fact, a dual role in systems science. In one of the roles, it is a methodological tool for dealing with systems problems. In the other role, it serves as a laboratory for experimenting with systems. The purpose of this experimentation is to discover or validate laws of systems science. In contrast to laws of nature, laws of systems science characterize properties of various categories of systems rather than categories of real-world objects. We perform experiments of some kind on the computer with many different systems of the same category. The aim of this experimentation is to discover useful properties characterizing the category of systems or, alternatively, to validate some conjectures regarding the category.

Experimentation with systems on computers to expand systems knowledge is only one of two sides of systems methodology. The other side consists of methods developed for dealing with various systems problems. These are problems that involve primarily systemhood properties.

An expertise in systems knowledge and systems problem-solving methodology may, in principle, be implemented on a computer in the form of an expert system. Such an expert system is complementary to the usual expert system, which is predominantly oriented to thinghood expertise in some special area of classical science, engineering, or some other profession. The two types of expert systems together should form a far better computer support for dealing with overall problems than either of them alone.

The *primacy of problems* in systems methodology is in sharp contrast with the *primacy of methods* in applied mathematics. It is the most fundamental commitment of systems methodology to develop methods for solving genuine systems problems in their natural formulation. Simplifying assumptions, if unavoidable, are introduced carefully, for the purpose of making the problem manageable, and yet distort it as little as possible. The methodological tools for dealing with a problem are of secondary importance; they are chosen in such a way as to best fit the problem rather than the other way around. Moreover, the tools need not be only mathematical in nature. They may consist of a combination of mathematical, computational, heuristic, experimental, or any other desirable methodological traits.

In order to choose an appropriate method for a specific problem, relevant characteristics of prospective methods must be determined. This is a subject of *systems* metamethodology — the study of systems methods as well as methodologies (integrated collections of methods). Let any particular study whose aim is to determine some specific characteristics of a method (or methodology) be called a *metamethodological inquiry*. Examples of the most fundamental characteristics of methods, which are relevant to virtually all problems, are computational complexity, performance, and generality of the methods involved.

Computational complexity is a characterization of the time or space (memory) requirements for solving a problem by a particular method. Either of these requirements is usually expressed in terms of a single parameter that represents the size of the problem, e.g., the number variables in the given systems. The dependence of the required time or memory space on the problem size is usually called a *time complexity function* or *space complexity function*, respectively. Either of these functions can be used for comparing different methods for dealing with the same problem type.

Performance of a method is characterized by the degree of its success in dealing with the class of problems to which it is applicable. It can be expressed in various ways, typically by the percentage of cases in which the desirable solution is reached, by the average closeness to the desirable solution, or by a characterization of the worst-case solution. Methods whose performance is not perfect are usually called heuristic methods. They are employed as a means for reducing computational complexity.

Generality of a method is determined by the assumptions under which it operates, e.g., by the axioms of a mathematical theory upon which the method is based. A particular set of assumptions, upon which several different methods may be based, is often referred to as a *methodological paradigm*. Each assumption contained in a methodological paradigm restricts the applicability of the associated methods and, consequently, restricts the set of possible solutions in some specific way.

In some instances, characteristics of methods or methodologies can be obtained mathematically. For example, worst-case complexity functions have been determined for many methods involved in systems problem solving. In many cases, however, mathematical treatment is not feasible. For example, it is often impossible to determine mathematically the performance of a heuristic method or the complexity function of a method for typical (average) problems of a given type. One way of obtaining the desired characteristics in these cases is to perform experimental investigations of the methods involved. That is, the application of the investigated method or methodology is simulated on a computer for a set of typical problems of a given type. Results obtained for these problems are then summarized in a desirable form to characterize the method or methodology and, possibly, compare it with its various competitors.

1.2.6 Complexity and Simplification of Systems

Our most troubling long-range problems, such as economic forecasting and trade balance, defense systems, and genetic modeling, center on systems of extraordinary complexity. The systems that host these problems — computer networks, economics, ecologies, and immune systems — appear to be as diverse as the problems. Humans as complex, intelligent systems have the ability to anticipate the future and learn and adapt in ways that are not yet fully understood. Engineers and scientists, who study or design systems, have to deal with complexity more often than ever, hence the interest in the field of complexity. Understanding and modeling system complexity can be viewed as a pretext for solving complex scientific and technological problems, such as finding a cure for the acquired immune deficiency syndrome (AIDS) or solving long-term environmental issues or using genetic engineering safely in agricultural products. The study of complexity led to, for example, chaos and catastrophe theories. Even if complexity theories did not produce solutions to problems, they can still help us to understand complex systems and perhaps direct experimental studies. Theory and experiment go hand in glove, therefore providing opportunities to make major contributions.

Complexity is perhaps as important a concept for systems science as the concept of a system. It is a difficult concept, primarily because it has many possible meanings. While various specific meanings of complexity have been proposed and discussed on many occasions, there is virtually no sufficiently comprehensive study that attempts to capture its general characteristics.

To begin with a broad perspective, a dictionary (*Webster's Third International Dictionary*) defines the quality or state of being complex as follows:

- "Having many varied interrelated parts, patterns, or elements and consequently hard to understand fully"
- "Marked by an involvement of many parts, aspects, details, notions, and necessitating earnest study or examination to understand or cope with"

These commonsense characterizations of complexity do not contain any qualification regarding the kind of entities to which they are applicable. As such, the entities could be material or abstract, natural or human-made, products of art or science. Regardless of their types, the degree of complexity according to the commonsense characterizations is associated with the number of recognized parts as well as the extent of their interrelationship; in addition, complexity is given a somewhat subjective connotation since it is related to the ability to understand or cope with the thing under consideration. The commonsense characterizations of complexity assume an interaction between an object (a part of the world that may have "many varied interrelated parts") and a human being (or, perhaps, a computer), for whom it may be difficult "to understand or cope with" the object. This means that the complexity of an object for a particular human being depends on the way he interacts with it (i.e., on his interests and capabilities). More poetically, we may say that the complexity of an object is in the eyes of the observer.

In most cases, there is virtually an unlimited number of ways one can interact with an object. As a consequence, the interaction is almost never complete. It is based on a limited (and usually rather small) number of attributes that the observer is capable of distinguishing on the object and that are relevant to his interests. These attributes are not available to the observer directly, but only in terms of their abstract images, which are results of perception or some specific measurement procedures. These abstract images are usually called variables. When a set of variables is established as a result of our interaction with an object of interest, we say that a system (or, more precisely, a source system) is distinguished on the object.

Since we deal with systems distinguished on objects and not with the objects themselves (in their totality), it is not operationally meaningful to view complexity as an intrinsic property of objects. While complexities of objects may exist in the ontological sense, they are epistemologically and methodologically vacuous, in contrast to complexities of systems.

Two *general principles of systems complexity* can be recognized; they are applicable to any of the systems types and can thus be utilized as guidelines for a comprehensive study of systems complexity. According to the *first general principle*, the complexity of a system (of any type) should be proportional to the amount of information required to describe the system. Here, the term *information* is used solely in a syntactic sense; no semantic or pragmatic aspects of information are employed. One way of expressing this *descriptive complexity* is to define it by the size of the shortest description of the system in some standard language or, alternatively, the size of the smallest program in a standard language by which the system can be simulated on a canonical universal computer. The primary advantage of this definition of descriptive complexity is that it is theoretically sound and applicable to all systems, regardless of their classification. Its primary weakness is methodological: it is rather difficult to determine in many cases the shortest description of a system. According to the *second general principle*, systems complexity should be proportional to the amount of the information needed to resolve any uncertainty associated with the system involved for predictive, retrodictive, or prescriptive purposes. Here, again, syntactic information is used, but information that is based on a measure of uncertainty.

Uncertainty is an inherent property of nondeterministic systems. Such systems describe situations that offer multiple choices. Several mathematical theories are now available within which various types of uncertainty can be formalized and measured.

Both descriptive complexity and uncertainty-based complexity are connected with information, i.e., information needed to describe a system (descriptive or algorithmic information) and information needed to resolve uncertainty embedded in it (uncertainty-reducing information). Simplifying a system, then, should be based on reducing both the complexity based on descriptive information and the complexity based on the uncertainty information. Unfortunately, these two complexities conflict with each other. In general, when we reduce one, the other increases or, at best, remains unchanged. Hence, a general problem of simplification is multiobjective criteria optimization.

One way of reducing the descriptive complexity of a system of any type is to exclude some variables from the system. *Excluding variables* from any relation reduces the relation in two ways. First, its dimension is reduced since some sets in its Cartesian product are excluded. Second, its cardinality is reduced since overall states that were distinguished only by the excluded variables become equivalent.

Reducing the uncertainty-based complexity involves an inverse procedure. That is, it requires adding input variables to the system, which contribute, at least potentially, some information that in turn reduces the uncertainty regarding the output variables.

Another way of reducing descriptive complexity of a system is to partition states of some variables of the system into equivalence classes and to replace each equivalence class with one state. This simplification strategy is usually referred to as *coarsening* or *quantizing* of variables; it reduces cardinalities of relations involved, but it leaves their dimensions intact. While descriptive complexity is always reduced by coarsening of variables, uncertainty-based complexity may be affected by coarsening of variables in either way.

An important strategy for reducing descriptive complexity of a system is to break the system down into appropriate subsystems. That is, a particular overall system is approximated with a structure system. The key issue in employing this strategy is to minimize the increase in uncertainty (or loss of information) while achieving the desired reduction of descriptive complexity. This process is called *reconstructability analysis* (Klir, 1985).

Conceptualizing systems as structure systems, possibly of higher orders, is certainly an efficient way of managing complexity. Such systems are organized hierarchically, with each system consisting of a network of interconnected subsystems, each of which consists again of a network of its own subsystems, and so on, until some ultimate subsystems are reached that are not further divided into more primitive subsystems. The power of organizing systems hierarchically has been recognized and utilized with great success in the sciences of the artificial. This organizing principle has undoubtedly been one of the basic tools of good designers, artists, and managers.

It has also played a key role in the process of developing efficient mass production by allowing the division of labor in manufacturing complex products.

The significance of hierarchically organized systems has also been recognized for a long time in the sciences of the natural; e.g., virtually all complex systems that we recognize in the real world (that is, models of the real world) have the tendency to organize hierarchically. Thus, for example, biological cells seem to group naturally into organs, while organs group into organisms, organisms group into populations of animals, and the animals group into ecosystems. The fact that we tend to perceive the world as hierarchically organized might have some ontological significance, but it may as well be solely of epistemological nature, reflecting the way the human brain and mind have evolved to deal with the complexity of the real world. Regardless of its ontological significance, it is undeniable that hierarchically organized systems play an important pragmatic role in our comprehension and management of reality, be it natural or human-made.

Another way of dealing with very complex systems, perhaps the most significant one, is to allow imprecision in describing properly aggregated data. Here, the imprecision is not of a statistical nature, but rather of a more general modality, even though the possibility of imprecise statistical descriptions is included as well. The mathematical frameworks for this new modality are, as already mentioned, the theories of fuzzy sets and nonadditive measures.

1.2.7 Computational Complexity and Limitations

The two types of complexity introduced thus far, the descriptive complexity and the uncertainty-based complexity, pertain to systems. Yet another face of complexity exists, a complexity that pertains to systems problems. This complexity, which is usually referred to as *computational complexity*, is a characterization of the time or space (memory) requirements for solving a problem by a particular algorithm. Either of these requirements is usually expressed by a function, f, of a single parameter, n, that represents the size of the problem. This function is called a *time* (or *space*) *complexity function*. The main distinction is between algorithms whose complexity function can be expressed in terms of a polynomial as follows:

$$f(n) = a_k n^k + a_{k-1} n^{k-1} + \dots + a_1 n + a_0 \tag{1.9}$$

for some positive integer k, and algorithms for which $f(n)$ is expressed by an exponential form, e.g., 2^n, 10^n, $2^{e(n)}$, where $f(n)$ is an exponential function of n. Due to the essential differences between polynomial and exponential time complexity functions, polynomial time algorithms are considered *efficient*, while exponential time algorithms are considered *inefficient*. As a consequence, problems for which it can be proven that they are not solvable by polynomial time algorithms are viewed as *intractable*, while problems for which polynomial time algorithms are known are viewed as *tractable*. Computational complexity has been extensively investigated since the early 1970s. Many important results regarding tractability of various systems problems are covered in the classic book on computational complexity by Garey and Johnson (1979).

From a broader and more realistic perspective, the size of a problem instance is not the only determinant of its computational complexity. That is, problem instances of the same type and size may have very different computational demands. Most studies in the area of computational complexity are oriented primarily to the characterization of the worst-case problem instances. Although this orientation is theoretically sound, it usually results in estimates that are rarely reached in practice and are therefore too pessimistic. To ameliorate this situation, the worst-case estimates are sometimes supplemented with average-case estimates. However, such estimates are based on the assumption that all problem instances are equally likely, which does not necessarily reflect the actual probability distribution of problem instance encountered in practice. The problem of determining the actual distribution for various problem types is predominantly an empirical problem. This problem can be studied, in principle, by monitoring and analyzing problem instances requested by users of the various systems problem packages. Any such study is an example of a metamethodological inquiry.

Complexity can also be classified into two broad categories: (1) complexity with structure and (2) complexity without structure. The complexity with structure was termed *organized complexity* by Weaver (1948). Organized complexity can be observed in a system that involves nonlinear differential equations with a lot of interactions among a large number of components and variables that define the system, such as in life, behavioral, social, and environmental sciences. Such systems are usually nondeterministic in their nature. Advancements in computer technology and numerical methods have enhanced our ability to obtain solutions of these problems effectively and inexpensively. As a result, engineers design complex systems in simulated environments and operations, such as a space mission to a distant planet, and scientists can conduct numerical experiments involving, for example, nuclear blasts. In the area of simulation-based design, engineers are using parallel computing and physics-based modeling to simulate fire propagation in engineering systems, or the turbulent flow of a jet engine using molecular motion and modeling (Garey and Johnson, 1979). These computer and numerical advancements are not limitless, as the increasing computational demands lead to what is termed *transcomputational problems* capped by the *Bremermann's limit* (Bremermann, 1962). This Bremermann's limit, which was derived on the basis of quantum theory, is expressed by the following proposition (Bremermann, 1962):

> No data processing systems, whether artificial or living, can process more than 1.36×10^{47} bits per second per gram of its mass.

Here, data processing is defined as transmitting bits over one or several of a system's channels. Bremermann (1962) provides additional information on the theoretical basis for this proposition. Considering a hypothetical computer that has the entire mass of the Earth (6×10^{27} g) operating for a time period equal to an estimated age of the Earth (3.14×10^{17} seconds), this imaginary computer would be able to process 2.56×10^{92} bits, or rounded to the nearest power of 10, 10^{93} bits, defining the Bremermann's limit. Many scientific and engineering problems defined with a lot of details can exceed this limit. Klir and Folger (1988) provide the examples of

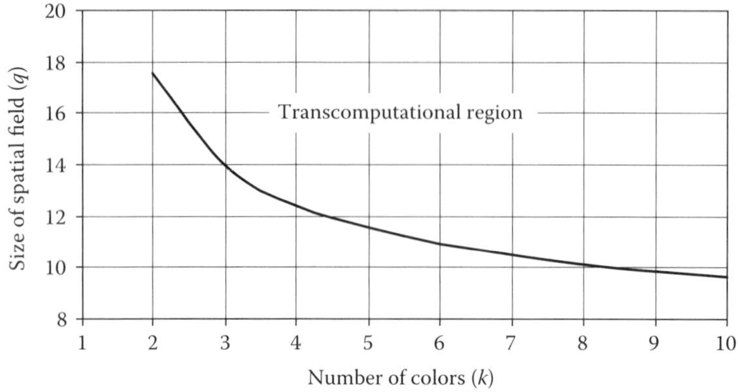

FIGURE 1.11 The Bremermann's limit for pattern recognition.

pattern recognition and human vision that can easily reach transcomputational levels. In pattern recognition, consider a square $q \times q$ spatial array defining $n = q^2$ cells that partition the recognition space. Pattern recognition often involves color. Using k colors, as an example, the number of possible color patterns within the space is k^n. In order to stay within the Bremermann's limit, the following inequality must be met:

$$k^n \leq 10^{93} \tag{1.10}$$

Figure 1.11 shows a plot of this inequality for values of $k = 2$ to 10 colors. For example, using only two colors, a transcomputational state is reached at $q = 18$ cells. These computations in pattern recognition can be directly related to human vision and the complexity associated with processing information by the retina of a human eye. According to Klir and Folger (1988), if we consider a retina of about 1 million cells, with each cell having only two states of *active* and *inactive* in recognizing an object, modeling the retina in its entirety would require the processing of

$$2^{1,000,000} \approx 10^{300,000} \tag{1.11}$$

bits of information, far beyond the Bremermann's limit.

Organized complexity in nature offers another interesting aspect of complexity in that it can be decomposed into an underlying repeated unit (Flake, 1998). For example, economic markets that defy prediction, the pattern recognition capabilities of any of the vertebrates, the human immune system's response to viral and bacterial attack, and the evolution of life on our planet are emergent in that they contain simple units that, when combined, form a more complex whole. They are examples of the whole of the system being greater than the sum of the parts, which is a fair definition of holism — the very opposite of reductionism. They are similar to an ant colony. Although a single ant exhibits a simple behavior that includes a very small number of tasks, depending on its caste, such as foraging for food, caring for the queen's brood, tending to the upkeep of the nest, defending against enemies, or, in the case of the queen, lay eggs, the behavior of the ant colony as a whole is very complex. The ant colony includes millions of workers that can sweep whole regions clean of animal life, and the fungus-growing ants that collect vegetable matter as food for symbiotic fungi and then harvest a portion of the fungi as food for the colony. The physical structure of the colony that ants build often contains thousands of passageways and appears mazelike to human eyes but are easily navigated by the inhabitants. The point herein is that an ant colony is more than just a bunch of ants. An organized complexity exists that is challenging to scientists. Knowing how each caste in an ant species behaves would not enable a scientist to magically infer that ant colonies possess so many sophisticated patterns of behavior.

Another dimension to this complexity is that agents that exist on one level of understanding are very different from agents on another level; for example, cells are not organs, organs are not animals, and animals are not species. The interactions on one level of understanding are often very similar to the interactions on other levels, as illustrated by finding:

- Self-similar structures in biology, such as trees, ferns, leaves, and twigs
- Self-similarity in inanimate objects, such as snowflakes, mountains, and clouds

Figure 1.12 shows an example of a plantlike structure constructed of mid-length branching fractals using a constant angle. The figure illustrates the progression of self-similarity to produce structures that can be observed in nature. Flake (1998) provides many other examples similar to Figure 1.12. Examining this self-similarity might enhance our understanding of complexity, and might help us to unravel complexities associated with predicting the stock market and the weather. Is the challenge posed herein due to limited knowledge, or is it somehow inherent in these systems?

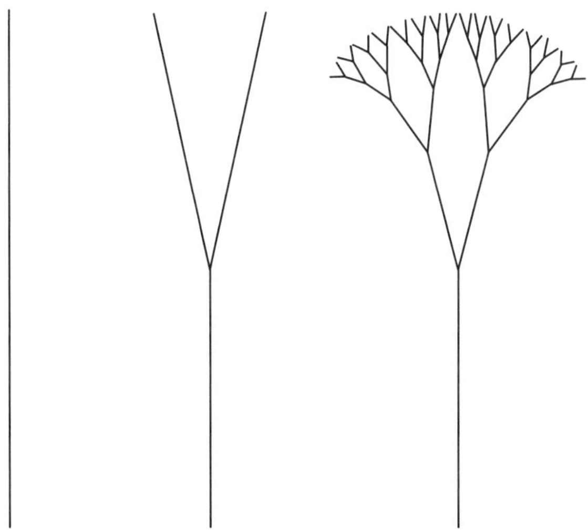

FIGURE 1.12 Plantlike branching fractals based on self-similarity.

This complexity is also self-organization, such as the collectives of ant colonies, human brains, and economic markets self-organizing to create enormously complex behavior that is much richer than the behavior of the individual component units. The complexity also exhibits evolution, learning, and the adaptation found in social systems.

Flake (1998) concludes that nature is frugal in that of all the possible rules that could be used to govern the interactions among agents, scientists are finding that nature often uses the simplest. Moreover, the same rules are repeatedly used in very different places.

Generally, an engineering system needs to be modeled with a portion of its environment that interacts significantly with it in order to assess some system attributes of interest. The level of interaction with the environment can only be subjectively assessed. By increasing the size of the environment and level of details in a model of the system, the complexity of the system model increases, possibly in a manner that does not have a recognizable or observable structure. This complexity without structure is more difficult to model and deal with in engineering and sciences. By increasing the complexity of the system model, our ability to make *relevant* assessments of the system's attributes can diminish. Therefore, there is a trade-off between relevance and precision in system modeling in this case. Our goal should be to model a system with a sufficient level of detail that can result in sufficient precision and can lean to relevant decisions in order meet the objective of the system assessment.

Living systems show signs of these trade-offs between precision and relevance in order to deal with complexity. The survival instincts of living systems have evolved and manifest themselves as processes to cope with complexity and information overload. The ability of a living system to make relevant assessments diminishes with the increase in information input, as discussed by Miller (1978). Living systems commonly need to process information in a continuous manner in order to survive. For example, a fish needs to process visual information constantly in order to avoid being eaten by another fish. When a school of larger fish rushes toward the fish, presenting it with images of threats and dangers, the fish might not be able to process all the information and images and becomes confused. Considering the information processing capabilities of living systems as input–output black boxes, the input and output to such systems can be measured and plotted in order to examine such relationships and any nonlinear characteristics that they might exhibit. Miller (1978) described these relationships for living systems using the following hypothesis, which was analytically modeled and experimentally validated:

> As the information input to a single channel of a living system — measured in bits per second — increases, the information output — measured similarly — increases almost identically at first but gradually falls behind as it approaches a certain output rate, the channel capacity, which cannot be exceeded. The output then levels off at that rate, and finally, as the information input rate continues to go up, the output decreases gradually towards zero as breakdown or the confusion state occurs under overload.

The above hypothesis was used to construct families of curves to represent the effects of information input overload, as shown schematically in Figure 1.13. Once

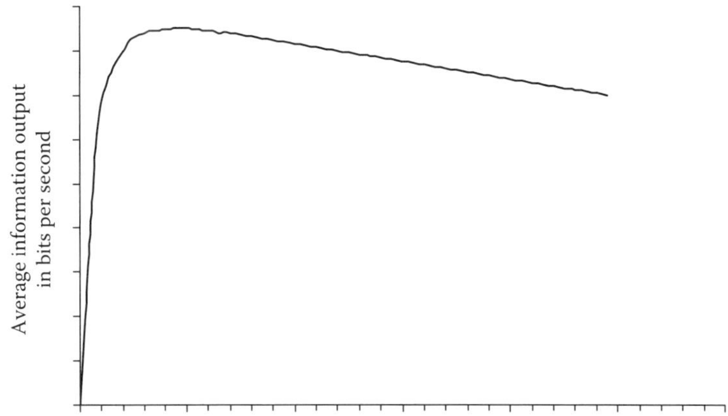

FIGURE 1.13 A schematic relationship of input and output information transmission rates for living systems.

the input overload is removed, most living systems recover instantly from the overload and the process is completely reversible; however, if the energy level of the input is much larger than the channel capacity, a living system might not fully recover from this input overload. Living systems also adjust the way they process information in order to deal with an information input overload using one or more of the following processes by varying degrees, depending on the level of a living system in terms of complexity:

1. Omission by failing to transmit information
2. Error by transmitting information incorrectly
3. Queuing by delaying transmission
4. Filtering by giving priority in processing
5. Abstracting by processing messages with less than complete details
6. Multiple channel processing by simultaneously transmitting messages over several parallel channels
7. Escape by acting to cut off information input
8. Chunking by transforming information into meaningful chunks

These actions can also be viewed as simplification means to cope with complexity or an information input overload.

1.3 KNOWLEDGE

1.3.1 Terminology and Definitions

Philosophers have concerned themselves with the study of knowledge, truth and reality, and knowledge acquisition since the Greek era of the early days of the Greek philosophers, such as Thales (c. 585 B.C.), Anaximander (611 to 547 B.C.), and

TABLE 1.9
Selected Knowledge and Epistemology Terms

Term	Definition
Philosophy	The fundamental nature of the world, the grounds for human knowledge, and the evaluation of human conduct
Epistemology	A branch of philosophy that investigates the possibility, origins, nature, and extent of human knowledge
Metaphysics	The investigation of ultimate reality; a branch of philosophy concerned with providing a comprehensive account of the most general features of reality as a whole, and the study of being as such; questions about the existence and nature of minds, bodies, God, space, time, causality, unity, identity, and the world are all metaphysical issues
• Ontology	A branch of metaphysics concerned with identifying, in the most general terms, the kinds of things that actually exist
• Cosmology	A branch of metaphysics concerned with the origin of the world
• Cosmogony	A branch of metaphysics concerned with the evolution of the universe
Ethics	A branch of philosophy concerned with the evaluation of human conduct
Aesthetics	A branch of philosophy that studies beauty and taste, including their specific manifestations in the tragic, the comic, and the sublime, where beauty is the characteristic feature of things that arouse pleasure or delight, especially to the senses of a human observer, and sublime is the aesthetic feeling aroused by experiences too overwhelming (i.e., awe) in scale to be appreciated as beautiful by the senses
Knowledge	A body of propositions that meet the conditions of justified true belief
Priori	Knowledge derived from reason alone
Posteriori	Knowledge gained by reference to the facts of experience
Rationalism	Inquiry based on *priori* principles, or knowledge based on reason
Empiricism	Inquiry based on *posteriori* principles, or knowledge based on experience

Anaximenes (c. 550 B.C.), who first proposed a rational explanation of the natural world and its powers. This section provides a philosophical introduction to knowledge, epistemology, their development, and related terminology to form a basis for analyzing and understanding ignorance and uncertainty.

Philosophy (*philosophia*) is a Greek term and literally means "love of wisdom." It deals with the careful thought about the fundamental nature of the world, the grounds for human knowledge, and the evaluation of human conduct. Philosophy, as an academic discipline, has chief branches that include logic, metaphysics, epistemology, and ethics. Selected terms related to knowledge and epistemology are defined in Table 1.9.

Philosophers defined knowledge, its nature, and methods of acquisition that evolved over time producing various schools of thought. Appendix A, which is adapted from Ayyub (2001), briefly summarizes these developments along a historical timeline, referring only to what was subjectively assessed as primary departures from previous schools. As new schools were introduced, they could be treated as new alternatives, since in some cases they could not invalidate previous ones.

1.3.2 Absolute Reality and Absolute Knowledge

The absolute reality of things is investigated in a branch of philosophy called *metaphysics*, which is concerned with providing a comprehensive account of the most general features of reality as a whole. The term *metaphysics* is believed to have originated in Rome about 70 B.C. by the Greek Peripatetic philosopher Andronicus of Rhodes in his edition of the works of Aristotle (384 to 322 B.C.).

Metaphysics typically deals with issues such as the ultimate nature of things, identification of objects that actually exist, things that compose the universe, the ultimate reality, the nature of mind and substance, and the most general features of reality as a whole. On the other hand, epistemology is a branch of philosophy that investigates the possibility, origins, nature, and extent of human knowledge. Metaphysics and epistemology are very closely linked and, at times, indistinguishable, as the former speculates about the nature of reality and the latter speculates about the knowledge of it. Metaphysics is often formulated in terms of three modes of reality, the *mind* (or *consciousness*), the *matter* (or *physical substance*), and a *higher nature* (one that transcends *both mind and matter*), according to three specific philosophical schools of thought: *idealism*, *materialism*, and *transcendentalism*, respectively.

Idealism is based on a theory of reality, derived from *Plato's Theory of Ideas* (427 to 347 B.C.), that attributes to *consciousness*, or the *immaterial mind*, a primary role in the constitution of the world. Metaphysical idealism contends that reality is mind dependent and that true knowledge of reality is gained by relying upon a spiritual or conscious source.

The school of *materialism* is based on the notion that all existence is resolvable into matter, or into an attribute or effect of matter. Accordingly, matter is the ultimate reality, and the phenomenon of consciousness is explained by physiochemical changes in the nervous system. In metaphysics, materialism is the antithesis of idealism, in which the supremacy of mind is affirmed and matter is characterized as an aspect or objectification of mind. The world is considered to be entirely mind independent, composed only of physical objects and physical interactions. Extreme or absolute materialism is known as *materialistic monism*, the theory that all reality is derived from one substance. Modern materialism has been largely influenced by the *theory of evolution*.

Plato developed the school of transcendentalism by arguing for a higher reality (metaphysics) than that found in sense experience, and for a higher knowledge of reality (epistemology) than that achieved by human reason. Transcendentalism stems from the division of reality into a realm of spirit and a realm of matter, and affirms the existence of absolute goodness characterized as something beyond description and as knowable ultimately only through intuition. Later, religious philosophers applied this concept of transcendence to divinity, maintaining that God can be neither described nor understood in terms that are taken from human experience. This doctrine was preserved and advanced by Muslim philosophers, such as Al-Kindi (800 to 873), Al-Farabi (870 to 950), Ibn Sina (980 to 1037), and Ibn Rushd (1128 to 1198), and adopted and used by Christian and Jewish philosophers, such as Aquinas (1224 to 1274), in the medieval period, as described in Appendix A.

The interest in knowledge and its theory resulted in the creation of a branch of philosophy called epistemology that investigates the possibility, origins, nature, and extent of human knowledge. It deals with issues such as the definition of knowledge and related concepts, the sources and criteria of knowledge, the kinds of knowledge possible and the degree to which each is certain, and the exact relation between the one who knows and the object known. Knowledge can be based on *priori*, knowledge derived from reason alone, and *posteriori*, knowledge gained by reference to the facts of experience. Epistemology can be divided into *rationalism*, inquiry based on *priori* principles or knowledge based on reason, and *empiricism*, inquiry based on a *posteriori* principles or knowledge based on experience.

In Section 1.2 realism and constructivism are discussed for system definition purposes. According to realism, a system that is obtained by applying correctly the principles and methods of science actually *represents* some aspect of the real world. According to constructivism, all systems are artificial abstractions and are not necessarily existent in reality. That is, systems are not presented to us to be discovered, but we construct them by our perceptual and mental capabilities within the domain of our experiences.

Humans perceive reality as a continuum in its composition of objects, concepts, and propositions. Humans construct knowledge in quanta to meet constraints related to their cognitive abilities and limitations, producing what can be termed as *quantum knowledge*. This quantum knowledge leads to and contains ignorance — manifested in two forms: (1) ignorance of some state of ignorance and (2) incompleteness or inconsistency, as discussed in detail in Section 1.4. The ignorance of a state of ignorance is called *blind ignorance*. The incompleteness form of ignorance stems from quantum knowledge that does not cover the entire domain of inquiry. The inconsistency form of ignorance rises from specialization and focusing on a particular science or discipline or phenomenon without accounting for interactions with or from other sciences or disciplines or phenomena.

1.3.3 Knowledge, Information, and Opinions

Many disciplines of engineering and the sciences rely on the development and use of predictive models that in turn require knowledge and information, and sometimes subjective opinions of experts. Working definitions for knowledge, information, and opinions are provided in this section.

Knowledge can be based on *evolutionary epistemology* (Honderich, 1995) using an *evolutionary model*. Knowledge can be viewed to consist of two types, *nonpropositional* and *propositional* knowledge. The nonpropositional knowledge can be further broken down into *know-how and concept knowledge* and *familiarity knowledge* (commonly called *object knowledge*). The know-how and concept knowledge requires someone to know how to do a specific activity, function, procedure, etc., such as riding a bicycle. The concept knowledge can be empirical in nature. In evolutionary epistemology the know-how knowledge is viewed as a historical antecedent to propositional knowledge. The object knowledge is based on a direct acquaintance with a person, place, or thing; for example, Mr. Smith knows the president of the U.S. The propositional knowledge is based on propositions that can be either true or false; for example, Mr. Smith knows that the Rockies are in North America (Sober, 1991; di Carlo, 1998). This proposition can be expressed as

$$\textit{Mr. Smith} \text{ knows that } \textit{the Rockies are in North America} \tag{1.12a}$$

$$S \text{ knows } P \tag{1.12b}$$

where S is the subject, i.e., Mr. Smith, and P is the claim "the Rockies are in North America." Epistemologists require the following three conditions for making this claim in order to have a true proposition:

- S must believe P.
- P must be true.
- S must have a reason to believe P; i.e., S must be justified in believing P.

The justification in the third condition can take various forms; however, simplistically it can be taken as justification through rational reasoning or empirical evidence. Therefore, propositional knowledge is defined as a body of propositions that meet the conditions of *justified true belief* (JTB). This general definition does not satisfy a class of examples called the *Gettier problem*, initially revealed in 1963 by Edmund Gettier (Austin, 1998). Gettier showed that we can have highly reliable evidence and still not have knowledge (Ayyub, 2001). Also, someone can skeptically argue that as long as it is possible for S to be mistaken in believing P (i.e., not meeting the third condition), the proposition is false. This argument, sometimes called a Cartesian argument, undermines empirical knowledge. In evolutionary epistemology, this high level of scrutiny is not needed, and need not be satisfied in the biological world. According to evolutionary epistemology, true beliefs can be justified causally from reliably attained law-governed procedures, where law refers to a natural law. Sober (1991) noted that there are very few instances, if ever, where we have perfectly infallible evidence. Almost all of our commonsense beliefs are based on evidence that is not infallible, even though some may have overwhelming reliability. The presence of a small doubt in meeting the justification condition does not make our evidence infallible, but only reliable. Evidence reliability and infallibility arguments form the bases of the *reliability theory of knowledge*. Figure 1.14 shows a breakdown of knowledge by types, sources, and objects that was based on a summary provided by Honderich (1995).

In engineering and the sciences, knowledge can be defined as a body of justified true beliefs (JTB), such as laws, models, objects, concepts, know-how, processes, and principles, acquired by humans about a system of interest, where the justification condition can be met based on the *reliability theory of knowledge*. The most basic knowledge category is called cognitive knowledge (*episteme*), that which can be acquired by human senses. The next level is based on correct reasoning from hypotheses, such as mathematics (*dianoi*). The third category moves us from intellectual categories to categories that are based on the realm of appearances and deception, and are based on propositions. The third category is belief (*pistis*). *Pistis*, the Greek word for faith, denotes intellectual or emotional acceptance of a propo-

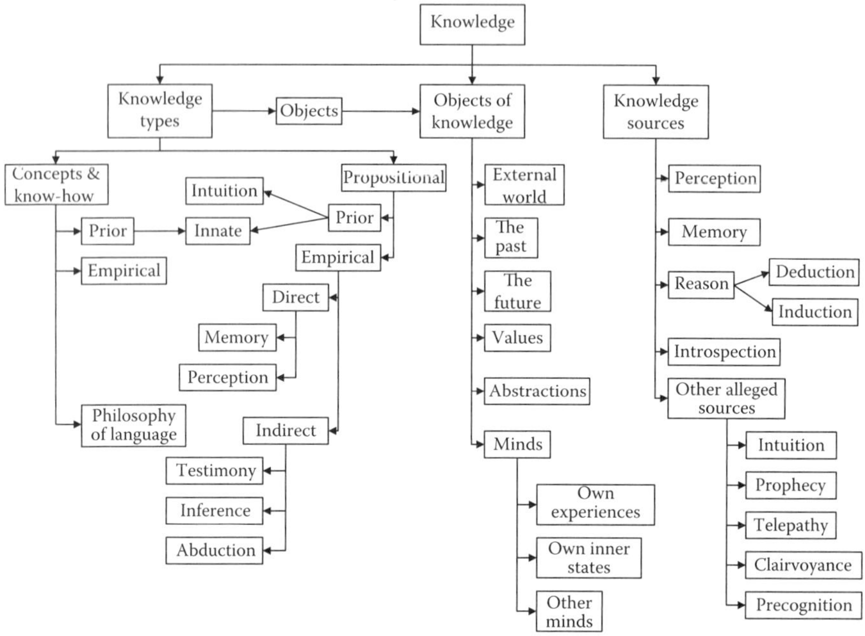

FIGURE 1.14 Knowledge types, sources, and objects. (From Ayyub, B.M., *Elicitation of Expert Opinions for Uncertainty and Risks: Theory, Applications, and Guidance*, CRC Press, Boca Raton, FL, 2001. With permission.)

sition. It is followed by conjecture (*eikasia*), where knowledge is based on inference, theorization, or prediction based on incomplete or unreliable evidence. The four categories are shown in Figure 1.15. They also define the knowledge box in Figure 1.16. These categories constitute the human cognition of human knowledge that might be different from a future state of knowledge achieved by an evolutionary process, as shown in Figure 1.16. The *pistis* and *eikasia* categories are based on expert judgment and opinions regarding system issues of interest. Although the *pistis* and *eikasia* knowledge categories might by marred with uncertainty, they are certainty sought after in many engineering disciplines and the sciences, especially by decision and policy makers.

Information can be defined as sensed objects, things, places, processes, and thoughts and knowledge communicated by language and multimedia. Information can be viewed as a preprocessed input to our intellect system of cognition, and knowledge acquisition and creation. Information can lead to knowledge through investigation, study, and reflection. However, knowledge and information about a system might not constitute the eventual evolutionary knowledge state about the system, as a result of not meeting the justification condition in JTB or the ongoing evolutionary process, or both. Knowledge is defined in the context of the humankind, evolution, language and communication methods, and social and economic dialectic processes, and cannot be removed from them. As a result, knowledge would always reflect the imperfect and evolutionary nature of humans that can be attributed to

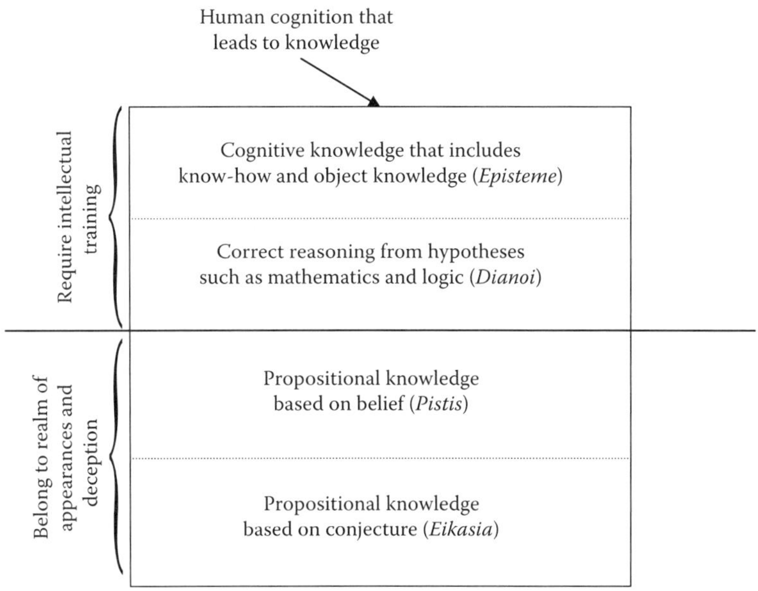

FIGURE 1.15 Knowledge categories and sources. (From Ayyub, B.M., *Elicitation of Expert Opinions for Uncertainty and Risks: Theory, Applications and Guidance*, CRC Press, Boca Raton, FL, 2001. With permission.)

their reliance on their senses for information acquisition; their dialectic processes; and their mind for extrapolation, creativity, reflection, and imagination, with associated biases as a result of preconceived notions due to time asymmetry, specialization, and other factors. An important dimension in defining the state of knowledge and truth about a system is nonknowledge or ignorance.

Opinions rendered by experts, which are based on information and exiting knowledge, can be defined as preliminary propositions with claims that are not fully justified or justified with adequate reliability but are not necessarily infallible. Expert opinions are seeds of propositional knowledge that do not meet one or more of the conditions required for the JTB with the reliability theory of knowledge. They are valuable, as they might lead to knowledge expansion, but decisions made based on them sometimes might be risky propositions, since their preliminary nature might lead to proving them false by others or in the future.

The relationships among knowledge, information, opinions, and evolutionary epistemology are schematically shown in Figure 1.16. The dialectic processes include communication methods such as languages, visual and audio formats, and other forms. Also, they include economics, classes, schools of thought, and political and social dialectic processes within peers, groups, colonies, societies, and the world.

1.3.4 Reasoning, Science, and Uncertainty

Philosophers and social scientists have been trying to define approaches for knowledge construction through mathematical formulations and scientific theories, based

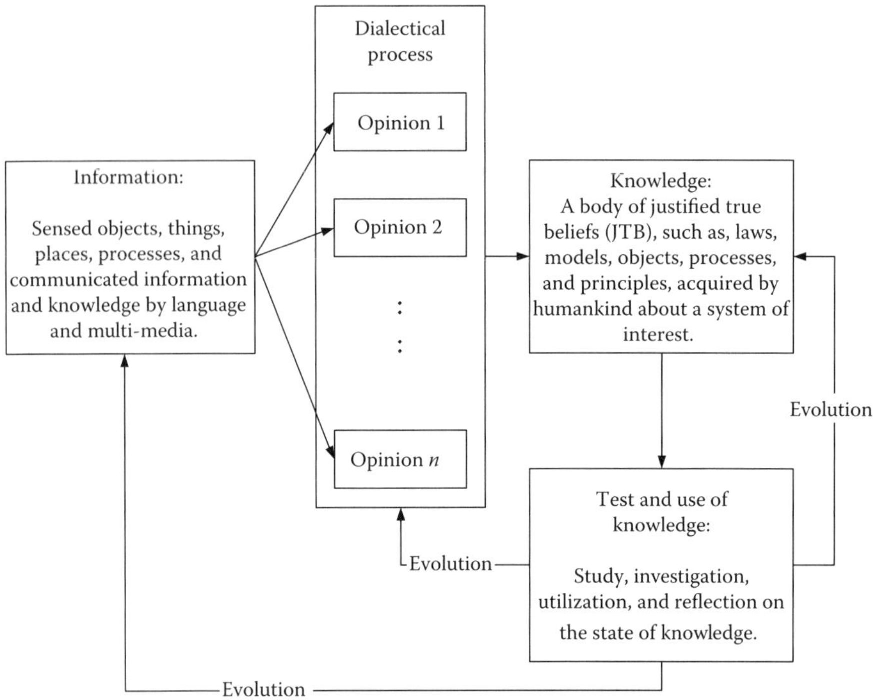

FIGURE 1.16 Knowledge, information, opinions, and evolutionary epistemology. (From Ayyub, B.M., *Elicitation of Expert Opinions for Uncertainty and Risks: Theory, Applications, and Guidance*, CRC Press, Boca Raton, FL, 2001. With permission.)

on human decisions under uncertain circumstances and outcomes. As C.S. Peirce said, "The object of reasoning is to find out, from the consideration of what we already know, something else which we do not know" (Ramsey, 1931). Emile Borel (1871 to 1956) developed the first effective theory of measure for a set of points, and the systematic theory for a divergent series in 1899 (New School University, 2003c). This work, along with the works of mathematicians Rene Baire and Henri Lebesgue, marked the beginning of the modern theory of functions of real variables.

Based on the theory of measure, J.M. Keynes (1883 to 1946) defined probability as being subject to human interpretations by stating that "probability is relative in a sense to the principles of human reason. The degree of probability, which it is rational for us to entertain, does not presume perfect logical insight, and is relative in part to the secondary propositions which we in fact know." Analysts make probable inferences, for which they claim objective validity, and proceed from full belief in one proposition to partial belief in another and claim the procedure is objectively right. Taking two propositions, one as a premise and another as a conclusion, a probability relation exits between the two with only some degree of belief based on a full belief in the premise. Rationally, someone should proceed from the premise to the conclusion only with the limited degree of belief. A fundamental criticism of this theory is the difficulty to perceive these degrees of belief connecting premises and conclusions. In his book *Treatise on Probability* (1921), Keynes provided an explanation for probability as constituting a "part of our knowledge which we obtain by argument."

Keynes's work aroused the interest of logician F.P. Ramsey (1903 to 1930) to outline his own *subjective theory of probability*. K. Popper (1902 to 1994) introduced induction in his book *The Logic of Scientific Discovery* in 1934 (Popper, 2002 release) and stated that scientific theories can never be proven, merely tested and corroborated. Scientific discovery is distinguished from all other types of investigation by the falsifiability of its theories, where falsifiability is analogous to testability, while irrefutability is associated only with a notion being untestable. In this context, therefore, untestable theories are unscientific. B. De Finetti (1906 to 1985) contended that probability does not exist objectively independent of the human mind (Nau, 2001). The subjective theory of probability is jointly attributed to de Finetti, Ramsey, and Leonard Savage (1917 to 1971) by incorporating it in the von Neumann and Morgenstern expected utility theory.

Science was an integral part of philosophy until recently, up to the start of the 20th century, since intellectual enterprise of science is essentially the progressive improvement of understanding nature. This relationship formed what can be termed *natural philosophy*, as distinguished from moral philosophy and metaphysical philosophy. As a result of specialization in natural sciences in the last century, the philosophy of science became recognized as a separate discipline. The philosophy of science "attempts first to elucidate the elements involved in scientific inquiry (i.e., observational procedures, patterns of argument, methods of representation and calculation, metaphysical presuppositions), and then to evaluate the grounds of their validity from the points of view of formal logic, practical methodology, and metaphysics" (Nau, 2001). The philosophy of science found roots in the 1890s and early 1900s when serious doubts grew up about the finality of the Newtonian synthesis in the writings of Ernst Mach, Heinrich Hertz, Max Planck, Henri Poincaré, Pierre Duhem, and others. The philosophy of science had three branches based on how uncertainty and unknowns were supposed to be handled and interpreted: *empiricism*, *qualified realism*, and *conventionalist position*. Mach and Avenarius expounded a sensationalist form of empiricism and stated that all ideas must be traceable to *impressions*, i.e., *sensations*, where theoretical concepts were intellectual fictions, introduced only to achieve economy in the intellectual organization of sensory impressions or observations (Pléh, 2003). Theories had cognitive validity only so long as they could be grounded in sensory impressions. This thinking resulted in skepticism about the reality of atoms. On the other hand, Planck, who developed the quantum theory, argued for a qualified realism where an ideal exists toward which all conceptual developments in physics should proceed. Without such a belief in the enduring reality of external nature, all motives for theoretical improvement in science would vanish. Poincaré and Duhem argued for an intermediate position of the so-called conventionalist position, which expressed reservations about the arbitrary elements in theory construction while avoiding the radical doubt about the status of theoretical entities.

Quantum mechanics emerged in the early part of the 20th century to explain three important problems in physics that could not be solved by the methods of *classical physics*: the problem of *black body radiation* (explained by Max Planck by his *theory of electromagnetic radiation and energy quanta*), the *photoelectric effect* (explained by Albert Einstein in 1905 using Planck's formula and the principle of conservation of energy), and the formula for the *spectral lines of the hydrogen atom*, provided by Rydberg in 1890 (the first acceptable explanation was provided by Niels Bohr in 1913) (Schaefer, 2003). These works were followed by Louis de Broglie's particle waves (1924), Erwin Schroedinger's equation governing the motions of electrons and protons (1926), Paul Dirac's incorporation of relativity into quantum mechanics (1927), and Werner Heisenberg's uncertainty principle (1927). The views of the developers of the theory of quantum mechanics, however, differed about the existence of uncertainty (Pearcey and Thaxton, 1994). Einstein, Planck, and de Broglie considered uncertainty in quantum mechanics to be merely a statement of human ignorance; for example, Einstein resisted a probabilistic interpretation of quantum mechanics by stating that "God does not play dice with the universe." Niels Bohr, however, maintained that uncertainty is not a result of transient ignorance, solvable by further research, but a fundamental and unavoidable limitation on human knowledge. Werner Heisenberg, on the other hand, stated that nature is not deterministic, as classical physics assumed; it is indeterminate since when a scientist intrudes a measuring device into any system, a particular outcome is forced to be actualized from what was before a fuzzy realm of potentialities. According to quantum mechanics, the results of a particular experiment cannot be predicted with absolute certainty; however, the probability of obtaining a specific result can be precisely computed. The theory of relativity evolved as an extension of the quantum mechanics theory by emphasizing that physical quantities must be defined from the perspective of an observer using a specific set of measuring instruments. Therefore, quantities like length and speed cannot be defined in absolute terms. Moreover, units and dimensions of measurement, which are commonly believed to be independent, are actually connected, particularly in conditions that are extreme to ordinary experience (Nau, 2001).

1.3.5 Cognition and Cognitive Science

Cognition can be defined as mental processes of receiving and processing information for knowledge creation and behavioral actions. Cognitive science is the interdisciplinary study of mind and intelligence (Stillings, 1995). Cognitive science deals with philosophy, psychology, artificial intelligence, neuroscience, linguistics, and anthropology. The intellectual origins of cognitive science started in the mid-1950s when researchers in several fields began to develop theories on how the mind works based on complex representations and computational procedures.

The origin of cognitive science can be taken as the theory of knowledge and the theory of reality of the ancient Greeks, when philosophers such as Plato and Aristotle tried to explain the nature of human knowledge. The study of the mind remained the province of philosophy until the 19nth century, when experimental psychology was developed by Wilhelm Wundt and his students by initiating laboratory methods for studying systematically mental operations. A few decades later, experimental psychology became dominated by behaviorism, by which the existence of the mind was virtually denied. Behaviorists, such as J.B. Watson, argued that psychology should restrict itself to examining the relation among observable stimuli and observable behavioral responses, and should not deal with consciousness and mental representations. The intellectual landscape began to change dramatically in 1956, when George Miller summarized numerous studies showing that the capacity of human thinking is limited, with short-term memory, for example, limited to around seven items. He proposed that memory limitations are compensated for by humans through their ability to recode information into chunks, and mental representations that require mental procedures for encoding and decoding the information. Although at this time primitive computers had been around for only a few years, pioneers such as John McCarthy, Marvin Minsky, Allen Newell, and Herbert Simon were founding the field of artificial intelligence. Moreover, Noam Chomsky rejected behaviorist assumptions about language as a learned habit and proposed instead to explain language comprehension in terms of mental grammars consisting of rules.

Cognitive science is based on a central hypothesis that thinking can best be understood in terms of representational structures in the mind and computational procedures that operate on those structures (Johnson-Laird, 1988). The nature of the representations and computations that constitute thinking are not fully understood. The central hypothesis is general enough to encompass the current range of thinking in cognitive science, including connectionist theories that model thinking using artificial neural networks. This hypothesis assumes that the mind has mental representations analogous to computer data structures, and computational procedures similar to computational algorithms. The mind is considered to contain such mental representations as logical propositions, rules, concepts, images, and analogies. It uses mental procedures such as deduction, search, matching, rotating, and retrieval for interpretation, generation of knowledge, and decision making. The dominant mind–computer analogy in cognitive science has taken on a novel twist from the use of another analog, that is, the brain. Cognitive science then works with a complex three-way analogy among the mind, brain, and computers. Connectionists have proposed a brainlike structure that uses neurons and their connections as inspirations for data structures, and neuron firing and spreading activation as inspirations for algorithms. There is not a single computational model for the mind, since different kinds of computers and programming approaches suggest different ways in which the mind might work, ranging from serial processors, such as the commonly used computers that perform one instruction at a time, to parallel processors, such as some recently developed computers that are capable of doing many operations at once.

Cognitive science claims that the human mind works by representation and computation using empirical conjecture. Although the computational-representational approach to cognitive science has been successful in explaining many aspects of human problem solving, learning, and language use, some philosophical critics argue that it is fundamentally flawed based on the following limitations (Thagard, 1996; Von Eckardt, 1993):

- *Emotions*: Cognitive science neglects the important role of emotions in human thinking.

- *Consciousness*: Cognitive science ignores the importance of consciousness in human thinking.
- *Physical environments*: Cognitive science disregards the significant role of physical environments on human thinking.
- *Social factors*: Human thought is inherently social and has to deal with various dialectic processes in ways that cognitive science ignores.
- *Dynamical nature*: The mind is a dynamical system, not a computational system.
- *Quantum nature*: Researchers argue that human thinking cannot be computational in the standard sense, so the brain must operate differently, perhaps as a quantum computer.

These open issues need to be considered by scientists and philosophers in developing new cognitive theories and a better understanding of how the human mind works.

1.4 IGNORANCE

1.4.1 KNOWLEDGE AND IGNORANCE

Generally, engineers and scientists, and even almost all humans, tend to focus on and emphasize what is known, and not what is unknown. Even the English language lends itself for this emphasis. For example, we can easily state that Expert A *informed* Expert B, whereas we cannot directly state the contrary. We can only state it by using the negation of the earlier statement: Expert A *did not inform* Expert B. Statements such as "Expert A *misinformed* Expert B" or "Expert A *ignored* Expert B" do not convey the same (intended) meaning. Another example is "John *knows* David," for which a meaningful direct contrary statement does not exist. The emphasis on knowledge and not on ignorance can also be noted in sociology by having a field of study called the *sociology of knowledge* and not having a *sociology of ignorance* field, although Weinstein and Weinstein (1978) introduced the *sociology of nonknowledge* and Smithson (1985) introduced the *theory of ignorance*.

Engineers and scientists tend to emphasize knowledge and information, and sometimes intentionally or unintentionally brush aside uncertainty and not acknowledge ignorance. In addition, information (or knowledge) can be misleading in some situations because it does not have the truth content that was assigned to it — leading potentially to overconfidence. In general, knowledge and ignorance can be classified as shown in Figure 1.17, using squares with crisp boundaries for the purpose of illustration. The shapes and boundaries can be made multidimensional, irregular, or fuzzy. The evolutionary infallible knowledge (EIK) about a system is shown as the top-right square in the figure and can be intrinsically unattainable due to the fallacy of humans and the evolutionary nature of knowledge. The state of reliable knowledge (RK) is shown using another square, i.e., the bottom-left square, for illustration purposes. The reliable knowledge represents the present state of knowledge in an evolutionary process, i.e., a snapshot of knowledge as a set of know-how, objects, and propositions that meet justifiable true beliefs within reasonable reliability levels.

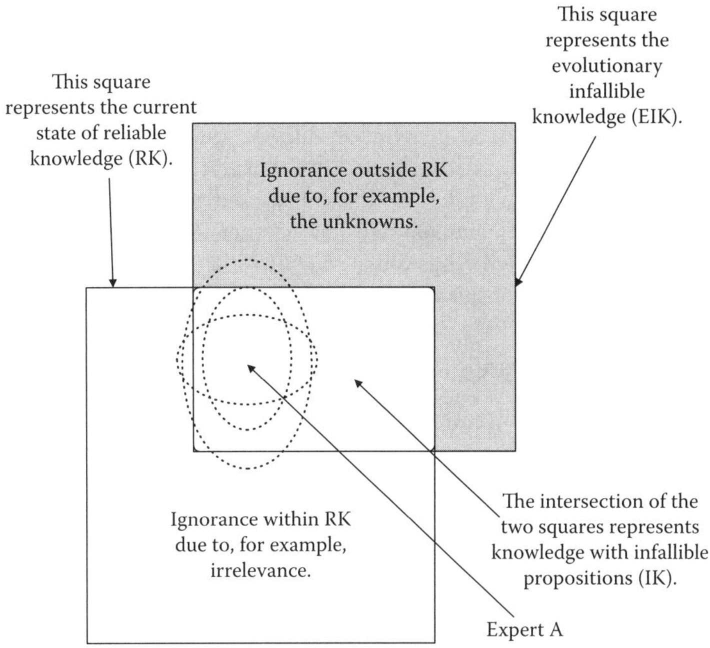

FIGURE 1.17 Human knowledge and ignorance.

At any stage of human knowledge development, this knowledge base about the system is a mixture of truth and fallacy. The intersection of EIK and RK represents the knowledge base with the infallible knowledge components (i.e., know-how, objects, and propositions). Therefore, the following relationship can be stated using the notations of set theory:

$$\text{Infallible knowledge (IK)} = \text{EIK} \cap \text{RK} \tag{1.13}$$

where $\cap$ means intersection. Infallible knowledge is defined as knowledge that can survive the dialectic processes of humans and societies, and passes the test of time and use. This infallible knowledge can be schematically defined by the intersection of these two squares of EIK and RK. Based on this representation, two primary types of ignorance can be identified: (1) ignorance within the knowledge base RK, due to factors such as irrelevance, and (2) ignorance outside the knowledge base, due to unknown objects, interactions, laws, dynamics, and know-how.

Expert A of some knowledge about the system can be represented as shown in Figure 1.17, using ellipses for illustrative purposes. Three types of ellipses can be identified: (1) a subset of the evolutionary infallible knowledge (EIK) that the expert has learned, captured, or created; (2) self-perceived knowledge by the expert; and (3) perception by others of the expert's knowledge. The EIK of the expert might be smaller than the self-perceived knowledge by the expert, and the difference between the two types is a measure of overconfidence that can be partially related to the expert's ego. Ideally, the three ellipses should be the same, but commonly they are not. They are greatly affected by communication skills of experts and their successes in dialectic processes that with time might lead to evolutionary knowledge marginal advances or quantum leaps. Also, their relative sizes and positions within the infallible knowledge (IK) base are unknown. It can be noted from Figure 1.17 that the expert's knowledge can extend beyond the reliable knowledge base into the EIK area as a result of creativity and imagination of the expert. Therefore, the intersection of the expert's knowledge with the ignorance space outside the knowledge base can be viewed as a measure of creativity and imagination. Another expert (i.e., Expert B) would have her or his own ellipses that might overlap with the ellipses of Expert A, and might overlap with other regions by varying magnitudes.

1.4.2 Ignorance Classification and Hierarchy

1.4.2.1 Ignorance Classification

The state of *ignorance* for a person (or a society) can be unintentional or deliberate, due to an erroneous cognition state and not knowing relevant information, or ignoring information and deliberate inattention to something for various reasons, such as limited resources or cultural opposition, respectively. The latter type is a state of *conscious ignorance*, which is not intentional, and once recognized, evolutionary species try to correct for that state for survival reasons, with varying levels of success. The former ignorance type belongs to the *blind ignorance* category. Therefore, ignoring means that someone can either *unconsciously* or *deliberately* refuse to acknowledge or regard, or leave out an account or consideration for relevant information (di Carlo, 1998). These two states should be treated in developing a hierarchal breakdown of ignorance.

Using the concepts and definitions from evolutionary knowledge and epistemology, ignorance can be classified based on the three knowledge sources as follows:

- *Know-how ignorance*: It can be related to the lack of know-how knowledge or having erroneous know-how knowledge. Know-how knowledge requires someone to know how to do a specific activity, function, procedure, etc., such as riding a bicycle.
- *Object ignorance*: It can be related to the lack of object knowledge or having erroneous object knowledge. Object knowledge is based on a direct acquaintance with a person, place, or thing; for example, Mr. Smith knows the president of the U.S.
- *Propositional ignorance*: It can be related to the lack of propositional knowledge or having erroneous propositional knowledge. Propositional knowledge is based on propositions that can be either true or false; for example, Mr. Smith knows that the Rockies are in North America.

The above three ignorance types can be cross-classified against two possible states for a knowledge agent, such as a person, of knowing their state of ignorance. These two states are as follows:

- *Nonreflective* (or *blind*) *state*: The person does not know of self-ignorance, a case of ignorance of ignorance.
- *Reflective state*: The person knows and recognizes self-ignorance. Smithson (1985) termed this type of ignorance *conscious ignorance*, and the blind ignorance was termed *meta-ignorance*. As a result, in some cases the person might formulate a proposition but still be ignorant of the existence of a proof or disproof, i.e., *ignoratio elenchi*. A knowledge agent's response to reflective ignorance can be either passive acceptance or a guided attempt to remedy one's ignorance that can lead to four possible outcomes:
    1. A successful remedy that is recognized by the knowledge agent to be a success, leading to fulfillment
    2. A successful remedy that is not recognized by the knowledge agent to be a success, leading to searching for a new remedy
    3. A failed remedy that is recognized by the knowledge agent to be a failure, leading to searching for a new remedy
    4. A failed remedy that is recognized by the knowledge agent to be a success, leading to blind ignorance, such as *ignoratio elenchi* or irrelevant conclusion.

The cross-classification of ignorance is shown in Figure 1.18 in two possible forms that can be used interchangeably. Although the blind state does not feed directly into the evolutionary process for knowledge, it represents a becoming knowledge reserve. The reflective state has a survival value to evolutionary species; otherwise, it can be argued that it never would have flourished (Campbell, 1974). Ignorance emerges as a lack of knowledge relative to a particular perspective from which such gaps emerge. Accordingly, the accumulation of beliefs and the emergence of ignorance constitute a dynamic process resulting in old ideas perishing and new ones flourishing (Bouissac, 1992). According to Bouissac (1992), the process of scientific discovery can be metaphorically described as not only a cumulative sum (positivism) of beliefs, but also an activity geared toward relentless construction of ignorance (negativism), producing architecture of holes, gaps, and lacunae, so to speak.

Hallden (1986) examined the concept of evolutionary ignorance in decision theoretic terms. He introduced the notion of gambling to deal with blind ignorance or lack of knowledge, according to which there are times when, in lacking knowledge, gambles must be taken. Sometimes gambles pay off with success, i.e., continued survival, and sometimes they do not, leading to sickness or death.

According to evolutionary epistemology, ignorance has factitious, i.e., human-made, perspectives. Smithson (1988) provided a working definition of ignorance based on "Expert A is ignorant from B's viewpoint if A fails to agree with or show awareness of ideas which B defines as actually or potentially valid." This definition allows for self-attributed ignorance, and either Expert A or B can be attributer or perpetrator of ignorance. Our ignorance and claimed knowledge depend on our current historical setting, which is relative to various natural and cultural factors, such as language, logical systems, technologies, and standards that have developed

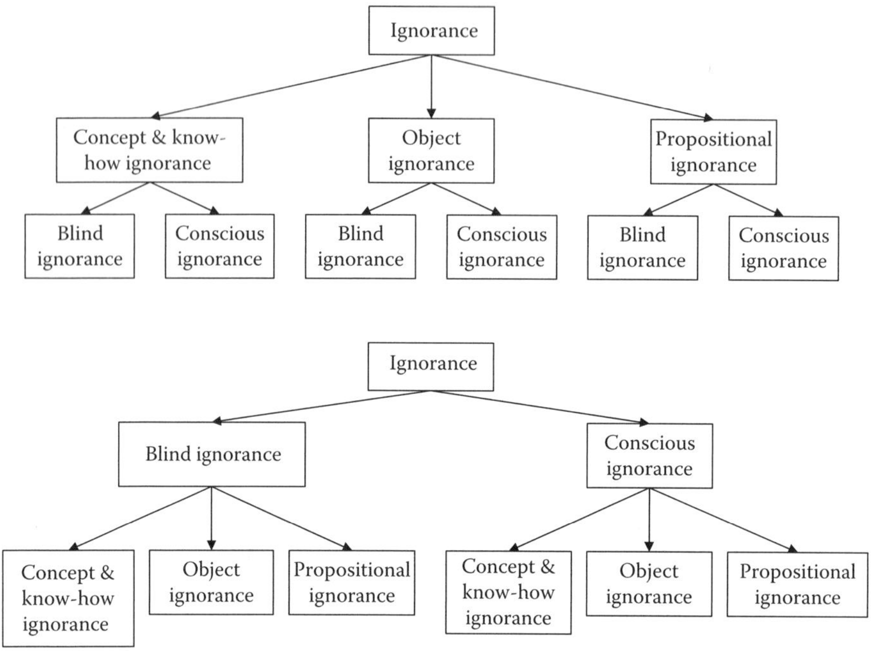

FIGURE 1.18 Classifying ignorance.

and evolved over time. Therefore, humans evolved from blind ignorance through gambles to a state of incomplete knowledge, with reflective ignorance recognized through factitious perspectives. In many scientific fields, the level of reflective ignorance becomes larger as the level of knowledge increases. Duncan and Weston-Smith (1977) stated in the *Encyclopedia of Ignorance* that compared to our pond of knowledge, our ignorance remains Atlantic. They invited scientists to state what they would like to know in their respective fields, and noted that the more eminent they were, the more readily and generously they described their ignorance. Clearly, before solving a problem, it needs to be articulated.

1.4.2.2 Ignorance Hierarchy

Figure 1.17 and Figure 1.18 express knowledge and ignorance in evolutionary terms as they are socially or factitiously constructed and negotiated. Ignorance can be viewed to have a hierarchal classification based on its sources and nature, as shown in Figure 1.19, with the brief definitions provided in Table 1.10. Ignorance can be classified into two types, blind ignorance (also called meta-ignorance) and conscious ignorance (also called reflective ignorance).

Blind ignorance includes *not knowing* relevant know-how, objects-related information, and relevant propositions that can be justified. The unknowable knowledge can be defined as knowledge that cannot be attained by humans based on current evolutionary progressions, or cannot be attained at all due to human limitations, or can only be attained through quantum leaps by humans. Blind ignorance also

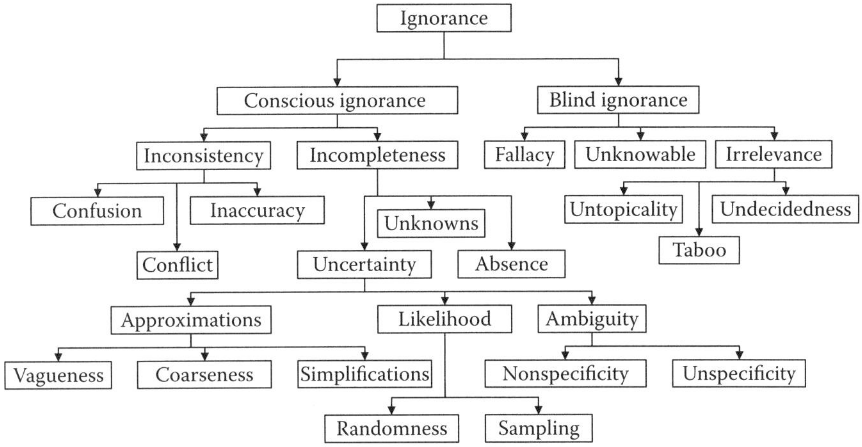

FIGURE 1.19 Ignorance hierarchy. (From Ayyub, B.M., *Elicitation of Expert Opinions for Uncertainty and Risks: Theory, Applications, and Guidance*, CRC Press, Boca Raton, FL, 2001. With permission.)

includes irrelevant knowledge that can be of two types: (1) relevant knowledge that is dismissed as irrelevant or ignored and (2) irrelevant knowledge that is believed to be relevant through nonreliable or weak justification, or as a result of *ignoratio elenchi*. The irrelevance type can be due to untopicality, taboo, and undecidability. Untopicality can be attributed to intuitions of experts that could not be negotiated with others in terms of cognitive relevance. Taboo is due to socially reinforced irrelevance. Issues that people must not know, deal with, inquire about, or investigate define the domain of taboo. The undecidedness type deals with issues that cannot be designated true or false because they are considered insoluble, or solutions that are not verifiable, or as a result of *ignoratio elenchi*. A third component of blind ignorance is fallacy that can be defined as erroneous beliefs due to misleading notions. In its 2003 quotes of the year, *Newsweek* magazine (December 29, 2003) selected a quote by U.S. Secretary of Defense Donald Rumsfeld, used to clarify the U.S. policy on the war on terror at a Pentagon briefing; it includes elements related to Figure 1.19:

> There are known knowns. These are things that we know. There are known unknowns. That is to say, there are things that we know we don't know. But there are also unknown unknowns. There are things we don't know we don't know.

Thomas Sowell, a senior fellow on public policy at the Hoover Institution (http://www-hoover.stanford.edu/bios/sowell.html) said:

> It takes considerable knowledge to realize the extent of your ignorance.

Kurt Gödel (1906 to 1978) showed that a logical system could not be both consistent and complete, and could not prove itself complete without proving itself

TABLE 1.10
Taxonomy of Ignorance

Term	Meaning
1. Blind ignorance	Ignorance of self-ignorance, or called meta-ignorance
1.1. Unknowable	Knowledge that cannot be attained by humans based on current evolutionary progressions, or cannot be attained at all due to human limitations, or can only be attained through quantum leaps by humans
1.2. Irrelevance	Ignoring something
1.2.1. Untopicality	Intuitions of experts that could not be negotiated with others in terms of cognitive relevance
1.2.2. Taboo	Socially reinforced irrelevance; issues that people must not know about, deal with, inquire about, or investigate
1.2.3. Undecidedness	Issues that cannot be designated true or false because they are considered insoluble, or solutions that are not verifiable, or *ignoratio elenchi*
1.3. Fallacy	Erroneous belief due to misleading notions
2. Conscious ignorance	A recognized self-ignorance through reflection
2.1. Inconsistency	Inconsistency in knowledge can be attributed to distorted information as a result of inaccuracy, conflict, contradiction, or confusion
2.1.1. Confusion	Wrongful substitutions
2.1.2. Conflict	Conflicting or contradictory assignments or substitutions
2.1.3. Inaccuracy	Bias and distortion in degree
2.2. Incompleteness	Lacking or nonwhole knowledge in its extent due to absence or uncertainty
2.2.1. Absence	Incompleteness in kind
2.2.2. Unknowns	The difference between the *becoming* knowledge state and *current* knowledge state
2.2.3. Uncertainty	Knowledge incompleteness due to inherent deficiencies with acquired knowledge
2.2.3.1. Ambiguity	The possibility of having multioutcomes for processes or systems
a) Unspecificity	Outcomes or assignments that are incompletely defined
b) Nonspecificity	Outcomes or assignments that are improperly or incorrectly defined
2.2.3.2. Approximations	A process that involves the use of vague semantics in language, approximate reasoning, and dealing with complexity by emphasizing relevance
a) Vagueness	Noncrispness of belonging and nonbelonging of elements to a set or a notion of interest
b) Coarseness	Approximating a crisp set by subsets of an underlying partition of the set's universe that would bound the set of interest
c) Simplifications	Assumptions needed to make problems and solutions tractable
2.2.3.3. Likelihood	Defined by its components of randomness, statistical and modeling
a) Randomness	Nonpredictability of outcomes
b) Sampling	Samples vs. populations

inconsistent, and vice versa. Also, he showed that there are problems that cannot be solved by any set of rules or procedures; instead, for these problems one must always extend the set of axioms. This philosophical view of logic can be used as a basis for classifying the conscious ignorance into *inconsistency* and *incompleteness*. This classification is also consistent with the concept of *quantum knowledge*, discussed in Section 1.3.2.

Inconsistency in knowledge can be attributed to distorted information as a result of inaccuracy, conflict, contradiction, or confusion, as shown in Figure 1.19. Inconsistency can result from assignments and substitutions that are wrong, conflicting, or biased, producing confusion, conflict, or inaccuracy, respectively. The confusion and conflict results from in-kind inconsistent assignments and substitutions, whereas inaccuracy results from a level bias or error in these assignments and substitutions.

Incompleteness is defined as lacking or nonwhole knowledge in its extent. Knowledge incompleteness consists of (1) absence and unknowns as incompleteness in kind and (2) uncertainty. The unknowns or unknown knowledge can be viewed in evolutionary epistemology as the difference between the *becoming* knowledge state and *current* knowledge state. The knowledge absence component can lead to one of the following scenarios: (1) no action and working without the knowledge, (2) unintentionally acquiring irrelevant knowledge, leading to blind ignorance, or (3) acquiring relevant knowledge that can be with various uncertainties and levels. The fourth possible scenario of deliberately acquiring irrelevant knowledge is not listed since it is not realistic.

Uncertainty can be defined as knowledge incompleteness due to inherent deficiencies with acquired knowledge. Uncertainty can be classified based on its sources into three types: ambiguity, approximations, and likelihood. The ambiguity comes from the possibility of having multioutcomes for processes or systems. Recognizing some of the possible outcomes creates uncertainty. The recognized outcomes might constitute only a partial list of all possible outcomes, leading to *unspecificity*. In this context, unspecificity results from outcomes or assignments that are *incompletely* defined. The *improper* or *incorrect* definition of outcomes, i.e., error in defining outcomes, can be called *nonspecificity*. In this context, nonspecificity results from outcomes or assignments that are improperly defined. The unspecificity is a form of knowledge absence and can be treated similarly to the absence category under incompleteness. The nonspecificity can be viewed as a state of blind ignorance.

The human mind has the ability to perform approximations through reduction and generalizations, i.e., induction and deduction, respectively, in developing knowledge. The process of approximation can involve the use of vague semantics in language, approximate reasoning, and dealing with complexity by emphasizing relevance. Approximations can be viewed to include vagueness, coarseness, and simplification. Vagueness results from the *imprecise* nature of belonging and nonbelonging of elements to a set or a notion of interest, whereas coarseness results from approximating a set by subsets of an underlying partition of the set's universe that would bound the crisp set of interest. Simplifications are assumptions introduced to make problems and solutions tractable.

The likelihood can be defined in the context of chance, odds, and gambling. Likelihood has primary components of randomness and sampling. Randomness stems from the nonpredictability of outcomes. Engineers and scientists commonly use samples to characterize populations, hence the last type.

1.4.3 Uncertainty Theories and Classifications

1.4.3.1 Ignorance Types, Mathematical Theories, and Applications

Systems analysis provides a general framework for modeling and solving various problems and making appropriate decisions, as discussed in Section 1.2. Such a framework should identify what is known about a system and associated uncertainties and unknowns. Any identified ignorance types according to Figure 1.19 and Table 1.10 would require the use of a mix of mathematical theories appropriately selected to effectively model this ignorance content. Table 1.11 provides for illustrative purposes a matrix of applications and mathematical theories and methodologies that are introduced in Chapters 2 and 3. For example, classical sets theory can effectively deal with ambiguity by modeling nonspecificity, whereas fuzzy and rough sets can be used to model vagueness, coarseness, and simplifications. The theories of probability and statistics are commonly used to model randomness and sampling uncertainty applied to quality control and reliability analysis. Bayesian methods can be used to combine randomness or sampling uncertainty with subjective information that can be viewed as a form of simplification and can be applied to reliability analysis. Ambiguity, as an ignorance type, forms a basis for randomness and sampling, as shown in the table, with classical sets, probability, statistics, Bayesian, evidence, and interval analysis methods. Inaccuracy, as an ignorance type, can be present in many problems, such as forecasting, risk analysis, and validation. The theories of evidence, possibility, and monotone measures can be used to model confusion and conflict in diagnostics, and vagueness in control. Interval probabilities and interval analysis can be used to model inaccuracy in risk analysis and validation, vagueness, and simplification in risk analysis.

1.4.3.2 Aleatory and Epistemic Uncertainties

Uncertainty in engineering analysis and design is commonly defined as knowledge incompleteness due to inherent deficiencies in acquired knowledge. It can also be used to characterize the state of a system as being unsettled or in doubt, such as the uncertainty of the outcome. Uncertainty is an important dimension in the analysis of risks. In this case, uncertainty can be present in the definition of the hazard threats and threat scenarios, the asset vulnerabilities and their magnitudes, failure consequence types and magnitudes, prediction models, underlying assumptions, effectiveness of countermeasures and consequence mitigation strategies, decision metrics, and appropriateness of the decision criteria. Traditionally, uncertainty in risk analysis processes is classified as follows:

- *Inherent randomness* (i.e., *aleatory uncertainty*): Some events and modeling variables are perceived to be inherently random and are treated to

TABLE 1.11
Theories and Example Applications to Model and Analyze Ignorance Types

Selected Theories and Methodologies	Ignorance Type: Confusion and Conflict	Inaccuracy	Ambiguity	Randomness and Sampling	Vagueness	Coarseness	Simplification
Classical sets			Modeling				
Probability		Forecasting	Modeling	Quality control			Modeling
Statistics			Analysis	Sampling			
Bayesian			Modeling	Reliability analysis			Modeling
Fuzzy sets					Control	Modeling	Modeling
Rough sets						Classification	Modeling
Evidence	Diagnostics		Modeling				
Possibility	Target tracking	Forecasting			Control		
Monotone measure							
Interval probabilities	Risk analysis	Risk analysis	Modeling		Risk analysis		Risk analysis
Interval analysis	Risk analysis	Validation	Analysis				Risk analysis

be nondeterministic in nature. The uncertainty in this case is attributed to the physical world because it cannot be reduced or eliminated by enhancing the underlying knowledge base. This type of uncertainty is sometimes referred to as aleatory uncertainty. An example of this uncertainty type is strength properties of materials such as steel and concrete, and structural load characteristics such as wave loads on an offshore platform. For a probability or consequence parameter of interest, the aleatory uncertainty is commonly represented probabilistically by a random variable $\bar{P}$.

- *Subjective* (or *epistemic*) *uncertainty*: In many situations, uncertainty is also present as a result of a lack of complete knowledge. In this case, the uncertainty magnitude could be reduced as a result of enhancing the state of knowledge by expending resources. Sometimes, this uncertainty cannot be reduced due to resource limitations, technological infeasibility, or sociopolitical constraints. This type of uncertainty, sometimes referred to as epistemic uncertainty, is the most dominant type in risk analysis. For example, the probability of an event can be computed based on many assumptions. A subjective estimate of this probability can be used in risk analysis; however, the uncertainty in this value should be recognized. With some additional modeling effort, this value can be treated as a random variable bounded using probability intervals or percentile ranges. By enhancing our knowledge base about this potential event, these ranges can be updated. For a probability or consequence parameter of interest, the epistemic uncertainty is commonly represented probabilistically by a random variable $\hat{P}$.

When uncertainty is recognizable and quantifiable, the framework of probability can be used to represent it. Objective or frequency-based probability measures can describe uncertainties associated with the aleatory uncertainty, and subjective probability measures (based on expert opinion, as provided in Chapter 7) can describe uncertainties associated with the epistemic uncertainty. Sometimes, however, uncertainty is recognized, but cannot be quantified in statistical terms. Examples include risks far into the future, such as those for radioactive waste repositories, where risks are computed over design periods of 1000 or 10,000 years, or risks aggregated across sectors and over the world, such as the cascading effects of a successful terrorist attack on a critical asset, including consequent government changes and wars (National Research Council, 1995).

The two primary uncertainty types of aleatory and epistemic can be combined for a parameter of interest as follows (Ang and Tang, 1975, 1984):

$$P = \bar{P}\,\hat{P} \tag{1.14}$$

where P is a random variable representing both uncertainty types, i.e., the combined uncertainty. $\bar{P}$ is a random variable to represent the aleatory uncertainty, and $\hat{P}$ is a random variable to represent the epistemic uncertainty. For example, the following lognormal distributions can be used for this purpose:

$$\hat{P} = LN[1.0, COV(\hat{P})] \tag{1.15}$$

where COV is the coefficient of variation of $\hat{P}$ and LN means lognormal. In Equation 1.15, the random variable is assumed to be an unbiased estimate of the true value. The aleatory uncertainty can be represented in a similar manner using $COV(\bar{P})$. The total COV (P) can be computed as follows:

$$COV(P) = \sqrt{[COV(\bar{P})]^2 + [COV(\hat{P})]^2} \tag{1.16}$$

However, it is often important to treat the aleatory uncertainty separately from the epistemic uncertainty; for example, in light of the epistemic uncertainty, the pertinent result, such as the true or correct expected value of P, will also be a random variable. If the two types of uncertainties were combined as indicated in Equation 1.16, however, then the expected value of P would be a deterministic value. In the case where P is the risk R, it is important that the decision maker be able to select or specify a risk-aversive value, such as the 90% value of the risk. This latter aspect can be provided only if the two types of uncertainty are treated separately.

1.4.3.3 Uncertainty in System Abstraction

1.4.3.3.1 Abstraction for System Modeling

Engineers use information for the purpose of system definition, analysis, and design. Information in this case is classified, sorted, analyzed, and used to predict system behavior and performances; however, classifying, sorting, and analyzing uncertainty in this information, and using it to assess uncertainties in our predictions, is a far more difficult task. Uncertainty in engineering was traditionally classified into objective and subjective types, i.e., aleatory and epistemic uncertainty. This classification was still deficient in completely capturing the nature of uncertainty and covering all its aspects. This difficulty stems from its complex nature and its invasion of almost all epistemological levels of a system by varying degrees.

Analysis of an engineering project commonly starts with a definition of a system that can be viewed according to realism (or constructivism) as an abstract representation of an object of interest (or a construction from the experimental domain). The abstraction is performed at different epistemological levels (Ayyub, 1992, 1994). The process of abstraction can be graphically represented as shown in Figure 1.20. A resulting model from this abstraction depends largely on the engineer (or analyst) who performed the abstraction, hence on the subjective nature of this process. During the process of abstraction, the engineer needs to make decisions regarding what aspects should or should not be included in the model. These aspects are shown in Figure 1.20. Aspects that are abstracted and not abstracted include the previously identified uncertainty types. In addition to the abstracted and nonabstracted aspects, unknown aspects of the system can exist due to blind ignorance, and they are more difficult to deal with because of their unknown nature, sources, extents, and impact on the system.

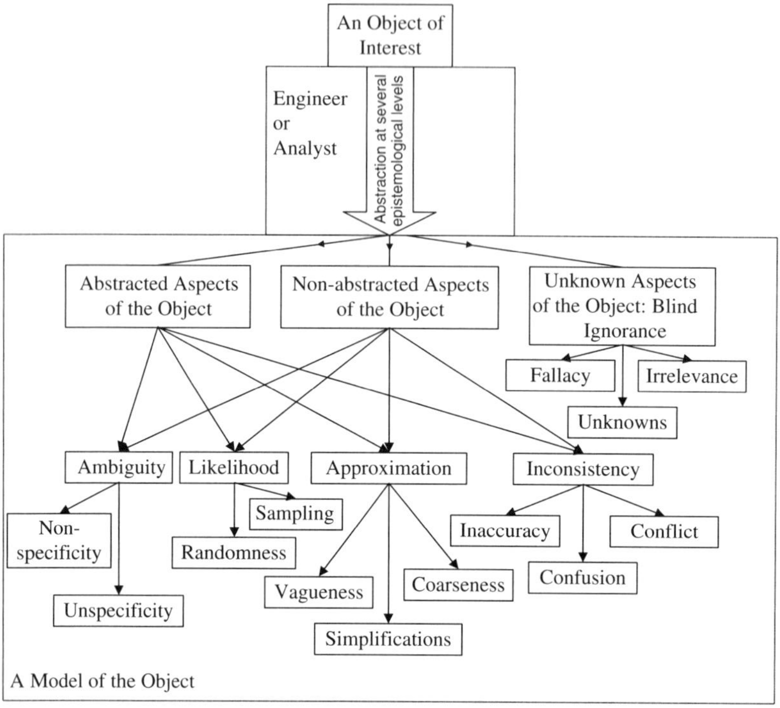

FIGURE 1.20 Abstraction and ignorance in modeling an object.

In engineering, uncertainty modeling and analysis are performed on the abstracted aspects of the system with some consideration of the nonabstracted aspects of a system. The division between abstracted and nonabstracted aspects can be a division of convenience that is driven by the objectives of the system modeling, or simplification of the model; however, the unknown aspects of the systems are due to blind ignorance that depends on the knowledge of the analyst, and the state of knowledge about the system in general. The effects of the unknown aspects on the ability of the system model to predict the behavior of the object of interest can range from none to significant.

1.4.3.3.2 Ignorance and Uncertainty in Abstracted System Aspects

Engineers and researchers dealt with the ambiguity and likelihood types of uncertainty in predicting the behavior and designing engineering systems using the theories of probability and statistics and Bayesian methods. Probability distributions were used to model system parameters that are uncertain. Probabilistic methods that include reliability methods, probabilistic engineering mechanics, stochastic finite element methods, reliability-based design formats, and other methods were developed and used for this purpose. In this treatment, however, a realization was established of the presence of the approximation's type of uncertainty. Subjective probabilities were used to deal with this type that are based on mathematics used for the frequency type of probability. Uniform and triangular probability distributions were used to model this type of uncertainty for some parameters. The Bayesian techniques were also used, for example, to deal with combining empirical and subjective information about these parameters. The underlying distributions and probabilities were therefore updated. Regardless of the nature of uncertainty in the gained information, similar mathematical assumptions and tools were used that are based on probability theory.

Approximations arise from human cognition and intelligence. They result in uncertainty in mind-based abstractions of reality. These abstractions are therefore subjective and can lack crispness, or they can be coarse in nature, or they might be based on simplifications. The lack of crispness, called vagueness, is distinct from ambiguity and likelihood in source and natural properties. The axioms of probability and statistics are limiting for the proper modeling and analysis of this uncertainty type and are not completely relevant, nor completely applicable. The vagueness type of uncertainty in engineering systems can be dealt with appropriately using fuzzy set theory, as discussed in Chapter 2. In engineering, this theory was proven to be a useful tool in solving problems that involve the vagueness type of uncertainty. To date, many applications of the theory in engineering have been developed, such as (Ayyub, 1991; Brown, 1979, 1980; Brown and Yao, 1983; Blockley, 1975, 1979a, 1979b, 1980; Blockley et al., 1983; Furuta et al., 1985, 1986; Ishizuka et al., 1981, 1983; Itoh and Itagaki, 1989; Kaneyoshi et al., 1990; Shiraishi and Furuta, 1983; Shiraishi et al., 1985; Yao, 1979, 1980; Yao and Furuta, 1986):

1. Strength assessment of existing structures and other structural engineering applications
2. Risk analysis and assessment in engineering
3. Analysis of construction failures, scheduling of construction activities, safety assessment of construction activities, decisions during construction, and tender evaluation
4. The impact assessment of engineering projects on the quality of wildlife habitat
5. Planning of river basins
6. Control of engineering systems
7. Computer vision
8. Optimization based on soft constraints

Coarseness in information can arise from approximating an unknown relationship or set by partitioning the universal space with associated belief levels for the partitioning subsets in representing the unknown relationship or set (Pawlak, 1991). Such an approximation is based on *rough sets* as described in Chapter 2. Pal and Skowron (1999) provide background and detailed information on rough set theory, its applications, and hybrid fuzzy–rough set modeling. Simplifying assumptions are common in developing engineering models. Errors resulting from these simplifications are commonly dealt with in engineering using bias random variables that are assessed empirically. A system can also be simplified by using knowledge-based if–then rules to represent its behavior based on fuzzy logic and approximate reasoning.

1.4.3.3.3 Ignorance and Uncertainty in Nonabstracted System Aspects

In developing a model, an analyst needs to decide at the different levels of modeling a system upon the aspects of the system that need to be abstracted, and the aspects that need not be abstracted. This division is for convenience and to simplify the model, and is subjective depending on the analysts, as a result of their background, and the general state of knowledge about the system. The abstracted aspects of a system and their uncertainty models can be developed to account for the nonabstracted aspects of the system to some extent. Generally, this accounting process is incomplete. Therefore, a source of uncertainty exists due to the nonabstracted aspects of the system. The ignorance categories and uncertainty types in this case are similar to the previous case of abstracted aspects of the system. These categories and types are shown in Figure 1.20. The ignorance categories and uncertainty types due to the nonabstracted aspects of a system are more difficult to deal with than the uncertainty types due to the abstracted aspects of the system. The difficulty can stem from a lack of knowledge or understanding of the effects of the nonabstracted aspects on the resulting model in terms of its ability to mimic the object of interest. Poor judgment or human errors about the importance of the nonabstracted aspects of the system can partly contribute to these uncertainty types, in addition to contributing to the next category, uncertainty due to the unknown aspects of a system.

1.4.3.3.4 Ignorance due to Unknown System Aspects

Some engineering failures have occurred because of failure modes that were not accounted for in the design stage of these systems. Failure modes were not accounted for due to various reasons, including (1) blind ignorance, negligence, using irrelevant information or knowledge, human errors, or organizational errors, or (2) a general state of knowledge about a system that is incomplete. These unknown system aspects depend on the nature of the system under consideration, the knowledge of the analyst, and the state of knowledge about the system in general. Not accounting for these aspects in the models results in varying levels of impact on the ability of these models to mimic the behavior of the systems. The effects of the unknown aspects on these models can range from none to significant. In this case, the ignorance categories include wrong information and fallacy, irrelevant information, and unknowns, as shown in Figure 1.20.

Engineers dealt with nonabstracted and unknown aspects of a system by assessing what is commonly called the modeling uncertainty, defined as the ratio of a predicted system's variables or parameter (based on the model) to the value of the parameter in the object of interest. This empirical ratio, which is called the bias, is commonly treated as a random variable that can consist of objective and subjective components. Factors of safety are intended to safeguard against failures. This approach of bias assessment is based on two implicit assumptions: (1) the value of the variable or parameter for the object of interest is known or can be accurately assessed from historical information or expert judgment, and (2) the state of knowledge about the object of interest is complete and reliable. For some systems, the first assumption can be approximately examined through verification and validation, whereas the second assumption generally cannot be validated.

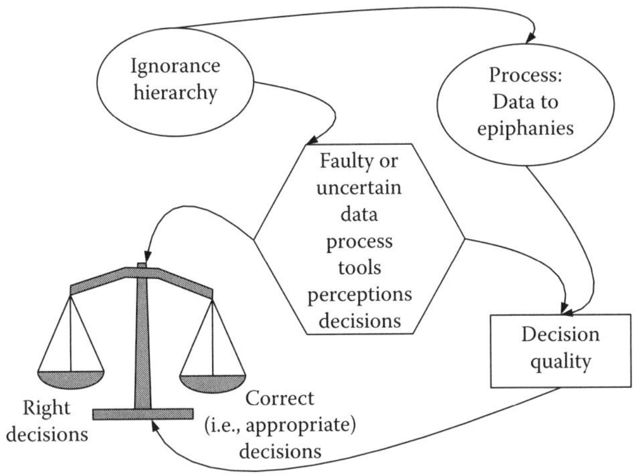

FIGURE 1.21 A decision-making process.

1.5 FROM DATA TO KNOWLEDGE FOR DECISION MAKING

A decision-making process can be represented as shown Figure 1.21 to include an ignorance hierarchy of Figure 1.19, a set of processes that transform data to epiphanies, resulting in decisions whose quality may be characterized in a spectrum of appropriate to inappropriate, and the goal of achieving optimum decisions, which are deemed either right (the optimum decision under conditions of certainty — the God's-eye view) or correct (the optimum decision under conditions of uncertainty). While there are many definitions for concepts such as information and knowledge, including technical definitions provided in previous sections, the following brief definitions are convenient for the purposes of Figure 1.21:

- Data: Unconnected numbers or symbols (e.g., names, dates, positions) representing objects and entities with appropriate levels of reliability or belief.
- Facts: Connected data.
- Information: Facts in context.
- Knowledge: A particular assemblage of information that forms justified true beliefs, information in context, and actionable information.
- Experience: Primarily from self-directed interaction with the real world, including internalization of knowledge and subsequent beliefs.
- Shared visions: Philosophical and emotional collective understandings founded on our universality and not individuality, and producing motivating forces in organizations that give purpose needed by leaders.
- Epiphanies: Level of perception beyond logic and intuition, and the rare creative brilliance.

These definitions imply a special cognitive process or operation (denoted as *). Data form the basis for this hierarchy of knowing, with the following sets of relationships:

- (data) * (order) = facts
- (facts) * (context) = information
- (information) * (synthesis) = knowledge
- (knowledge) * (perspective) = experience
- (experience) * (unifying principles) = shared vision
- (shared vision) * (meta-logic) = epiphanies

Order, appropriate to the problem at hand, imposed on data generates facts. Facts in the appropriate context create information. Information that is aggregated and synthesized properly, with respect to the situation of interest, leads to knowledge. Knowledge that is contemplated in the perspective of other relevant knowledge, past and present, provides experience, i.e., experience in the active sense of acquiring broad knowledge, not simply in the passive sense of observing or living through an event. Unifying principles, such as developed through social induction, lead to shared vision, i.e., imagining what is possible. Meta-logic in conjunction with shared vision can lead to epiphanies and new solutions to problems.

Emerging from this process, as shown in Figure 1.21, are decisions of varying quality. The ignorance hierarchy impinges on the hierarchy of knowing and contributes to the generation of faulty data, processes that act on data and the hierarchy of knowing, tools used as decision aids, perceptions that arise from the hierarchy of knowing, and the ultimate decisions. The outcomes of this decision-making process are of two kinds: right decisions and correct decisions. Right decisions are optimum decisions given a God's-eye view of the situation, where all of the relevant data are known. Correct decisions, however, are the best decisions that can be expected of most mortals — optimum decisions where only some of the relevant data are known to the decision maker, whether due to blind or conscious ignorance, as previously described in Section 1.4. For example, assume there is a decision to send a military unit to an objective over a certain path. If the path is optimum in terms of selected metrics (e.g., safety, timeliness, etc.) and the unit arrives safely, then the decision may be considered to be a right decision. If the unit were ambushed and destroyed along the path, the decision selecting the path could still be a correct decision if the decision maker did not (or could not) possess relevant data about the prospective ambush. If a faulty sensor, for example, did not detect an ambush, and the decision maker had no reason to suspect that the sensor was faulty (which should cause a prudent decision maker to take compensating actions to gather the critical data), then the decision was still correct. The decision maker should not be pilloried for making the best, most prudent decision possible given the data available, regardless of whether the outcome is fortunate or unfortunate. This no-fault decision making also assumes that the decision maker has taken prudent action to gather all relevant data, has processed the data appropriately into relevant information or knowledge, and has used the best human associates and tools available as decision aids. Decision makers who are consistently able to make decisions that are either right or correct have achieved the exalted state of wisdom.

EXERCISE PROBLEMS

1.1. Describe three engineering systems that can be modeled using the state-based method. What are the states for each system?
1.2. Build an information-based hierarchical system definition for a residential building by defining the source system, data system, generative system, structure system, and metasystem.
1.3. Repeat Problem 1.2 for a highway bridge.
1.4. Repeat Problem 1.2 for a residential house.
1.5. Provide engineering examples of structured and unstructured complexity.
1.6. Provide examples in science of structured and unstructured complexity.
1.7. Provide two cases of transcomputational problems. Why are they transcomputational in nature?
1.8. Using Plato's theory of reality, provide three examples of forms or ideas.
1.9. What is skepticism? Describe its origin and progression through the times.
1.10. Write an essay of about 400 words on the book *Tuhafut al-Tuhafut* by Ibn Rushd, summarizing primary arguments in it, its significance, and its effect on Europe.
1.11. What is positivism? Describe its origin and progression.
1.12. What is the theory of meaning?
1.13. What are the differences between knowledge, information, and opinions?
1.14. What is ignorance?
1.15. What are knowledge types and sources? Provide examples.
1.16. What are the primary differences between appropriate decisions and right decisions. Provide engineering examples of appropriate decisions and right decisions.
1.17. Provide engineering examples of the various ignorance types in the hierarchy provided in Figure 1.19.
1.18. Provide examples from the sciences of the various ignorance types in the hierarchy provided in Figure 1.19.
1.19. What are the differences between an unknown and an unknowable? Provide examples.

2 Encoding Data and Expressing Information

2.1 INTRODUCTION

This chapter provides background information, mathematical methods, and analytical tools that can be used to encode data and express information for the purpose of creating some structure or order needed to solve problems in engineering and the sciences. The various ways of encoding and expressing data and information are important components of uncertainty modeling and analysis, and sometimes are termed as *formalized languages*.

Encoding data also includes the expression of an opinion of an expert that can be defined in a numeric or nonnumeric manner, including a representation in natural language or a picture, or a figure representing or symbolizing the opinion. The expression in this case might be sensitive to the choice of a particular word, phrase, sentence, symbol, or picture. It can also include a show of feeling or character. It can be in the form of a symbol or a set of symbols expressing some mathematical or analytical relationship, as a quantity or operation.

In this chapter, we present formalized languages and selected functions of formalized languages that include the fundamentals of classical set theory, fuzzy sets, generalized measures, rough sets, and gray systems. These languages can be used to express and encode data and information. Basic operations for these theories are defined and demonstrated. Operations relating to these theories are presented for the purpose of combining collected information or data for solving problems or system modeling. Relations and operations can be used to express and combine collected information. In addition, methods for dealing with system complexity and simplification are provided in this chapter. Examples are used in this chapter to demonstrate the various methods and concepts. The level of coverage detail of a particular method was set based on the maturity and potentials of the method.

2.2 IDENTIFICATION AND CLASSIFICATION OF THEORIES

Defining a system, as discussed in Chapter 1, commonly constitutes a starting point for solving many problems in engineering and the sciences. System definition commonly involves data collection and encoding, and expressing information in a suitable format or manner for a problem or the system under consideration. The process of encoding data and information expression needs to be performed for each aspect of the system in the context of a universe or a universal set. In probability theory, the universal set is called the sample space. A universal set can be defined as the

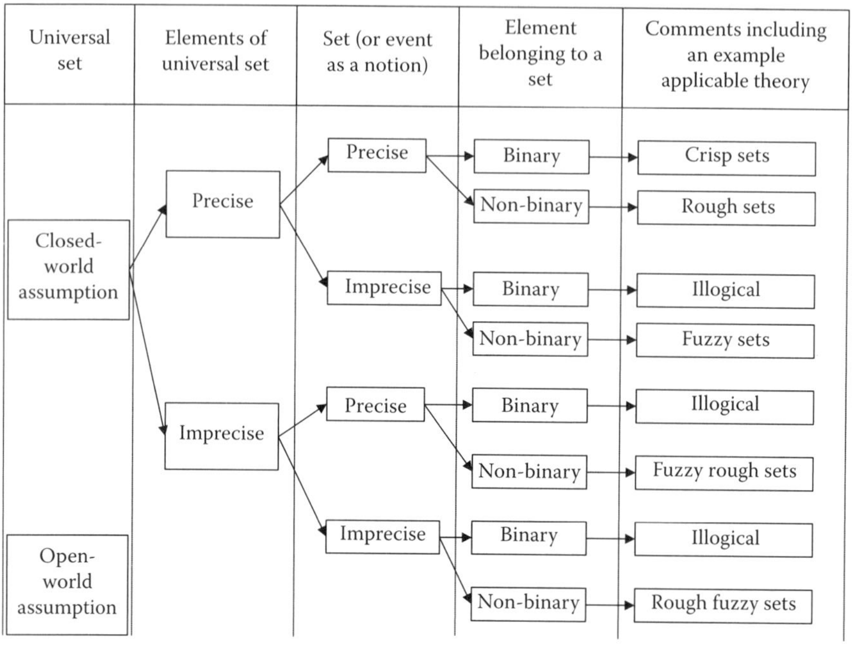

FIGURE 2.1 Identification and classification of theories.

totality of all the things that exist pertaining to the domain of interest. Mathematically, a universal set is defined as the set of all objects or elements considered in a given problem or for a given system. The universal set is commonly treated as a complete set that is known with absolute certainty, termed in this case the closed-world assumption. This assumption can be relaxed to allow for cases of an uncertain universal set definition that involve models based on an open-world assumption. This latter case is considered in Chapter 5. In this chapter, all modeling cases involve theories that are based on the closed-world assumption, as shown in the first column of Figure 2.1.

The elements of the universal set (X) are commonly assumed as precise objects without any uncertainty in defining such objects. The meaning of the term *precisely defined elements* might vary by application. It could mean that the elements are (1) strictly described, (2) accurately stated, (3) definite, (4) distinctly defined with no variation, or (5) strictly conform to usage and rules. This case of precise elements defining X is shown in Figure 2.1 as the first branching of a tree representing several cases that are discussed in this section. The second branch in the first level of branching in the second column is the case of imprecise elements. Imprecisely defined elements carry a contrary meaning to precisely defined elements. This term could mean, depending on the application, (1) not clearly, precisely, or definitely expressed or stated; (2) indefinite in shape, form, or character; (3) hazily or indistinctly seen or sensed; (4) not sharp, certain, or precise in thought, feeling, or expression; (5) imprecisely determined or known; or (6) uncertain in nature. In this case, the elements of the universal set cannot be defined precisely, and are defined nevertheless in meaningful terms. Examples of the precise elements are integer numeric values or letters of the alphabet. For the case of vague elements, an example is the illnesses and diseases that could infect humans that are of varying imprecision levels. Some diseases are characterized by viruses, others by bacteria, and others by only symptoms attributed to many factors, including genetic disorder.

The third column of Figure 2.1 addresses a notion of interest that can be expressed by a set or an event that is defined herein as a collection of elements from a universal set of interest. Such a notion can be precisely or imprecisely expressed. This second level of branching should be viewed distinctly from the precise or imprecise nature of the element comprising the universal set, and provides an added layer of classification to the first branching level to produce the next tree segment in Figure 2.1. The next column addresses the two cases related to belonging (i.e., membership) of an element to a set (or notion or event) of interest. Two cases are considered: the case of binary belonging (i.e., 0 for nonbelonging and 1 for belonging to a set) and the case of nonbinary belonging (e.g., graded belonging, where it is assigned a membership value in the continuous interval 0 to 1). Adding this belonging branching to the tree produces the eight cases shown in Figure 2.1, with branches corresponding to various theories that are built on the assumptions enumerated along each branch. The top branch of precisely defined elements in a universal set with precisely defined notions and binary belonging forms the basis for crisp set theory. For example, the set of integers {0, 1, 2, ..., 10} from the universal set of integers is a case of precisely defined notions and binary belonging. In cases where a set is not fully known in terms of what elements belong to it, the set can be approximated using bounding by rough sets (Pawlak, 1991). For example, the Milky Way Galaxy, which is the home of our solar system together with at least 200 to 400 billion other stars and their planets, and thousands of clusters and nebulae can be considered to form the universal set of interest. The notion of planets that have or had water is a precisely defined notion and uncertain belonging that can be represented by graded membership, whereas the notion of planets that were physically probed by humans is a precisely defined notion and binary belonging.

Cases involving imprecise notions with graded membership can be modeled by fuzzy sets (Zadeh, 1965). For example, the imprecise, but meaningful, notion of high-quality products produced by a manufacturing plant is an example of this case. The branch of precisely defined elements in a universal set with imprecisely defined notions and binary belonging is illogical since imprecise notions lead to graded membership, and therefore is disregarded. Imprecisely defined elements in a universal set with precisely defined notions and graded belonging form the basis for fuzzy rough sets. Making the notions in this case imprecise also leads to rough fuzzy sets. For example, the set of fatal human diseases is a crisp notion, but with imprecise elements. The remaining two cases under imprecisely defined elements in a universal set are illogical.

This chapter covers the fundamentals of these theories as bases for introducing other theories, uncertainty measures, and computational methods in subsequent chapters. Examples to illustrate these types of sets and measures are provided in subsequent sections.

2.3 CRISP SETS AND OPERATIONS

2.3.1 A UNIVERSE AND ITS ELEMENTS

A universe or universal set (X) can be defined as the totality of all the things that exist pertaining to the attribute of interest for a system. A universal set can be mathematically defined as the set of all objects or elements considered in a given problem or for modeling an attribute of interest for the system. The elements of the universe can be either precisely or vaguely defined as shown in Figure 2.1. Some fundamental properties of these elements need to be defined and introduced. The number of elements is either *finite* or *infinite*. The universal set can be either *bounded* or *unbounded*. Also, the set can be either *discrete* (i.e., countable) or *continuous* (i.e., noncountable). The elements can be of the following types:

1. Unordered, such as a nominal scale for data collection that could include, for example, gender and political party affiliation
2. Partially ordered, such as an ordinal scale for data collection that could include, for example, military ranks and education level
3. Ordered, such as interval and ratio scales used in data collection that could include weight and height of a person

The unordered type represents elements or measurements at the lowest level, because there is no order to the information. Measurements consist of simply identifying the elements as individual objects or categories. Nominal measurement scales are commonly both discrete and qualitative. However, numbers may be assigned to the categories for the purpose of coding. Frequently used examples of variables measured on a nominal scale include (1) gender (female or male), (2) political affiliation (Republican, Democrat, Independent, or other), and (3) college major (engineering, sciences, physical education, or others). Engineering data are sometimes provided using a nominal scale, for example, (1) project failed or did not fail, (2) fatal and nonfatal accidents, and (3) land use (urban, rural, forest, institutional, commercial, or others).

The partially ordered type represents elements or measurements at a higher scale than the nominal scale because it has the added property that there is order among the elements. However, the magnitude of the differences between elements is not meaningful. For example, military ranks are measured on an ordinal scale. The major is above the sergeant, and the sergeant is above the private, but we cannot say that a major is two or three times higher than a sergeant. Variables of interest in engineering that are measured on an ordinal scale include the infiltration potential of soil texture classes and hazard classifications for dam design (high hazard, moderate hazard, low hazard). Soils are classified into one of several categories, such as sand, sandy loam, clay loam, and clay. In this respect, soil texture is measured on a nominal scale. However, if we consider the infiltration potential of the soil, then we can put the soil textures in order according to the infiltration potential, high to low.

The ordered type represents elements or measurements that have the characteristics of the ordinal scale, in addition to having a meaningful separation between any two numbers on the scale. Temperature is defined on the interval scale. We recognize that a difference in temperature of 5°C is less than a difference of 10°C. Values on an interval scale may be treated with arithmetic operators. For example, the mean value of a set of test grades requires addition and division. Engineering data are frequently recorded on an interval scale. The yield strength of steel, the compressive strength of concrete, and the shear stress of soil are variables measured on an interval scale. The annual number of traffic fatalities and the number of lost worker-hours on construction sites due to accidents are also engineering variables recorded on an interval scale. A special case of ordered elements is *linearly* ordered elements, where the relationship between two elements of the universe is governed by a linear equation, i.e., first-order polynomial. Linearly ordered elements are commonly used in engineering and the sciences. The ratio scale is used in cases where the zero value has a physical meaning, not a notational meaning, such as in the case of error measurement.

As was previously discussed, the elements of the universe are commonly assumed as precise objects without any uncertainty in defining such objects. In engineering and the sciences, cases of imprecisely or vaguely defined objects could be encountered. In this case, the elements of the universe cannot be defined precisely, and are defined in vague terms that are nevertheless meaningful. Examples of precise elements are the number of automobiles crossing a highway bridge and the number of cars waiting to make a left turn at an intersection. The vague elements are exemplified by the conditions of aging structures in the inventory of some legal district or an organization. A universal set consisting of vaguely defined elements introduces uncertainty at a fundamental level that impacts subsequent models constructed based on this universal set.

2.3.2 Classical (Crisp) Sets and Events

Sets constitute a fundamental concept needed for uncertainty analysis. Any collection of distinct individuals is called a *set*, the individuals being termed its *elements* (or *members*). A set can be defined as a collection of elements or members from a universe of interest. If it is possible to determine uniquely whether any given individual is or is not a member of a given set, the set is called *classical* or *crisp*; otherwise, it is called *fuzzy*. If members of a set are sets, the set is called a *family of sets*.

Capital letters are usually used to denote sets, e.g., A, B, X, Y, etc. Small letters are commonly used to denote their members, e.g., a, b, x, y, etc., respectively. An individual member is expressed as belonging to a crisp set A as follows:

$$x \in A \tag{2.1}$$

The opposite case (x is not a member of A) is written as

$$x \notin A \tag{2.2}$$

A crisp set can be described either by naming all its members (the *list method*) or by specifying certain well-defined properties of the members (the *rule method*).

In either case, the description is written within braces. Thus, the crisp set A whose members are $a_1, a_2, \ldots, a_n$ is written as

$$A = \{a_1, a_2, \ldots, a_n\} \tag{2.3}$$

and the set B whose members satisfy the properties $p_1, p_2, \ldots, p_m$ is written as

$$B = \{b \mid b \text{ has properties } p_1, p_2, \ldots, p_m\} \tag{2.4}$$

The symbol $\mid$ stands for the phrase "such that." The set B is thus defined as the set of all individuals b such that each b satisfies the properties $p_1, p_2, \ldots, p_m$. If no logical connectives between the listed properties are specified, it is assumed that all the properties must be satisfied by each member of the set. Alternatively, appropriate logical connectives are used. A family of sets can be defined in the form $\{A_i \mid i \in I\}$, where i and I are called the *set identifier* and the *identification* (or *index*) *set*, respectively.

Example 2.1 Selected Crisp Sets

The following are example sets:

$$A = \{2, 4, 6, 8, 10\} \tag{2.5}$$

$$B = \{b \mid b \text{ is a real number} > 0\}, \text{ where } \mid \text{ means "such that"} \tag{2.6}$$

$$C = \{\text{Maryland, Virginia, Washington, D.C.}\} \tag{2.7}$$

$$D = \{\text{P, M, 2, 7, U, E}\} \tag{2.8}$$

$$F = \{1, 3, 5, 7, \ldots\}; \text{ the set of odd numbers} \tag{2.9}$$

In these example sets, each set consists of a collection of elements. In set A, 2 belongs to it and 12 does not belong to it. Using mathematical notations, this is usually expressed as $2 \in A$ and $12 \notin A$.

The set of integers greater than 2 and smaller than 10 can be written as $\{3, 4, 5, 6, 7, 8, 9\}$ or as $\{a \mid a \text{ is an integer}, a > 2, a < 10\}$. Similarly, the set $\{b \mid b$ is an integer greater than 1 and smaller than 20 and either b is divisible by 2 or b is divisible by 3$\}$ can also be written as $\{2, 3, 4, 8, 9, 10, 14, 15, 16\}$. Other examples of sets are {*, [, /, | |}, $\{a \mid a$ is a lowercase script letter of the Russian alphabet$\}$, and $\{b \mid b$ is a state of the U.S.$\}$. Specification of a set as $\{a, a, b\}$ is redundant and represents, essentially, the set $\{a, b\}$ (members of a set are required to be distinct individuals).

2.3.3 Properties of Sets and Subsets

A crisp set whose members can be labeled by the positive integers is called a *countable* set. If such labeling is not possible, the set is called *uncountable*. For instance, the set $\{a \mid a \text{ is a real number}, 0 \le a \le 1\}$ is uncountable. Every uncountable set is *infinite*; countable sets are classified into *finite* and *countably infinite* (also called denumerable). The singular set that does not contain any member is called the *empty set* and is denoted by ϕ.

For any finite set A, its number of elements is called a *cardinality* of A, and it is denoted by $|A|$. The cardinality of a set is a measure of its size. For a finite, discrete set A with n elements belonging to it, the cardinality of the set is n, i.e., $|A| = n$. For uncountable sets, such as $\{x \mid x \text{ is a real number}, a \le x \le b\}$, the size of the set can be quantified by the Lebesgue measure as $(b - a)$.

If every member of a crisp set A is also a member of crisp set B, i.e., if $a \in A$ implies that $a \in B$, A is called a *subset* of B and is expressed as

$$A \subseteq B \tag{2.10}$$

If both $A \subseteq B$ and $B \subseteq A$, the sets A and B contain the same members and are called *equal sets* expressed as

$$A = B \tag{2.11}$$

If A and B are *not equal*, the following expression is used:

$$A \neq B \tag{2.12}$$

If both $A \subseteq B$ and $A \neq B$, then B contains at least one individual that is not a member of B; A is called a *proper subset* of B, and this property is denoted as

$$A \subset B \tag{2.13}$$

Every set is considered to be a subset of itself. The null set Ø is considered to be a subset of every set.

Example 2.2 Example Subsets

For example, sets A, C, and D in Example 2.1 are finite sets, whereas sets B and F are infinite sets.

The following are example subsets:

$$A_1 = \{2, 4\} \text{ is a subset of } A = \{2, 4, 6, 8, 10\} \tag{2.14}$$

$$B_1 = \{b \mid 7 < b \le 200\} \text{ is a subset of } B = \{b \mid b \text{ is real value} > 0\} \tag{2.15}$$

$$F = \{1, 2, 3, 4, 5, 6, 7, \ldots\} \text{ is a subset of } F = \{1, 2, 3, 4, 5, 6, 7, \ldots\} \tag{2.16}$$

Other examples are $\{0, 1\} \subset \{0, 1, 2\} \neq \{1, 2, 3\} = \{2, 1, 3\} = \{3, 1, 2\} \subset \{a \mid a \text{ is a real number}, 1 \le a \le 3\}$.

2.3.4 Characteristic Function

If all sets under consideration within a certain context are subsets of a set X, then X is referred to as the *universal set* (within that context). Every universal set is required to be a crisp set. A crisp set of some elements of X, say set A, can be defined by a *characteristic function* (or a *discrimination function*) of the form

$$A: X \to \{0, 1\} \tag{2.17}$$

which assigns to each $x \in X$ a value $A(x) \in \{0, 1\}$, such that

$$A(x) = \begin{cases} 1 & \forall x \in A \\ 0 & \forall x \notin A \end{cases} \tag{2.18}$$

where $\forall$ means for all. This mapping is defined from the universe X to the integer values $\{0, 1\}$, where 0 means a value x does not belong to A and 1 means a value x belongs to A. The meaning of this membership function is that there are only two possibilities for an element x, either being a member of A, i.e., $A(x) = 1$, or not being a member of A, i.e., $A(x) = 0$. In this case the set A has sharp boundaries.

The difference between fuzzy sets and classical (crisp or nonfuzzy) sets is that the membership function takes on values in the interval [0, 1] instead of one of the two values $\{0, 1\}$. This assumption that elements belong or do not belong to sets in a binary manner constitutes a basis for the classical set theory. Relaxing this assumption to allow some form of degradation in belonging and nonbelonging to sets leads to fuzzy sets, as discussed in Section 2.4.

2.3.5 Sample Space and Events

In probability theory, the set of all possible outcomes (or events) constitutes the *universal set* or the *sample space*. A sample space consists of sample points that correspond to the possible outcomes. A subset of the sample space is called an event. These definitions form the set basis of probabilistic analysis. An event without sample points is an empty set and is called the impossible set ϕ. A set that contains all the sample points is called the certain event, i.e., the equivalent of the universe X. The certain event is equal to the sample space.

Example 2.3 Example Sample Spaces and Events

The following are example sample spaces:

$$A = \{a \mid a = \text{number of vehicles making a left turn at a specified traffic light within an hour}\} \tag{2.19}$$

$$B = \{b \mid b = \text{number of products produced by an assembly line within an hour}\} \tag{2.20}$$

$$C = \{\text{the strength of concrete delivered at a construction site}\} \tag{2.21}$$

Based on the sample space A, the following events can be defined:

$$A_1 = \{\text{number of cars making a left turn at the specified traffic light within an hour without waiting at the light}\} \quad (2.22)$$

$$A_2 = \{\text{number of vehicles making a left turn at the specified traffic light within an hour after waiting more than 2 minutes}\} \quad (2.23)$$

2.3.6 Euclidean Vector Space and Set Convexity

A frequently used universal set is the space defined by all n-tuples of real numbers as points in the n-dimensional Euclidean vector space R^n, where R is the set of real numbers. Sets belonging to this space are commonly required to possess the property of convexity. A set A is said to be *convex*, if for every pairs of points that belong to A, a straight line that connects these two point stays within the bounds of A. Convex sets are illustrated in Example 2.4.

2.3.7 Venn–Euler Diagrams

Events and sets can be represented using spaces that are bounded by closed shapes, such as circles or ellipses for plane geometry, i.e., two-dimensional spaces, or spheres or cubes for solid geometry, i.e., three-dimensional spaces. These shapes are called Venn–Euler (or simply Venn) diagrams. Belonging, nonbelonging, and overlaps between events and sets can be represented by these diagrams. For elements defined on n-dimensional Euclidean vector space R^n, sets can be classified as *convex* and *nonconvex*.

Example 2.4 A Four-Event Venn Diagram

In the Venn diagram shown in Figure 2.2, four sets, A, B, C, and D, belong to a two-dimensional Euclidean sample space S. The event $C \subset B$ and $A \neq B$. Also, the events A and B have an overlap in the sample space S. Events A, B, and C are convex sets, whereas D is not.

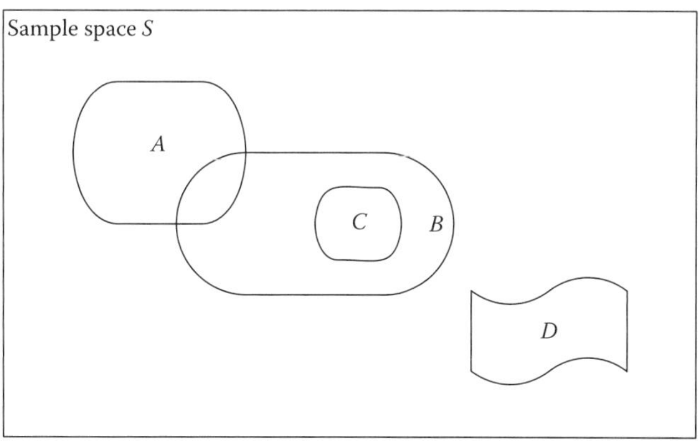

FIGURE 2.2 Crisp events or sets in two-dimensional Euclidean space.

2.3.8 Basic Operations on Sets

In this section, basic operations that can be used for sets are introduced. These operations can, of course, be applied to events as well.

The *union* of sets A and B is the set $A \cup B$ of all elements that belong to A or B, or both. It is denoted by $A \cup B$ and expressed as

$$A \cup B = \{x \mid x \in A \text{ or } x \in B\} \tag{2.24}$$

where the logical connective *or* is *inclusive*, standing for either-or-both. Several events are called *collectively exhaustive* events if the union of these events results in the sample space.

The *intersection* of sets A and B is the set $A \cap B$ of all elements that belong to both A and B. It is denoted by $A \cap B$ and expressed as

$$A \cap B = \{x \mid x \in A, x \in B\} \tag{2.25}$$

Events are called *mutually exclusive* if the occurrence of one event precludes the occurrence of other events. Two sets A and B are called *disjoint* if they have no common member, i.e., if

$$A \cap B = \emptyset \tag{2.26}$$

The *difference* of events A and B is the set of all elements that belong to A but that do not belong to B. It is also called the *relative complement* of a crisp set A with respect to a crisp set B; it is denoted by $B - A$, and defined formally as

$$B - A = \{x \mid x \in B, x \notin A\} \tag{2.27}$$

If the set X is the universal set in a certain discussion, the complement is called *absolute*, and it is usually denoted by $\bar{A}$ rather than $X - A$ for that discussion.

Fundamental rules of these basic operations on sets are shown in Table 2.1.

A collection of disjoint nonempty subsets $(A_i \mid i \in I)$ of A is called a partition on A, written as $\pi(A)$, if and only if the union of these subsets forms the original set A. Thus,

$$\pi(\mathrm{A}) = \{A_i \mid i \in I; A_i \subseteq A, A_i \neq \emptyset\} \tag{2.28}$$

is a partition on A if and only if

$$A_i \cap A_j = \emptyset \tag{2.29}$$

for each pair $i \neq j$ $(i, j \in I)$, and

TABLE 2.1
Other Operational Rules

Rule Type	Operations
Identity rules	$A \cup \emptyset = A$
	$A \cap \emptyset = \emptyset$
	$A \cup S = S$
	$A \cap S = A$
Idempotent rules	$A \cup A = A$
	$A \cap A = A$
Complement rules	$A \cup \bar{A} = S$
	$A \cap \bar{A} = \emptyset$
	$\bar{\bar{A}} = A$
	$\bar{S} = \emptyset$
	$\bar{\emptyset} = S$
Commutative rules	$A \cup B = B \cup A$
	$A \cap B = B \cap A$
Associative rules	$(A \cup B) \cup C = A \cup (B \cup C)$
	$(A \cap B) \cap C = A \cap (B \cap C)$
Distributive rules	$(A \cup B) \cap C = (A \cap C) \cup (B \cap C)$
	$(A \cap B) \cup C = (A \cup C) \cap (B \cup C)$
de Morgan's rules	$\overline{A \cup B} = \bar{A} \cap \bar{B}$
	$\overline{E_1 \cup E_2 \cup \ldots \cup E_n} = \bar{E}_1 \cap \bar{E}_2 \cap \ldots \cap \bar{E}_n$
	$\overline{A \cap B} = \bar{A} \cup \bar{B}$
	$\overline{E_1 \cap E_2 \cap \ldots \cap E_n} = \bar{E}_1 \cup \bar{E}_2 \cup \ldots \cup \bar{E}_n$
Combinations of rules	$\overline{A \cup (B \cap C)} = \bar{A} \cap \overline{(B \cap C)} = (\bar{A} \cap \bar{B}) \cup (\bar{A} \cap \bar{C})$

$$\bigcup_{i \in I} A_i = A \tag{2.30}$$

The Cartesian product of sets A and B is the set of all ordered pairs (a, b) such that $a \in A$ and $b \in B$. In mathematical notations, this can be expressed as

$$A \times B = \{(a, b): a \in A \text{ and } b \in B\} \tag{2.31a}$$

For example, if $A = \{a_1, a_2\}$ and $B = \{b_1, b_2\}$, $A \times B = \{(a_1, b_1), (a_1, b_2), (a_2, b_1), (a_2, b_2)\}$. Clearly, if $A \neq B$, then $A \times B \neq B \times A$. The Cartesian product of a family $\{A_1, A_2, \ldots, A_n\}$ of crisp sets is the set of all n-tuples $(a_1, a_2, \ldots, a_n)$ such that $a_i \in A_i$ $(i = 1, 2, \ldots, n)$. It is written as $A_1 \times A_2 \times \ldots \times A_n$. For finite sets,

$$|A_1 \times A_2 \times \cdots \times A_n| = |A_1|\,|A_2| \cdots |A_n| \tag{2.31b}$$

Cartesian products are commonly used to define spaces for the purpose of constructing relationships between the underlying sets.

Example 2.5 Operations on Sets

Let $A = \{1, 3, 5, 7\}$ and $B = \{b \mid b \text{ is an real number}, b > 3\}$, then

$$A - B = \{1, 3\}$$

$$B - A = \{x \mid x \text{ is a real number}, 3 < x < 5 \text{ or } 5 < x < 7 \text{ or } x > 7\}$$

$$A \cup B = \{x \mid x = 1 \text{ or } x \geq 3\}$$

$$A \cap B = \{5, 7\}$$

Let $A_1 = \{1, 3, 5, 7\}$, $A_2 = \{2, 3, 6, 7\}$, and $A_3 = \{4, 5, 6, 7\}$, then

$$\bigcup_{i=1,2,3} A_i = A_1 \cup A_2 \cup A_3 = \{1, 2, 3, 4, 5, 6, 7\}$$

$$\bigcap_{i=1,2,3} A_i = A_1 \cap A_2 \cap A_3 = \{7\}$$

Example 2.6 Partitions

Let $A = \{1, 2, 3, 4, 5\}$, then

$$\pi_1 = \{\{1, 3\}, \{2, 4, 5\}\}$$

is a partition on A. Another partition on A is

$$\pi_2 = \{\{1, 2, 3, 4\}, \{5\}\}$$

The set $\{\{1, 2, 3\}, \{3, 4, 5\}\}$ is not a partition because $\{1, 2, 3\} \cap \{3, 4, 5\} \neq \phi$. The set $\{\{2, 3\}, \{4, 5\}\}$ is also not a partition because $\{2, 3\} \cup \{4, 5\} \neq A$.

Example 2.7 Cartesian Product

Let $A = \{0, 1\}$ and $B = \{a, b, c\}$, then

$$A \times B = \{(0, a), (1, a), (0, b), (1, b), (0, c), (1, c)\}$$

$$A \times A = A^2 = \{(0, 0), (0, 1), (1, 0), (1, 1)\}$$

2.3.9 Power Sets

For a given set A, the set of all subsets of A is called the *power set* of A and is denoted P_A. For a finite, discrete set A with n elements belonging to it, its power set contains 2^n subsets. Each member of P_A is represented by an $|A|$-tuple of binary digits (values of the membership function A). There are $2^{|A|}$ such tuples; therefore,

$$|P_A| = 2^{|A|} \tag{2.32}$$

Notice that both the empty set Ø and the set A are members of P_A.

Example 2.8 Power Set and Cardinality

For example, the following set A is used to determine its power set and cardinality:

$$A = \{1, 2, 3\} \tag{2.33a}$$

The set A has the following power set:

$$P_A = \{\phi, \{1\}, \{2\}, \{3\}, \{1, 2\}, \{1, 3\}, \{2, 3\}, \{1, 2, 3\}\} \tag{2.33b}$$

These sets have the following respective cardinalities:

$$|A| = 3 \tag{2.34}$$

$$|P_A| = 8 \tag{2.35}$$

Another example, the power set of

$$A = \{2, 4, 6, 8, 10\} \tag{2.36a}$$

as provided in Equation 2.5, is

$$\begin{aligned} P_A = \{&\emptyset, \{2\}, \{4\}, \{6\}, \{8\}, \{10\}, \{2, 4\}, \{2, 6\}, \{2, 8\}, \{2, 10\}, \{4, 6\}, \\ &\{4, 8\}, \{4, 10\}, \{6, 8\}, \{6, 10\}, \{8, 10\}, \{2, 4, 6\}, \{2, 4, 8\}, \{2, 4, 10\}, \\ &\{2, 6, 8\}, \{2, 6, 10\}, \{2, 8, 10\}, \{4, 6, 8\}, \{4, 6, 10\}, \{4, 8, 10\}, \\ &\{6, 8, 10\}, \{2, 4, 6, 8\}, \{2, 4, 6, 10\}, \{2, 4, 8, 10\}, \{2, 6, 8, 10\}, \\ &\{4, 6, 8, 10\}, \{2, 4, 6, 8, 10\}\} \end{aligned} \tag{2.36b}$$

The power set contains $2^5 - 32$ subsets.

2.4 FUZZY SETS AND OPERATIONS

Classical sets are defined in terms of *characteristic functions* by which elements that belong to a given set are labeled by 1 and those that do not belong to it are labeled by 0. However, these numbers, 1 and 0, play in classical set theory a purely *symbolic role*. They are convenient, but may be as well replaced by any other pair of symbols (e.g., m for members and n for nonmembers, $\in$ for members and $\notin$ for nonmembers, etc.). Contrary to classical sets, fuzzy sets are not required to have sharp boundaries that distinguish their members from other objects. The membership in a fuzzy set is not a matter of affirmation or denial, as it is in any classical set, but it is a matter of degree. As stated earlier, the difference between fuzzy sets and classical (crisp or nonfuzzy) sets is that the membership function takes on values in the interval [0, 1] instead of one of the two values {0, 1}. A degenerate fuzzy set A in which the membership function takes values of either $A(x) = 0$ or $A(x) = 1$ for all $x \in X$ is called a *crisp set.*

2.4.1 Membership Function

The degree of membership of objects in fuzzy sets is most commonly expressed by real numbers in the unit interval [0, 1]. Fuzzy sets in which membership degree is expressed in this way are predominant in the literature and are called *standard fuzzy sets.*

Let X be a universe (classical, nonfuzzy) of all elements that are relevant in the context of a particular application. For each application, set X forms the universe of discourse of that application. Let A be a subset of X. Each element of X, x, is associated with a membership value to the subset A, $A(x)$. For a standard fuzzy set, the membership function is given by

$$A: X \rightarrow [0, 1] \tag{2.37}$$

as a mapping from the universe X to the interval of real values [0, 1], where a value in this range means the grade of membership of each element x of X to the set A; i.e., the value of $A(x)$ for each $x \in X$ can be viewed as a measure of the degree of compatibility of x with respect to the concept represented by A, where a value of 1 = the highest degree of compatibility and 0 = no compatibility. The number $A(x)$ represents the *grade of membership* of x in A. The larger this number, the higher the grade of membership of x in A (the more evident it is that x is a member of A, the more compatible is x with the concept represented by A). If $A(x) = 1$, then $x \in A$; if $A(x) = 0$, then $x \notin A$. Hence, crisp sets may be viewed as special cases of fuzzy sets. When they are viewed in this way, symbols 0 and 1 obtain their numerical significance. This is contrary to their strictly symbolic meaning in classical set theory. The assignment of values $A(x)$ to members x of the universal set X is called a *membership function.*

For a fuzzy set A consisting of m discrete elements, the membership function is often expressed as

$$A = \{x_1/A(x_1),\ x_2/A(x_2),\ \ldots,\ x_m/A(x_m)\} \tag{2.38}$$

in which = should be interpreted as "is defined to be" and / is a delimiter. For an infinite A with elements $x \in X$, the membership function of A can be expressed as

$$A = \{x/A(x), \text{ for all } x \in X\} \tag{2.39}$$

in which the function $A(x)$ takes values in the range [0, 1]. In the case of fuzzy sets, the boundaries of A are not sharp, and the membership of any x to A is fuzzy. The *support* of a fuzzy set is defined as all $x \in X$ such that $A(x) > 0$.

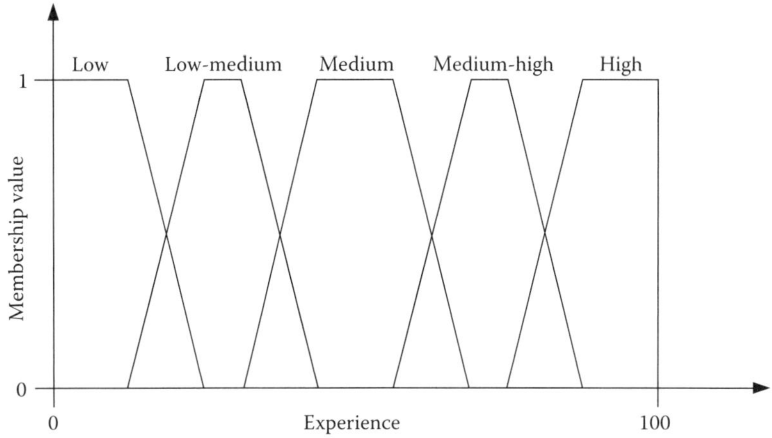

FIGURE 2.3 Experience levels as fuzzy sets.

Fuzzy sets are capable to express more realistically gradual transitions from membership to nonmembership. For example, experts sometimes might provide their opinions using vague terms, phrases, or words in natural languages, such as, *likely*, *large*, and *poor quality*. The meanings of these expressions, which are strongly context dependent, cannot be modeled by crisp sets. Membership functions can be constructed subjectively based on experience and judgment as described in subsequent sections.

Other examples of fuzzy sets include the set of fast swimmers, the set of beautiful women, and the set of large universities in the U.S.

Example 2.9 Fuzzy Sets to Represent Experience

As an example of fuzzy sets, let X be the universal set of experience of an individual to perform some job, such that $x \in X$ can take a real value in the range from $x = 0$, meaning *absolutely no experience in performing the job*, to $x = 100$, meaning *absolutely the highest level of experience*. The range 0 to 100 was selected for representation convenience. Other ranges could be used, such as 0 to 1. Five levels of experience are shown in Figure 2.3 using linguistic descriptors, such as *low*, *medium*, and *high experience*. These experience classifications are meaningful although vague. A fuzzy set representation offers a means of translating this vagueness into meaningful numeric expressions using membership functions.

Another method of presenting fuzzy sets can be based on dividing the range of experience into increments of 10. Therefore, a linguistic variable of the type "low or short experience," designated as A, can be expressed using the following illustrative fuzzy definition that does not correspond to Figure 2.3:

$$\begin{gathered}\text{Short experience, } A = \{x_1 = 100/A(x_1) = 0,\ x_2 = 90/A(x_2) = 0,\\ x_3 = 80/A(x_3) = 0,\ x_4 = 70/A(x_4) = 0,\ x_5 = 60/A(x_5) = 0,\ x_6 = 50/A(x_6) = 0,\\ x_7 = 40/A(x_7) = 0.1,\ x_8 = 30/A(x_8) = 0.5,\ x_9 = 20/A(x_9) = 0.7,\\ x_{10} = 10/A(x_{10}) = 0.9,\ x_{11} = 0/A(x_{11}) = 1.0\}\end{gathered} \tag{2.40}$$

This expression can be written in an abbreviated form by showing experience levels with only nonzero membership values as follows:

$$\text{Short experience, } A = \{40/0.1,\ 30/0.5,\ 20/0.7,\ 10/0.9,\ 0/1\} \tag{2.41a}$$

The fuzziness in the definition of short experience is obvious from Equation 2.40 or 2.41, as opposed to a definition in the form of Equation 2.17. Based on the fuzzy definition of short experience, different grades of experience have different membership values to the fuzzy set "short experience *A*." The membership values are decreasing as a function of increasing grade of experience. In this example, the values of x with nonzero membership values are 40, 30, 20, 10, and 0, and the corresponding membership values are 0.1, 0.5, 0.7, 0.9, and 1.0, respectively. Other values of x larger than 40 have zero membership values to the subset *A*. These membership values should be assigned based on subjective judgment with the help of experts and can be updated with more utilization of such linguistic measures in real-life applications. If a crisp set were used in this example of defining short experience, the value of x would be 0 with a membership value of 1.0. Similarly, long experience, *B*, can be defined as

$$\text{Long experience, } B = \{100/1,\ 90/0.9,\ 80/0.7,\ 70/0.2,\ 60/0.1\} \tag{2.41b}$$

It should be noted that Equations 2.40 and 2.41 show experience taking discrete values for convenience only, since values between these discrete values have membership values that can be computed using interpolation between adjacent values. In order to use fuzzy sets in practical problems, some operational rules similar to those used in classical set theory (Table 2.1) need to be defined.

Example 2.10 Membership Functions of Fuzzy Sets

Let *X* be the set of all real numbers, and let *A* be the fuzzy set of real numbers that are much greater than 1. Then one can give a precise, although subjective or problem oriented, characterization of *A* by defining its membership function $A(x)$, defined as

$$A(x) = \begin{cases} 0 & \text{for } x \le 10 \\ \dfrac{x-10}{90} & \text{for } 10 < x < 100 \\ 1 & \text{for } x \ge 100 \end{cases} \tag{2.42}$$

From Equation 2.42, the following can be stated: $A(12) = 0.22$, $A(20) = 0.11$, $A(50) = 0.44$, and $A(90) = 0.89$.

Example 2.11 Membership Functions of Real-Valued Sets

Examples of fuzzy sets (membership functions) are shown graphically in Figure 2.4. These fuzzy sets are defined on the set of real numbers, which is the universal set in

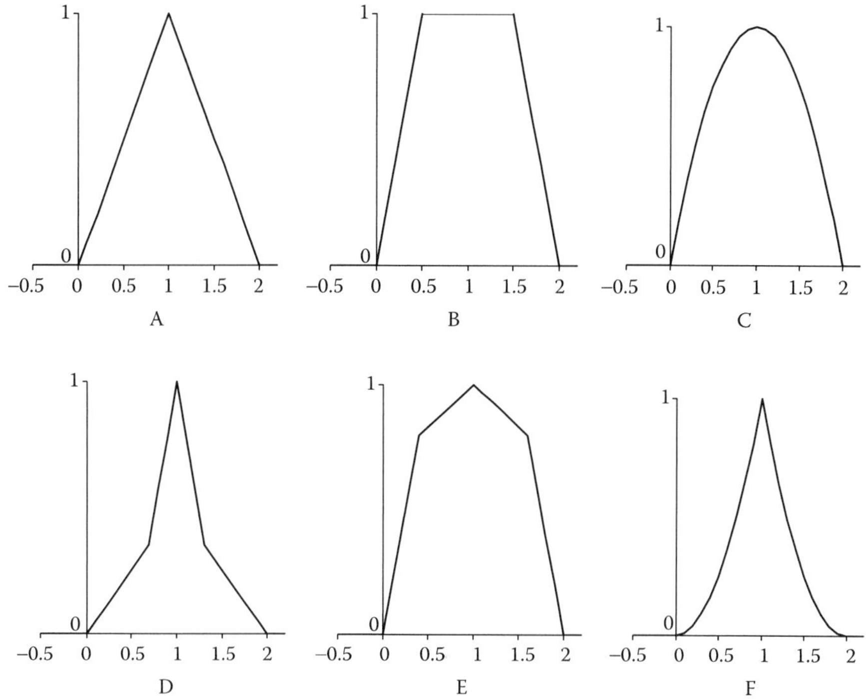

FIGURE 2.4 Example membership functions for the set “around 1.”

this case. They may represent the concept “around 1” (or “close to 1”). Which of these or other possible membership functions is actually an appropriate representation of the concept must be determined in the context of each particular application.

Membership functions in Figure 2.4 are quite typical. Each of them is defined by a set of formulas as follows:

$$A(x) = \begin{cases} x & \text{when } x \in [0,1] \\ 2-x & \text{when } x \in [1,2] \\ 0 & \text{otherwise} \end{cases} \tag{2.43a}$$

$$C(x) = \begin{cases} 2x - x^2 & \text{when } x \in [0,2] \\ 0 & \text{otherwise} \end{cases} \tag{2.43b}$$

$$F(x) = \begin{cases} x^2 & \text{when } x \in [0,1] \\ (2-x)^2 & \text{when } x \in [1,2] \\ 0 & \text{otherwise} \end{cases} \tag{2.43c}$$

2.4.2 α-Cut Representation of Sets

Fuzzy sets can be described effectively using an important concept called an *α-cut* that is needed also to facilitate the performance of fuzzy set operations. For a fuzzy set A defined on universe X and a number α in the unit interval of membership [0, 1], the α-cut of A, denoted by ${}^{\alpha}A$, is the *crisp* set that consists of all elements of A with membership degrees in A greater than or equal to α, i.e.,

$$ {}^{\alpha}A = \{x \mid A(x) \geq \alpha\} \tag{2.44}$$

This crisp set is a set of x values such that the membership value $A(x)$ is greater than or equal to α. A *strong* α-cut of A is denoted and defined as

$$ {}^{\alpha+}A = \{x \mid A(x) > \alpha\} \tag{2.45}$$

The α-cut of A, with $\alpha = 1.0$, is called the *core set* of the fuzzy set A. A fuzzy set with an empty core is called a *subnormal fuzzy set* since the largest value of its membership function is less than 1; otherwise, the fuzzy set is called a *normal set*. It follows directly from the definition provided in Equation 2.45 that by increasing α the next α-cut is always contained in the previous one. Hence, the set of all α-cuts of any given fuzzy set always forms a *nested family of sets,* which uniquely represents the fuzzy set where, for example, $A_1 \subseteq A_2 \subseteq A_3 \ldots \subseteq A_n$ is a family of n nested sets. The strong α-cut, with $\alpha = 0$, is called the *support set* of the fuzzy set A.

A set of nested α-cuts of A, i.e., ${}^{\alpha}A$, can be constructed by incrementally changing the value of α. A convenient representation of such nested sets at *quartile α values* is as follows:

$$\begin{gathered} \text{Nested } {}^{\alpha}A = \\ \{(\alpha_1, \underline{a}_1(\alpha_1), \overline{a}_1(\alpha_1)),\ (\alpha_2, \underline{a}_2(\alpha_2), \overline{a}_2(\alpha_2)),\ (\alpha_3, \underline{a}_3(\alpha_3), \overline{a}_3(\alpha_3)), \\ (\alpha_4, \underline{a}_4(\alpha_4), \overline{a}_4(\alpha_4)),\ (\alpha_5, \underline{a}_5(\alpha_5), \overline{a}_5(\alpha_5))\} \end{gathered} \tag{2.46}$$

where the five sets of triplets correspond to α values of 1, 0.75, 0.50, 0.25, and 0+, respectively, and the underlined and overlined a refer to lower and upper values of an interval, respectively, for each α value. Other quartile levels can be termed *upper quartile set*, *mid-quartile set*, and *lower quartile set*.

The significance of the α-cut representation of fuzzy sets is that it connects fuzzy sets with crisp sets. While each crisp set is a collection of objects that are conceived as a whole, each fuzzy set is a collection of nested crisp sets that are also conceived as a whole. Fuzzy sets are thus wholes of a higher category.

The α-cut representation of fuzzy sets allows us to extend the various properties of crisp sets, established in classical set theory, into their fuzzy counterparts. This is accomplished by requiring that the classical property be satisfied by all α-cuts of the fuzzy set concerned. Any property that is extended in this way from classical set theory into the domain of fuzzy set theory is called *cutworthy property*. For example, when convexity of fuzzy sets is defined by the requirement that all α-cuts of a fuzzy convex set be convex in the classical sense, this conception of fuzzy convexity is cutworthy. Other important examples are cutworthy definitions of fuzzy equivalence, fuzzy compatibility, and various kinds of fuzzy orderings. It is important to realize that many (perhaps most) properties of fuzzy sets, perfectly meaningful and useful, are not cutworthy. These properties do not have any counterparts in classical set theory.

Example 2.12 α-Cut of Experience

Figure 2.5a shows a fuzzy set representation or expression of medium experience based on the subjective assessment of an expert. The fuzzy set has a core set defined by the real range [40, 60], and a support defined by [20, 80]. The α-cuts of A are shown in Table 2.2 for all the quartile values and can be expressed as

$$\text{Nested } {}^{\alpha}A \text{ for medium experience} =$$
$$\{(1,40,60),(0.75,35,65),(0.5,30,70),(0.25,25,75),(0+,20,80)\}$$

Example 2.13 α-Cut of Fuzzy Sets

The α-cuts of all fuzzy sets in Figure 2.4 are closed intervals of real numbers. We can obtain them in this case by determining the inverse functions corresponding to the endpoints of the intervals. For example,

$${}^{\alpha}A = [\alpha, 2-\alpha] \tag{2.47a}$$

$${}^{\alpha}C = [1-\sqrt{1-\alpha}, 1+\sqrt{1-\alpha}] \tag{2.47b}$$

$${}^{\alpha}F = [\sqrt{\alpha}, 2-\sqrt{\alpha}] \tag{2.47c}$$

2.4.3 Fuzzy Venn–Euler Diagrams

Similar to crisp events, fuzzy events and sets can be represented using spaces that are bounded by closed shapes, such as circles with fuzzy boundaries showing the transitional stage from membership to nonmembership. Belonging, nonbelonging, and overlaps between events and sets can be represented by these diagrams. Figure 2.5b shows an example fuzzy event (or set) A with fuzzy boundaries. The various nested shapes can be considered similar to contours in topographical representations that correspond to the five-quartile α-cuts of A.

2.4.4 Operations on Fuzzy Sets

Each of the three basic operations of complement, intersection, and union is unique in classical set theory; however, the counterparts of these operations in fuzzy set

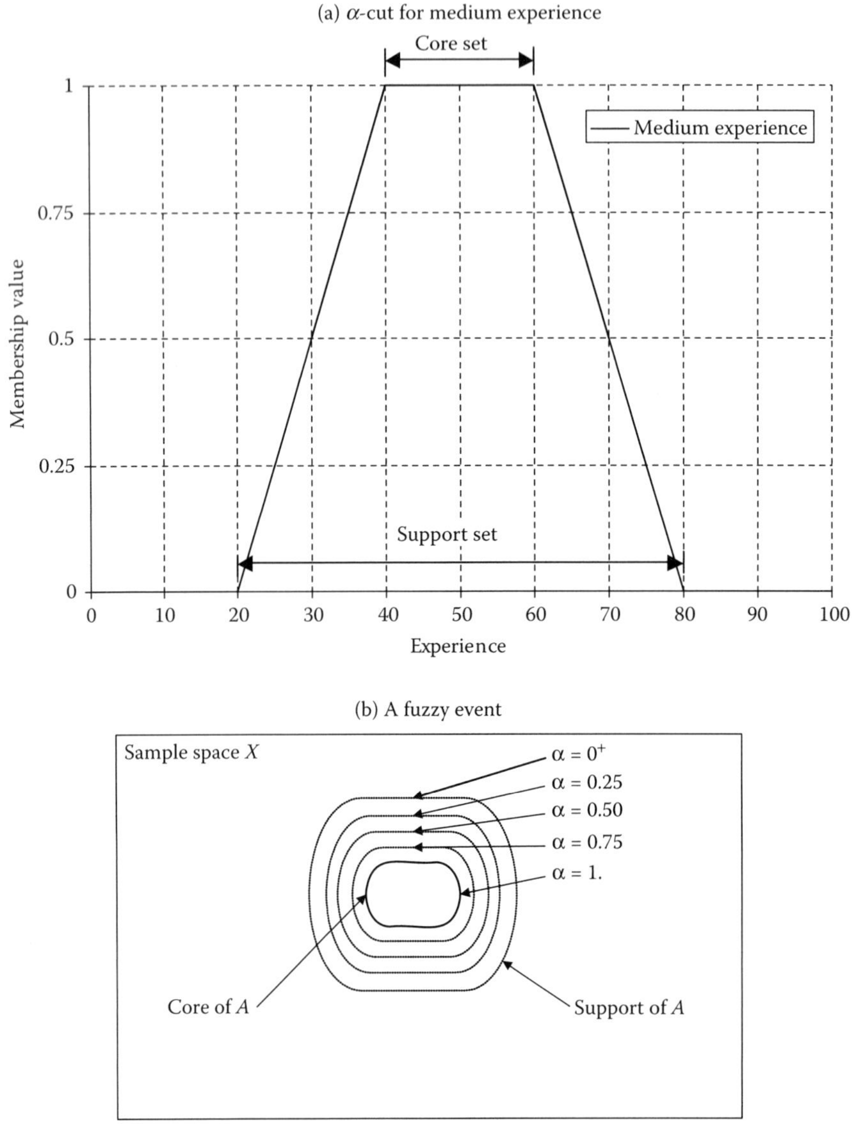

FIGURE 2.5 α-Cut for medium experience and a fuzzy event.

theory are not unique. Each of them consists of a class of functions that satisfy certain properties. In this section, only some notable features of these operations are described. A more comprehensive coverage can be found, for example, in the text by Klir and Yuan (1995a). These operations are defined in an analogous form to the corresponding operations of crisp sets in Section 2.3.7.

The standard *union* of sets $A \subset X$ and $B \subset X$ is the set $A \cup B \subset X$, which corresponds to the connective *or*, and its membership function is defined for each $x \in X$ as follows:

TABLE 2.2
α-Cuts of Medium Experience as Provided in Figure 2.5

α	Lower Limit $(\underline{a})$	Upper Limit $(\bar{a})$	Name of Set
1.	40	60	Core set
0.75	35	65	Upper quartile set
0.50	30	70	Mid-quartile set
0.25	25	75	Lower quartile set
0^+	20	80	Support set

$$(A \cup B)(x) = \max[A(x), B(x)] \tag{2.48}$$

This definition can be generalized to obtain what is called the *triangular conorms*, or for short, the *t-conorms*, such as the *Yager class* of fuzzy unions (Yager, 1980a), provided by

$$(A \cup B)(x) = \min[1, \sqrt[\beta]{(A(x))^{\beta} + (B(x))^{\beta}}\,] \tag{2.49}$$

where $\beta \in (0, \infty)$ is called the *intensity factor*. Equation 2.49 reduces to Equation 2.48 as $\beta \to \infty$. The union based on Equation 2.49 depends on β and can take any value in the following range with lower and upper limits that correspond to $\beta \to \infty$ and $\beta \to 0$, respectively:

$$\max(A(x), B(x)) \le (A \cup B)(x) \le \begin{cases} A(x) & \text{if } B(x) = 0 \\ B(x) & \text{if } A(x) = 0 \\ 1 & \text{otherwise} \end{cases} \tag{2.50}$$

The Yager class is one of several available classes for performing unions, all involving parameters similar to the intensity factor β used in the Yager class. The same notation β is used for these parameters in these classes, although they are fundamentally different and could require different values to achieve similar computational objectives. Table 2.3 provides a summary of some of these classes for computing the union of two fuzzy sets, including the Yager class.

The standard *intersection* of sets $A \subset X$ and $B \subset X$ is the set $A \cap B \subset X$, which corresponds to the connective *and*, and its membership function is defined by

$$(A \cap B)(x) = \min[A(x), B(x)] \tag{2.51}$$

This definition can be generalized to obtain what is called the *triangular norms*, or for short, the *t-norms*, such as the *Yager class* of fuzzy intersections (Yager, 1980a), provided by

TABLE 2.3
Classes of Fuzzy Unions of t-Conorms

Membership Function $(A \cup B)(x)$	Parameters (e.g., Intensity Factor)	Source
$\dfrac{A(x)+B(x)-(\beta-2)A(x)B(x)}{1+(\beta-1)A(x)B(x)}$	$\beta \in (0, \infty)$	Hamacher (1978)
$1-\log_\beta\left[1+\dfrac{(\beta^{1-A(x)}-1)(\beta^{1-B(x)}-1)}{\beta-1}\right]$	$\beta \in (0, \infty)$, $\beta \neq 1$	Frank (1979)
$\min\left[1, \sqrt[\beta]{(A(x))^\beta+(B(x))^\beta}\right]$	$\beta \in (0, \infty)$	Yager (1980a)
$1-\dfrac{(1-A(x))(1-B(x))}{\max(A(x),B(x),1-\beta)}$	$\beta \in [0, 1]$	Dubbois and Prade (1980)
$\left[1+\left[\left(\dfrac{1}{A(x)}-1\right)^{-\beta}+\left(\dfrac{1}{B(x)}-1\right)^{-\beta}\right]^{-1/\beta}\right]^{-1}$	$\beta \in (0, \infty)$	Dombi (1982)
$1-[\max(0,(1-A(x))^\beta+(1-B(x))^\beta-1)]^{1/\beta}$	$\beta \neq 0$	Schweizer and Sklar (1983)

$$(A \cap B)(x) = 1-\min[1, \sqrt[\beta]{(1-A(x))^\beta+(1-B(x))^\beta}\,] \tag{2.52}$$

where $\beta \in (0, \infty)$ is called the *intensity factor.* Equation 2.52 reduces to Equation 2.51 as $\beta \rightarrow \infty$. The intersection based on Equation 2.52 depends on β and can take any value in the following range with lower and upper limits that correspond to $\beta \rightarrow 0$ and $\beta \rightarrow \infty$, respectively:

$$\left.\begin{array}{ll} A(x) & \text{if } B(x)=1 \\ B(x) & \text{if } A(x)=1 \\ 0 & \text{otherwise} \end{array}\right\} \leq (A \cap B)(x) \leq \min(A(x), B(x)) \tag{2.53}$$

The Yager class for the intersection of two fuzzy sets is one of several available classes for performing intersections, all involving parameters similar to the intensity factor β used in the Yager class. The same notation, β, is used for these parameters in these classes, although they are fundamentally different and could require different values to achieve similar computational objectives. Table 2.4 provides a summary

TABLE 2.4
Classes of Fuzzy Intersections of t-Norms

Membership Function $(A \cap B)(x)$	Parameters (e.g., Intensity Factor)	Source
$\dfrac{A(x)B(x)}{\beta+(1-\beta)(A(x)+B(x)-A(x)B(x))}$	$\beta \in (0, \infty)$	Hamacher (1978)
$\log_\beta\left[1+\dfrac{(\beta^{A(x)}-1)(\beta^{B(x)}-1)}{\beta-1}\right]$	$\beta \in (0, \infty)$, $\beta \neq 1$	Frank (1979)
$1-\min\left[1, \sqrt[\beta]{(1-A(x))^\beta+(1-B(x))^\beta}\right]$	$\beta \in (0, \infty)$	Yager (1980a)
$\dfrac{A(x)B(x)}{\max(A(x),B(x),\beta)}$	$\beta \in [0, 1]$	Dubbois and Prade (1980)
$\left[1+\left[\left(\dfrac{1}{A(x)}-1\right)^\beta+\left(\dfrac{1}{B(x)}-1\right)^\beta\right]^{1/\beta}\right]^{-1}$	$\beta \in (0, \infty)$	Dombi (1982)
$[\max(0,(A(x))^\beta+(B(x))^\beta-1)]^{1/\beta}$	$\beta \neq 0$	Schweizer and Sklar (1983)

of some of these classes for computing the intersection of two fuzzy sets, including the Yager class.

The *difference* between sets A and B is the set A-B of all elements that belong to A but do not belong to B. The difference is mathematically expressed as

$$(A-B)(x)=\begin{cases} A(x) & \text{if } B(x)=0 \\ 0 & \text{if } B(x)\neq 0 \end{cases} \tag{2.54}$$

The membership function of the standard complement $\bar{A}$ of a fuzzy set A is defined by

$$\bar{A}(x)=1-A(x) \tag{2.55}$$

This definition has the property of involution expressed as

$$\bar{\bar{A}}(x)=A(x)$$

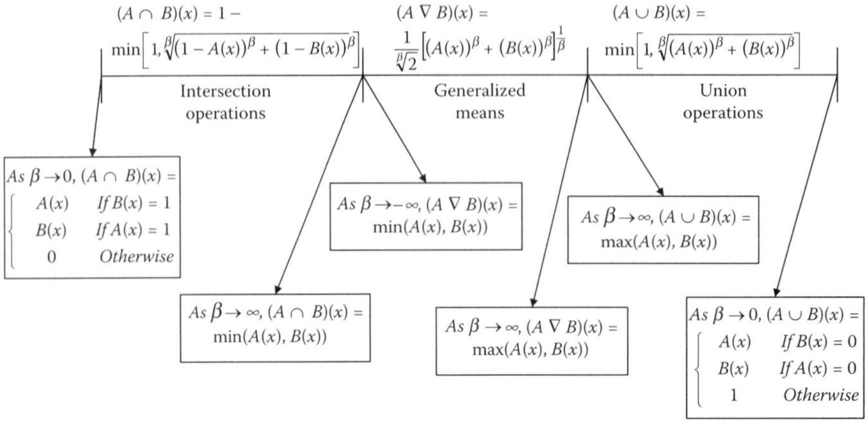

FIGURE 2.6 Examples of generalized fuzzy operations.

The definition can be generalized to obtain the *Yager class* of fuzzy complements (Yager, 1980a) as follows:

$$\bar{A}_\beta(x) = \sqrt[\beta]{1-(A(x))^\beta} \tag{2.56}$$

where $\beta \in (0, \infty)$ is called the *intensity factor.* Equation 2.56 reduces to Equation 2.55 for $\beta = 1$. The definition of the complement can also be generalized, for example, by the *Sugeno class* of fuzzy complements as follows:

$$\bar{A}_\beta(x) = \frac{1-A(x)}{1+\beta B(x)} \tag{2.57}$$

where $\beta \in (0, \infty)$ is called the *intensity factor.* Equation 2.57 reduces to Equation 2.55 as $\beta = 0$. Complements stand for logical negations of concepts represented by fuzzy sets. Which of possible complements to choose is basically an experimental question. The choice is determined in the context of each particular application by eliciting the meaning of negating a given concept by employing a suitable parameterized class of complementation functions.

The selection of a β value in the generalized definitions of fuzzy set operations requires the use of experts to calibrate these operations based on the context of use and application. The generalized definitions offer *softer* and *harder* forms of union, intersection, and complement by changing the value of β. The generalized unions and intersections for two fuzzy sets produce membership values as shown in Figure 2.6.

Figure 2.6 shows also the *averaging* operations that span the gap between the unions and intersections for the case of two arguments *A* and *B* (Klir and Folger, 1988; Klir and Yuan, 1995a). A special case of interest, herein, is the averaging operations for two arguments, $A\nabla B$, called the *generalized means*, for all $x \in X$, defined as

$$(A\nabla B)(x) = \frac{1}{\sqrt[\beta]{2}}[(A(x))^{\beta} + (B(x))^{\beta}]^{1/\beta} \tag{2.58}$$

where β is an *intensity factor* whose range is the set of all real numbers excluding 0. Averaging operations including the generalized means are not associative. They are generalized to more than two arguments in Equation 2.61. Equation 2.58 becomes the geometric mean

$$(A\nabla B)(x) = \sqrt{A(x)B(x)} \tag{2.59}$$

as $\beta \to 0$, and it becomes the arithmetic mean when $\beta = 1$. Equation 2.58 converges to the min and max operations as $\beta \to -\infty$ and $\beta \to \infty$, respectively. Other generalized operations and additional information on this subject are provided by Klir and Yuan (1995a), Klir and Folger (1988), Yager (1980a), Schweizer and Sklar (1983), Frank (1979), Dubois and Prade (1980), and Dombi (1982).

The following cases are examples of some common fuzzy intersections with their usual names (each defined for all $A(x)$ and $B(x) \in [0, 1]$):

Standard fuzzy intersection: $(A \cap B)(\text{x}) = \min[A(x), B(x)]$

Algebraic product: $(A \cap B)(x) = A(x)B(x)$

Bounded difference: $(A \cap B)(x) = \max(0, A(x) + B(x) - 1]$

Drastic intersection: $(A \cap B)(x) = \begin{cases} A(x) & \text{when } B(x) = 1 \\ B(x) & \text{when } A(x) = 1 \\ 0 & \text{otherwise} \end{cases}$

The following cases are examples of some common fuzzy unions with their usual names (each defined for all $A(x)$ and $B(x) \in [0, 1]$):

Algebraic sum: $(A \cup B)(x) = A(x) + B(x) - A(x)B(x)$

Bounded sum: $(A \cup B)(x) = \min(1, a + b) \min[1, A(x) + B(x)]$

Drastic union: $(A \cup B)(x) = \begin{cases} A(x) & \text{when } B(x) = 0 \\ B(x) & \text{when } A(x) = 0 \\ 1 & \text{otherwise} \end{cases}$

Averaging operations can be generalized to several arguments, i.e., membership values $A_1(x) = a_1$, $A_2(x) = a_2$, ..., $A_n(x) = a_n$. They are monotone nondecreasing and idempotent, but are not associative operations. Due to the lack of associativity, they must be defined as functions of n arguments for any $n \geq 2$. It is well known that any averaging operation, h, satisfies the following inequalities:

$$\min(a_1, a_2, \ldots, a_n) \leq h(a_1, a_2, \ldots, a_n) \leq \max(a_1, a_2, \ldots, a_n) \tag{2.60}$$

for any n-tuple $(a_1, a_2, \ldots, a_n) \in [0, 1]^n$. This means that the averaging operations fill the gap between intersection operations and union operations. This concept was introduced as the generalized means by Equation 2.58 for the case of $n = 2$. One class of averaging operations, h_λ, which covers the entire interval between min and max operations, is defined for each n-tuple $(a_1, a_2, \ldots, a_n)$ in $[0, 1]^n$ by the formula

$$h_\lambda(a_1, a_2, \ldots, a_n) = \left(\frac{a_1^\lambda + a_2^\lambda + \ldots + a_n^\lambda}{n} \right)^{1/\lambda} \tag{2.61}$$

where λ is a parameter whose range is the set of all real numbers except 0. For $\lambda = 0$, function h_λ is defined by the limit

$$\lim_{\lambda \to 0} h_\lambda(a_1, a_2, \ldots, a_n) = (a_1, a_2, \ldots, a_n)^{1/n} \tag{2.62}$$

which is the well-known geometric mean. Moreover,

$$\lim_{\lambda \to -\infty} h_\lambda(a_1, a_2, \ldots, a_n) = \min(a_1, a_2, \ldots, a_n) \tag{2.63a}$$

$$\lim_{\lambda \to \infty} h_\lambda(a_1, a_2, \ldots, a_n) = \max(a_1, a_2, \ldots, a_n) \tag{2.63b}$$

Therefore, the standard operations of intersection and union may also be viewed as extreme opposites in the range of averaging operations. Other classes of averaging operations are now available, some of which use weight factors to express relative importance of the individual fuzzy sets involved. For example, the averaging function can be expressed as

$$h(a_i, w_i \,|\, i = 1, 2, \ldots, n) = \sum_{i=1}^{n} w_i a_i \tag{2.64a}$$

where the weighting factors w_i usually take values in the unit interval [0, 1], and

$$\sum_{i=1}^{n} w_i = 1 \tag{2.64b}$$

$h(a_i, w_i | i = 1, 2, \ldots, n)$ expresses for each choice of values w_i the corresponding weighted average of values a_i ($i = 1, 2, \ldots, n$). Again, the choice is an experimental issue.

Additional types of operations applicable to fuzzy sets are modifiers. *Modifiers* are unary operations that are order preserving. Their purpose is to modify fuzzy sets to account for *linguistic hedges*, such as *very*, *fairly*, *extremely*, *more or less*, etc. The most common modifiers either increase or decrease all values of a given membership function $A(x)$. A convenient class of functions, m_λ, that qualify as increasing or decreasing modifiers is defined for each $A(x) \in [0, 1]$ by the formula

$$m_\lambda(A(x)) = (A(x))^\lambda \tag{2.65a}$$

or, denoting $A(x) = a$,

$$m_\lambda(a) = (a)^\lambda \tag{2.65b}$$

where $\lambda > 0$ is a parameter whose value determines which way and how strongly m_λ modifies a given membership function. Clearly, $m_\lambda(a) > a$ when $\lambda \in (0, 1)$, $m_\lambda(a) < a$ when $\lambda \in (1, \infty)$, and $m_\lambda(a) = a$ when $\lambda = 1$. The farther the value of λ from 1, the stronger the modifier m_λ. For example, to modify the set A "close to 1" provided in Figure 2.4 and Equation 2.43a to represent the concept "*very* close to 1," the values of $A(x)$ should be increased. This can be done by modifiers according to Equation 2.65a and b provided that $\lambda > 1$. Applying these modifiers to A results in a new membership function whose shape is exemplified by the function labeled as F in Figure 2.4, and provided by Equation 2.43c using $\lambda = 2$. The smaller the value of λ, the wider is the modified membership function.

Example 2.14 Operations on Fuzzy Experience Levels

Two experts provided the following assessments of long experience using fuzzy sets B and C:

$$\text{Long experience, } B = \{100/1, 90/0.9, 80/0.7, 70/0.2, 60/0.1\} \tag{2.66}$$

$$\text{Long experience, } C = \{100/1, 90/0.8, 80/0.6, 70/0.4, 60/0.2\} \tag{2.67}$$

The union and intersection of these two sets can be computed according to the maximum and minimum operators of Equations 2.48 and 2.51, respectively, to obtain the following:

$$B \cup C = \{100/1, 90/0.9, 80/0.7, 70/0.4, 60/0.2\} \tag{2.68}$$

$$B \cap C = \{100/1, 90/0.8, 80/0.6, 70/0.2, 60/0.1\} \tag{2.69}$$

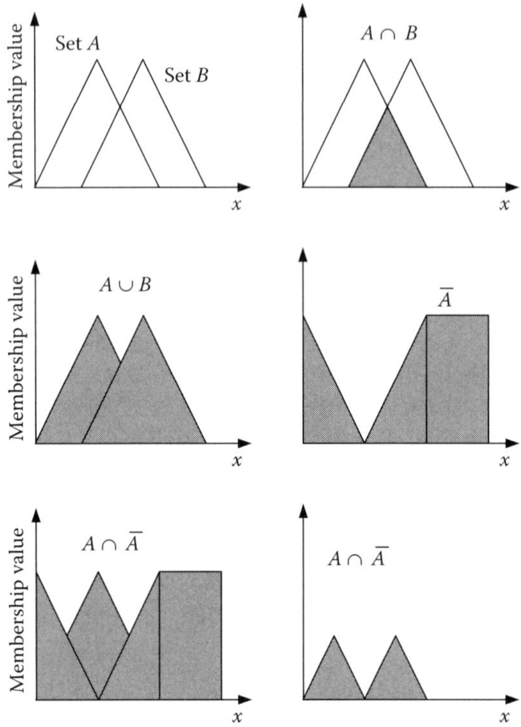

FIGURE 2.7 Examples of fuzzy operations.

The above definitions of the union and intersection of fuzzy sets are the hard definitions. The difference of the two sets is the empty set and can be stated as

$$B - C = \emptyset \tag{2.70}$$

The complement of B according to Equation 2.55 is given by

$$\overline{B}(x) = \{90/0.1,\ 80/0.3,\ 70/0.8,\ 60/0.9,\ 50/1,\ 40/1,\ 30/1,\ 20/1,\ 10/1,\ 0/1\} \tag{2.71}$$

The α-cut of B at $\alpha = 0.7$ according to Equation 2.43a is given by an interval of values as follows:

$$B_{0.7} = \{80,\ 90,\ 100\} \tag{2.72}$$

Example 2.15 Additional Operations on Fuzzy Sets

Figure 2.7 shows examples of fuzzy operations on two fuzzy events A and B. The intersection and union are shown using the min and max rules of Equations 2.51 and 2.48, respectively. The complement is also shown based on Equation 2.55. The last two figures show the unique properties of fuzzy sets with respect to standard operations $A \cup \bar{A} \neq S$ and $A \cap \bar{A} \neq \emptyset$.

Example 2.16 Failure Definition for Reliability and Safety Studies of Structures

Classical structural reliability assessment techniques are based on precise and crisp definitions of failure and survival of a structure in meeting a set of strength, functional, and serviceability criteria. Consider the following performance function:

$$Z = g(X_1, X_2, \ldots, X_n) \tag{2.73}$$

where $X_1, X_2, \ldots, X_n$ = basic random variables and Z = performance measure or safety margin as the difference between structural strength as a response (R) to applied loads (L), i.e., $Z = R - L$. Both R and L are functions of the basic random variables. Equation 2.73 is defined such that failure occurs where $Z < 0$, survival occurs where $Z > 0$, and the limit state equation is defined as $Z = 0$. The probability of failure can then be determined by computing the integral:

$$P_f = \int\int \cdots \int f_{\underline{X}}(x_1, x_2, \ldots, x_n) dx_1, dx_2, \ldots, dx_n \tag{2.74}$$

where $f_{\underline{X}}$ is the joint probability density function of $\{X_1, X_2, \ldots, X_n\}$ and the integration is performed over the region where $Z < 0$. Ayyub and McCuen (2003), Ayyub and Haldar (1984), and White and Ayyub (1985) provide additional information on this model.

The model for crisp failure consists of two basic, mutually exclusive events, i.e., complete survival and complete failure. The transition from one to another is abrupt rather than continuous. This model is illustrated in Figure 2.8, where R_f is the structural response

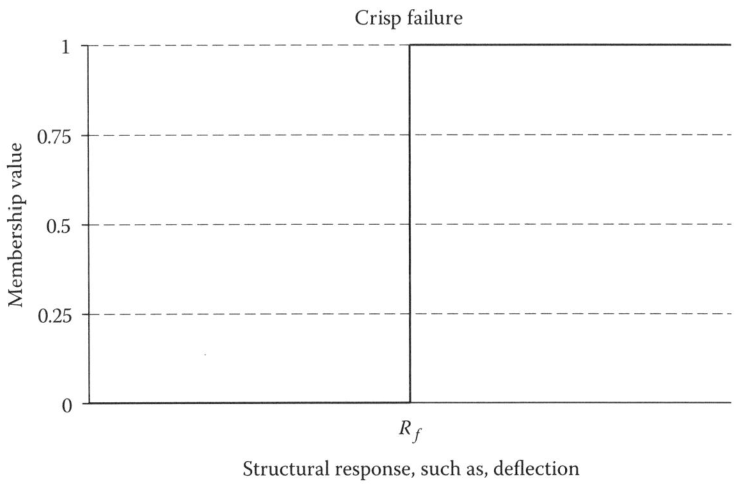

FIGURE 2.8 A crisp failure definition.

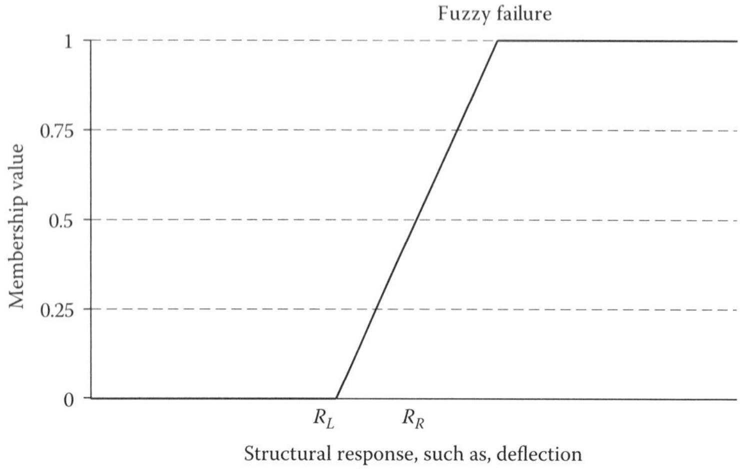

FIGURE 2.9 A fuzzy failure definition.

at the limiting state for a selected design criterion. If the structural response R is smaller than R_f, i.e., $R < R_f$, the complete survival state exists, and it is mapped to the zero failure level ($\alpha = 0$). If the structural response R is larger than R_f, i.e., $R > R_f$, the complete failure state occurs and it is mapped to $\alpha = 1$. The limit state is defined where $R = R_f$.

The fuzzy failure model is illustrated by introducing a subjective failure level index α as shown in Figure 2.9, where R_L and R_R are the left (lower) bound and right (upper) bound of structural response for the region of transitional or partial failure, respectively. The complete survival state is defined where $R \le R_L$, the response in the range ($R_L < R < R_U$) is the transitional state, and the response ($R \ge R_U$) is the complete failure state. In Figure 2.9, the structural response is mapped to the failure-level scale to model some performance event as follows:

$$\text{Performance event } A\text{: } R \rightarrow A = \{\alpha\text{: } \alpha \in [0, 1]\} \tag{2.75}$$

where 0 = failure level for complete survival, 1 = failure level for complete failure, and [0, 1] = all real values in the range of 0 to 1 for all failure levels.

The index α can also be interpreted as a measure of degree of belief in the occurrence of some performance condition. In this case, $\alpha = 0$ is interpreted as no belief of the occurrence of an event, and $\alpha = 1$ means absolute belief in the occurrence of the event.

A mathematical model for structural reliability assessment that includes both ambiguity and vagueness types of uncertainty was suggested by Alvi and Ayyub (1990), Ayyub and Lai (1992) and Lai and Ayyub (1994). The model results in the likelihood of failure over a damage spectrum. Since the structural reliability assessment is based on the fuzzy failure model, the probability of failure, in this case, is a function of α. Figure 2.10 shows various grades of failures expressed as fuzzy events as used by Ayyub and Lai (1992) to assess the structural reliability of ship hull girders.

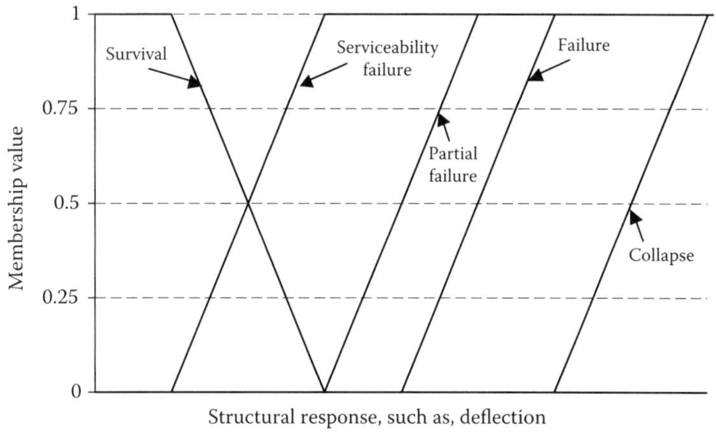

FIGURE 2.10 Fuzzy failure definitions for structural reliability assessment.

2.4.5 Cardinality of Fuzzy Sets

For crisp sets, the cardinality of a set is defined as a measure of its size. For example, a finite, discrete crisp set A with n elements has a cardinality of n, i.e., $|A| = n$.

For discrete fuzzy sets defined on a finite universe, two types of cardinality are used: (1) the scalar cardinality (also called the sigma count), denoted $|A|$ for a fuzzy set A, and (2) the cardinality as a fuzzy number, denoted C_A.

The scalar cardinality for a discrete, bounded set A is defined as

$$|A| = \sum_{x \in X} A(x) \tag{2.76}$$

In many applications using the sum of the membership grades is a good approximation for the cardinality; however, the resulting cardinality is not necessarily an integer. This scalar cardinality might not have an appropriate interpretation in some applications. For example, using a fuzzy subset A of high-quality items produced by a factory to determine how many high-quality items in a universal set X are used to define A could produce a value greater than 1, although the individual membership values are small numbers. In this case, the cardinality does not provide a reasonable count of high-quality items.

The fuzzy cardinality for a discrete, finite set A is a fuzzy nonnegative integer number defined as follows:

$$C_A(|{}^{\alpha}A|) = \alpha \tag{2.77}$$

where $|{}^{\alpha}A|$ is the cardinality of the α-cut of A.

Example 2.17 Cardinality of Fuzzy Sets

For example, the finite set $A = \{40/0.1, 30/0.4, 20/0.7, 10/0.9, 0/1.0\}$ has a scalar cardinality of $|A| = 0.1 + 0.4 + 0.7 + 0.9 + 1.0 = 3.1$. It has a fuzzy cardinality calculated at the given membership values as follows:

$$C_A(|{}^{1}A|) = 1 \text{ for } {}^{1}A = \{0\} \text{ and } |{}^{1}A| = 1$$

$$C_A(|{}^{0.9}A|) = 0.9 \text{ for } {}^{0.9}A = \{0, 10\} \text{ and } |{}^{0.9}A| = 2$$

$$C_A(|{}^{0.7}A|) = 0.7 \text{ for } {}^{0.7}A = \{0, 10, 20\} \text{ and } |{}^{0.7}A| = 3$$

$$C_A(|{}^{0.5}A|) = 0.5 \text{ for } {}^{0.5}A = \{0, 10, 20, 30\} \text{ and } |{}^{0.5}A| = 4$$

$$C_A(|{}^{0.1}A|)\ 0.1 \text{ for } {}^{0.1}A = \{0, 10, 20, 30, 40\} \text{ and } |{}^{0.1}A| = 5$$

Thus, the fuzzy cardinality is

$$C_A = |A| = \{5/0.1, 4/0.5, 3/0.7, 2/0.9, 1/1\}\}$$

It can be also computed as other membership values (e.g., quartiles) as follows:

$$C_A(|{}^{1}A|) = 1 \text{ for } {}^{1}A = \{0\} \text{ and } |{}^{1}A| = 1$$

$$C_A(|{}^{0.75}A|) = 0.75 \text{ for } {}^{0.75}A = \{0, 10\} \text{ and } |{}^{0.75}A| = 2$$

$$C_A(|{}^{0.5}A|) = 0.5 \text{ for } {}^{0.5}A = \{0, 10, 20\} \text{ and } |{}^{0.5}A| = 3$$

$$C_A(|{}^{0.25}A|) = 0.25 \text{ for } {}^{0.25}A = \{0, 10, 20, 30\} \text{ and } |{}^{0.25}A| = 4$$

$$C_A(|{}^{0+}A|)\ 0+ \text{ for } {}^{0+}A = \{0, 10, 20, 30, 40\} \text{ and } |{}^{0+}A| = 5$$

The result in this case can be expressed as

$$C_A = \{1/1, 2/0.75, 3/0.5, 4/0.25, 5/0\}$$

2.4.6 Fuzzy Subsets

A fuzzy set A is called to be a *subset* of or equal to a fuzzy set B, $A \subseteq B$, if and only if $A(x) \leq B(x)$ for all $x \in X$. A fuzzy set A is called to be equal to fuzzy set B, $A = B$, if and only if $A(x) = B(x)$ for all $x \in X$. The set of all fuzzy subsets of X is called the *fuzzy power set* of X. This set is crisp, even though its members are fuzzy sets. Moreover, this set is always infinite, even if X is finite.

In many situations involving two fuzzy sets A and B, neither of the sets is a subset of the other. In such cases, the extent to which one set, say A, is a subset of a set B can be measured by a scalar quantity called subsethood measure (*sub*) that is based on the scalar cardinality definition of Equations 2.76 and 2.77 as follows:

$$sub(A \subset B) = \frac{\sum_{x \in X} A(x) - \sum_{x \in X} \max[0, A(x) - B(x)]}{\sum_{x \in X} A(x)} \tag{2.78a}$$

Similarly, the subsethood measure of B in A is

$$sub(B \subset A) = \frac{\sum_{x \in X} B(x) - \sum_{x \in X} \max[0, B(x) - A(x)]}{\sum_{x \in X} B(x)} \tag{2.78b}$$

The negative term in the numerator describes the sum of the degrees to which the subset inequality $A(x) \leq B(x)$ is violated, the positive terms describe the largest possible violation of the inequality, the difference in the numerator describes the sum of the degrees to which the inequality is not violated, and the term in the denominator is a normalizing factor to obtain the range. This subsethood measure has the following property:

$$0 \leq sub\ (A \subset B) \leq 1 \tag{2.78c}$$

The minimum, $sub\ (A \subset B) = 0$, occurs where sets A and B do not intersect, and the maximum, $sub\ (A \subset B) = 1$, occurs where a set A is a complete subset of B. When sets A and B are defined on a bounded subset of real numbers (i.e., X is a closed interval of real numbers), the three Σ terms in Equation 2.78a and b are replaced with integrals over X.

2.4.7 Fuzzy Intervals, Numbers, and Arithmetic

Fuzzy sets that are defined on either the set of real numbers, R, or the set of integers, I, have a special significance in fuzzy set theory. Among them, the most important are cutworthy *fuzzy intervals* that are defined by requiring that each α-cut be a single closed and bounded interval of real numbers for all $\alpha \in [0, 1]$. A fuzzy interval, A, may conveniently by represented for each $x \in R$ by the canonical form

$$A(x)=\begin{cases} f_A(x) & \text{when } x\in[a,b] \\ 1 & \text{when } x\in[b,c] \\ g_A(x) & \text{when } x\in(c,d] \\ 0 & \text{otherwise} \end{cases} \tag{2.79}$$

where a, b, c, and d are specific real numbers such that $a \le b \le c \le d$, f_A is a real-valued function that is increasing, and g_A is a real-valued function that is decreasing. In most applications, functions f_A and g_A are continuous, but in general, they may be only semicontinuous from the right and left, respectively. When $A(x) = 1$ for exactly one $x \in R$ (i.e., $b = c$ in the canonical representation), A is called a *fuzzy number*. If the shape of the membership function is triangular, it is called a triangular fuzzy number.

For any fuzzy interval A expressed in the canonical form, the α-cuts of A are expressed for all $\alpha \in [0, 1]$ by the formula

$$^{\alpha}A=\begin{cases} [f_A^{-1}(\alpha), g_A^{-1}(\alpha)] & \text{when } \alpha\in(0,1) \\ [b,c] & \text{when } \alpha=1 \end{cases} \tag{2.80}$$

where f_A^{-1} and g_A^{-1} are the inverse functions of f_A and g_A, respectively. An α-cut of a fuzzy interval A can be expressed for an individual α as follows:

$$^{\alpha}A = {}^{\alpha}[\underline{a},\bar{a}] \tag{2.81}$$

where the interval range $^{\alpha}[\underline{a},\bar{a}]$, is a function at α.

A triangular fuzzy number A can be denoted as $A[a_L, a_m, a_R]$, where a_L = the left (i.e., the lowest) value of support, a_R = the right (i.e., the highest) value of support, and a_m = the middle value at the mode of the triangle. A trapezoidal fuzzy interval A can similarly be denoted as $A[a_L, a_{mL}, a_{mR}, a_R]$, where a_L = the left (i.e., the lowest) value of support, a_R = the right (i.e., the highest) value of support, and a_{mL} and a_{mR} = the left and right middle values at the mode range of the trapezoid. A triangular fuzzy number is commonly used to represent an approximate number, such as the weight of a machine is approximately 200 pounds, whereas a trapezoidal fuzzy interval is an approximate interval, such as the capacity of a machine is approximately 200 to 250 pounds. These examples of fuzzy numbers and fuzzy intervals are shown in Figure 2.11 with their crisp counterparts. A crisp number can be represented mathematically as

$$A = a_m \tag{2.82}$$

A fuzzy number $A[a_L, a_m, a_R]$ can be represented mathematically as

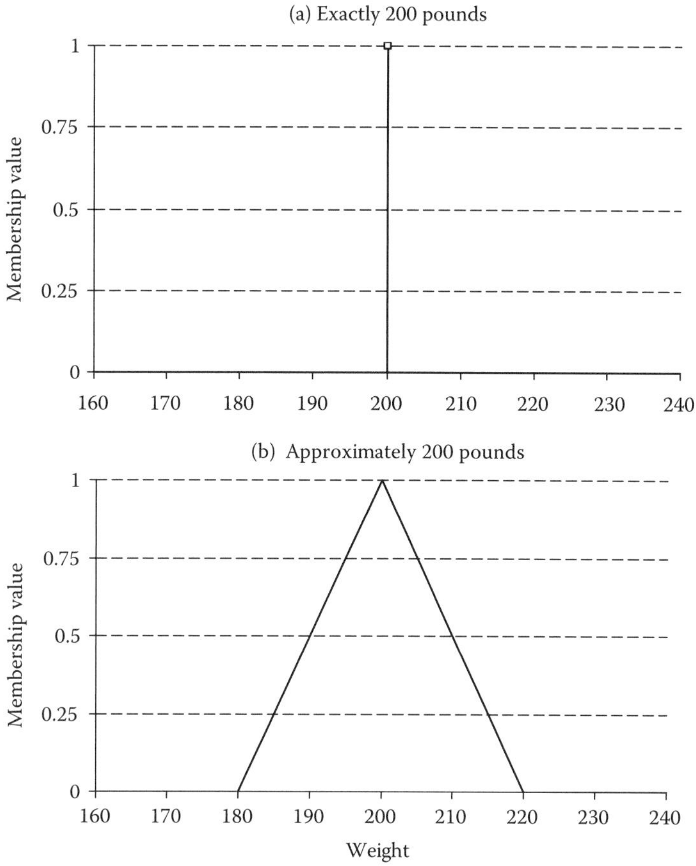

FIGURE 2.11 (a) Crisp number, (b) fuzzy number, (c) crisp interval, and (d) fuzzy interval. *Continued.*

$$A(x) = \begin{cases} \dfrac{x - a_L}{a_m - a_L} & \text{for } a_L \le x \le a_m \\ \dfrac{a_R - x}{a_R - a_m} & \text{for } a_m \le x \le a_R \\ 0 & \text{otherwise} \end{cases} \tag{2.83}$$

The α-cuts for this fuzzy number are

$$^{\alpha}A = [a_L + \alpha(a_m - a_L), a_R - \alpha(a_R - a_m)] \tag{2.84}$$

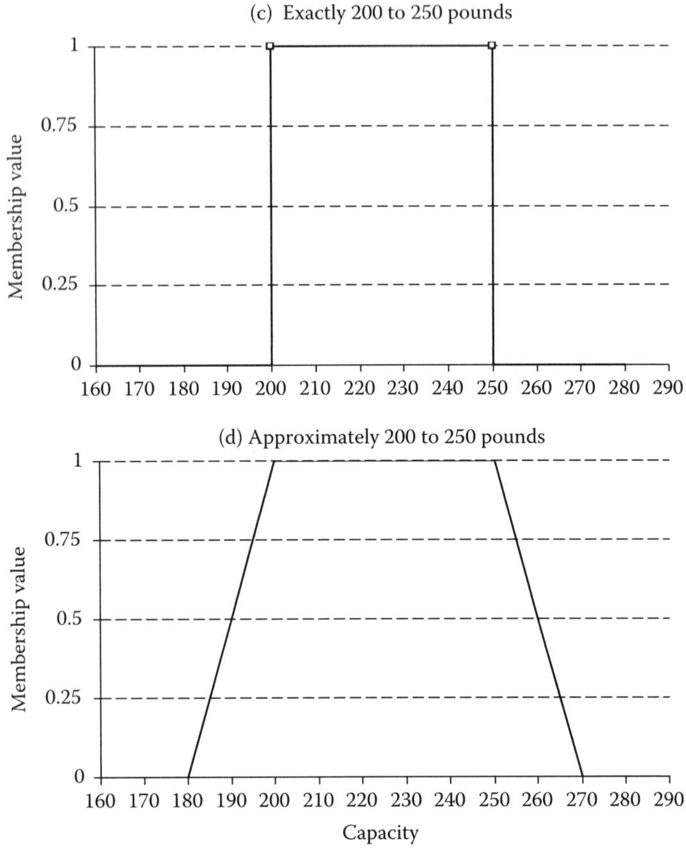

FIGURE 2.11 *Continued.*

A crisp interval can be represented mathematically as

$$A = [a_{mL}, a_{mR}] \tag{2.85}$$

A fuzzy interval $A[a_L, a_{mL}, a_{mR}, a_R]$ can be represented mathematically as

$$A(x) = \begin{cases} \dfrac{x - a_L}{a_{mL} - a_L} & \text{for } a_L \le x \le a_{mL} \\ 1 & \text{for } a_{mL} \le x \le a_{mR} \\ \dfrac{a_R - x}{a_R - a_{mR}} & \text{for } a_{mR} \le x \le a_R \\ 0 & \text{otherwise} \end{cases} \tag{2.86}$$

The α-cuts for this fuzzy interval are

$$^{\alpha}A = [a_L + \alpha(a_{mL} - a_L), a_R - \alpha(a_R - a_{mR})] \tag{2.87}$$

For two fuzzy numbers or intervals A and B, let $^{\alpha}A = {}^{\alpha}[\underline{a}, \overline{a}]$ and $^{\alpha}B = {}^{\alpha}[\underline{b}, \overline{b}]$, where a and b are real numbers on the lower and upper ends of the ranges for $\alpha \in [0, 1]$. The fuzzy arithmetic of addition, subtraction, multiplication, and division, respectively, can be defined as follows (Kaufmann and Gupta, 1985) based on interval-valued arithmetic (Moore, 1966, 1979):

$$^{\alpha}A + {}^{\alpha}B = {}^{\alpha}[\underline{a}, \overline{a}] + {}^{\alpha}[\underline{b}, \overline{b}] = {}^{\alpha}[\underline{a} + \underline{b}, \overline{a} + \overline{b}] \tag{2.88}$$

$$^{\alpha}A - {}^{\alpha}B = {}^{\alpha}[\underline{a}, \overline{a}] - {}^{\alpha}[\underline{b}, \overline{b}] = {}^{\alpha}[\underline{a} - \overline{b}, \overline{a} - \underline{b}] \tag{2.89}$$

$$^{\alpha}A \times {}^{\alpha}B = {}^{\alpha}[\underline{a}, \overline{a}] \times {}^{\alpha}[\underline{b}, \overline{b}] = {}^{\alpha}[\min(\underline{a}\underline{b}, \underline{a}\overline{b}, \overline{a}\underline{b}, \overline{a}\overline{b}), \max(\underline{a}\underline{b}, \underline{a}\overline{b}, \overline{a}\underline{b}, \overline{a}\overline{b})] \tag{2.90}$$

$$\begin{aligned} &^{\alpha}A / {}^{\alpha}B = \\ &^{\alpha}[\underline{a}, \overline{a}] / {}^{\alpha}[\underline{b}, \overline{b}] = {}^{\alpha}[\min(\underline{a}/\underline{b}, \underline{a}/\overline{b}, \overline{a}/\underline{b}, \overline{a}/\overline{b}), \max(\underline{a}/\underline{b}, \underline{a}/\overline{b}, \overline{a}/\underline{b}, \overline{a}/\overline{b})] \end{aligned} \tag{2.91}$$

Equation 2.91 requires that $0 \notin [\underline{b}, \overline{b}]$. The above equations can be used to propagate interval input into input–output models to obtain interval outputs using methods such as the *vertex method* (Dong and Wong, 1986a, 1986b, 1986c). Equations 2.88 to 2.91 can also be used to perform fuzzy arithmetic when one of the numbers is a real number (i.e., when either $\underline{a} = \overline{a}$ or $\underline{b} = \overline{b}$).

In order to use fuzzy arithmetic in numerical methods for the purpose of accommodating fuzzy coefficients, for example, solving a system of linear equations with fuzzy coefficients, the fuzzy subtraction, multiplication, and division of Equations 2.89 to 2.91, respectively, should be revised to the constrained type (Ayyub and Chao, 1998) as defined by Klir and Cooper (1996) and Klir and Pan (1998). For example, the definition of fuzzy division for a fuzzy number by another fuzzy number of the same magnitude can be different than the fuzzy division of a fuzzy number by itself. Such a difference for $^{\alpha}A/^{\alpha}A$ can be provided for $^{\alpha}A = [\underline{a}, \overline{a}]$ with $0 \notin [\underline{a}, \overline{a}]$, and for all $x \in {}^{\alpha}A$ and $y \in {}^{\alpha}A$ as follows:

1. For nonconstrained x and y, the unconstrained fuzzy division based on Equation 2.91 can be expressed as

$$\begin{aligned} &^{\alpha}A / {}^{\alpha}A = \\ &^{\alpha}[\underline{a}, \overline{a}] / {}^{\alpha}[\underline{a}, \overline{a}] = {}^{\alpha}[\min(\underline{a}/\underline{a}, \underline{a}/\overline{a}, \overline{a}/\underline{a}, \overline{a}/\overline{a}), \max(\underline{a}/\underline{a}, \underline{a}/\overline{a}, \overline{a}/\underline{a}, \overline{a}/\overline{a})] \end{aligned} \tag{2.92a}$$

2. For a constrained case where $x = y$, the fuzzy division is given by

$$ {}^{\alpha}A/{}^{\alpha}A\Big|_{x=y} = {}^{\alpha}[\underline{a},\bar{a}]/{}^{\alpha}[\underline{a},\bar{a}]\Big|_{x=y} = 1 \tag{2.92b}$$

For fuzzy subtraction, a similar definition for ${}^{\alpha}A - {}^{\alpha}A$ can be given for all $x \in {}^{\alpha}A$ and $y \in {}^{\alpha}A$ as follows:

1. For nonconstrained x and y, the unconstrained fuzzy subtraction based on Equation 2.89 is given by

$$ {}^{\alpha}A - {}^{\alpha}A = {}^{\alpha}[\underline{a},\bar{a}] - {}^{\alpha}[\underline{a},\bar{a}] = {}^{\alpha}[\underline{a} - \bar{a}, \bar{a} - \underline{a}] \tag{2.92c}$$

2. For a constrained case where $x = y$, the constrained fuzzy subtraction is

$$ {}^{\alpha}A - {}^{\alpha}A\Big|_{x=y} = {}^{\alpha}[\underline{a},\bar{a}] - {}^{\alpha}[\underline{a},\bar{a}]\Big|_{x=y} = {}^{\alpha}[\underline{a} - \underline{a}, \bar{a} - \bar{a}] = 0 \tag{2.92d}$$

For fuzzy multiplication, a similar definition for ${}^{\alpha}A \times {}^{\alpha}A$ can be given for all $x \in {}^{\alpha}A$ and $y \in {}^{\alpha}A$ as follows:

1. For nonconstrained x and y, the unconstrained fuzzy multiplication based on Equation 2.90 is given by

$$ {}^{\alpha}A \times {}^{\alpha}A = {}^{\alpha}[\underline{a},\bar{a}] \times {}^{\alpha}[\underline{a},\bar{a}] = {}^{\alpha}[\min(\underline{a}\underline{a}, \underline{a}\bar{a}, \bar{a}\underline{a}, \bar{a}\bar{a}), \max(\underline{a}\underline{a}, \underline{a}\bar{a}, \bar{a}\underline{a}, \bar{a}\bar{a})] \tag{2.92e}$$

2. For a constrained case where $x = y$, the constrained fuzzy multiplication is

$$ {}^{\alpha}A \times {}^{\alpha}A\Big|_{x=y} = {}^{\alpha}[\underline{a},\bar{a}] \times {}^{\alpha}[\underline{a},\bar{a}]\Big|_{x=y} = \begin{cases} {}^{\alpha}[\bar{a}^2, \underline{a}^2] & \text{where } \bar{a} < 0 \\ {}^{\alpha}[\underline{a}^2, \bar{a}^2] & \text{where } \underline{a} > 0 \\ {}^{\alpha}[0, \max(\underline{a}^2, \bar{a}^2)] & \text{where } 0 \in [\underline{a}, \bar{a}] \end{cases} \tag{2.92f}$$

Computing the square root and n^{th} power requires the use of constrained arithmetic. the square root is defined as follows, respectively:

$$ {}^{\alpha}[\sqrt{A}] = {}^{\alpha}[\sqrt{\underline{a}}, \sqrt{\bar{a}}] \quad \text{where } 0 \le \underline{a} \le \bar{a} \tag{2.93}$$

For fuzzy addition, the definition for ${}^{\alpha}A \times {}^{\alpha}A$ can be given for all $x \in {}^{\alpha}A$, and $y \in {}^{\alpha}A$ is the same for the constrained and nonconstrained types as follows:

$$ {}^{\alpha}A + {}^{\alpha}A = {}^{\alpha}[\underline{a},\bar{a}] + {}^{\alpha}[\underline{a},\bar{a}] = {}^{\alpha}[2\underline{a}, 2\bar{a}] \tag{2.94}$$

Many problems in engineering and the sciences can be modeled based on physical laws that would require constrained fuzzy arithmetic for propagating fuzzy input into these models. For example, a system of linear equations with fuzzy coefficients and other numerical problems can be constructed to solve such problems requiring constrained fuzzy arithmetic. General formulations of constrained fuzzy arithmetic are needed. The constraint does not need to be limited to $x = y$. The concept can be extended to any constraint, such as equalities of the type $x + y = 100$ and $x^2 + y^2 = 1$, or inequalities of the type $x < y$ and $x^2 + y^2 \leq 1$. The inequality constraints require the use of union operations to deal with numerical answers that can be produced by several x and y combinations, i.e., lack of uniqueness or mapping from many to one. Fuzzy arithmetic can be used to develop methods for aggregating expert opinions that are expressed in linguistic or approximate terms. This aggregation procedure retains uncertainties in the underlying opinions by obtaining a fuzzy combined opinion. Klir and Cooper (1996), Klir (1997a, 1997b), and Klir and Pan (1998) provide additional information on constrained arithmetic.

Example 2.18 Additional Operations on Fuzzy Sets

The following two fuzzy numbers are used to perform a series of arithmetic operations as provided below for demonstration purposes:

$$A = [1, 2, 3]\text{, a triangular fuzzy number}$$

$$B = [2, 4, 6]\text{, a triangular fuzzy number}$$

$$A + B = B + A = [3, 6, 9]\text{, a triangular fuzzy number}$$

$$A - B = [-5, -2, 1]\text{, a triangular fuzzy number}$$

The multiplication and division do not produce triangular numbers, and they need to be evaluated using α-cuts. The computations can also be performed using the α-cuts of Equations 2.84 and 2.87. For example, at $\alpha = 0$, 0.5, and 1, the intervals for the product and division are

$$\text{At } \alpha = 0,\ A \times B = [2, 18]$$

$$\text{At } \alpha = 0.5,\ A \times B = [4.5, 12.5]$$

$$\text{At } \alpha = 1,\ A \times B = [8, 8]$$

$$\text{At } \alpha = 0,\ B/A = [2/3, 6]$$

$$\text{At } \alpha = 0.5,\ B/A = [1.2, 10/3]$$

$$\text{At } \alpha = 1,\ B/A = [2, 2]$$

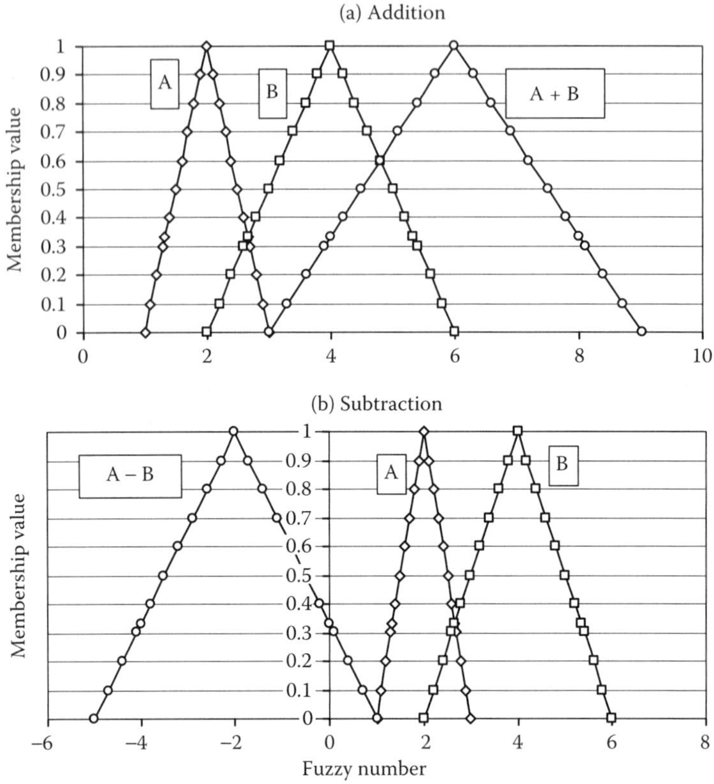

FIGURE 2.12 Example fuzzy arithmetic. *Continued.*

Figure 2.12 shows graphically the results of addition, subtraction, multiplication, division, constrained addition, and constrained multiplication at various α-cuts. As an example for the case of constrained addition, the following is provided:

$$\text{At } \alpha = 0,\ A + B = [1, 3] + [1, 3] = [2(1), 2(3)] = [2, 6]$$

$$\text{At } \alpha = 0.2,\ A \times B = [1.2, 2.8] \times [1.2, 2.8] = [1.2^2, 2.8^2] = [1.44, 7.84]$$

Figure 2.12e and f is not fully illustrative of constrained fuzzy arithmetic because the results are in these cases exactly the same as with standard fuzzy arithmetic. A simple example, such as $(A + B)/B$, would be very illustrative because this function is increasing in A and decreasing in B for this case, where A and B are positive for all α-cuts, and hence the result based on constraint fuzzy arithmetic can be expressed in terms of the endpoints of α-cuts of A and B. The standard and constrained fuzzy arithmetic give different results, but if the expression is rewritten as $A/B + 1$, the standard and constrained fuzzy arithmetic give the same results. The computations in this example were performed at discrete points for illustration purposes, but they can be performed to deal with both kinds of fuzzy arithmetic by working with analytic

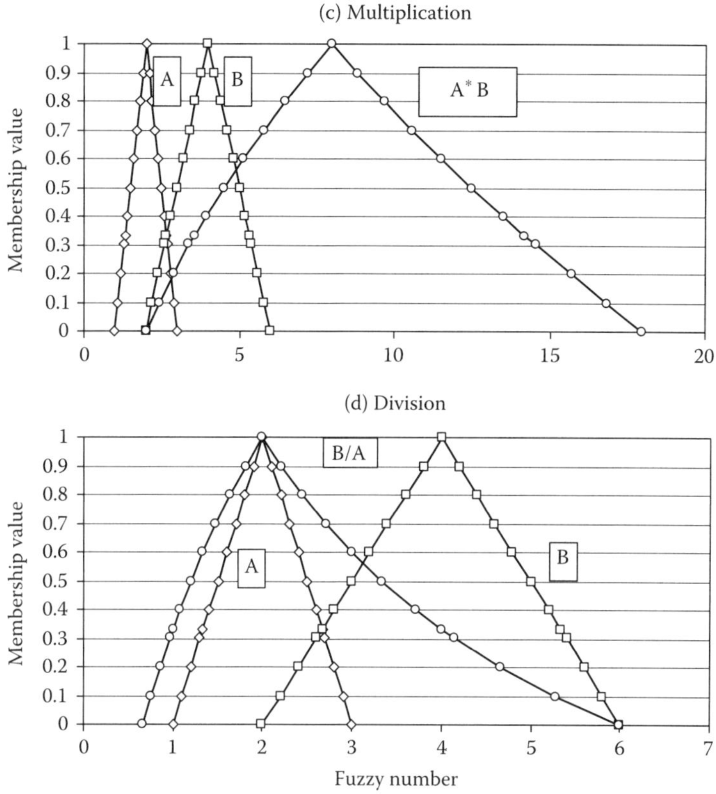

FIGURE 2.12 *Continued.*

forms of the membership functions (such as Equations 2.42 to 2.45), rather than with the discrete forms.

2.4.8 Fuzzy Relations

Fuzzy sets defined on a universal set in the form of a Cartesian product of two or more sets are called *fuzzy relations*. Individual sets in the Cartesian product of a fuzzy relation are called *dimensions* of the relation. When n-sets are involved in the Cartesian product, we call the relation n-dimensional ($n \geq 2$). Fuzzy sets may be viewed as degenerate, one-dimensional relations.

A relation between two or more sets is defined as an expression of association, interrelationship, interconnection, or interaction among these sets. The expression can be made in a crisp format indicating the presence or absence of such a relation, or it can be made in a fuzzy format indicating the strength of the relationship. An n-dimensional fuzzy relation R is thus defined by a membership function of the general form

$$R: X_1 \times X_2 \times \ldots \times X_n \rightarrow [0, 1] \tag{2.95}$$

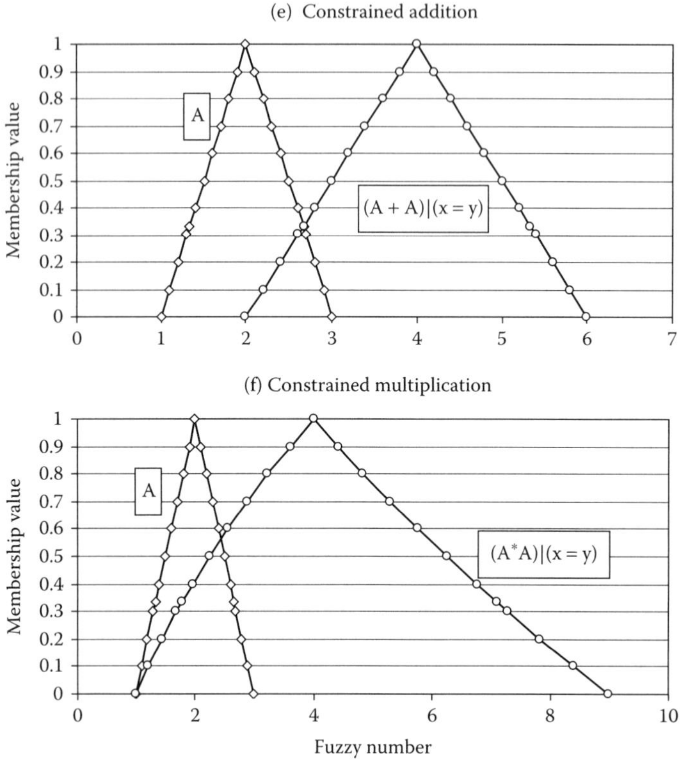

FIGURE 2.12 *Continued.*

The strength of a relationship R between $x \in X$ and $y \in Y$ is commonly expressed using $R(x, y) \in [0, 1]$. If $R(x, y) \in \{0, 1\}$, i.e., $R(x, y)$ can take one of two values 0 or 1, then R is considered to be a crisp relation, whereas if $R(x, y) \in [0, 1]$, i.e., $R(x, y)$ can take any real value in the range [0, 1], then R is considered to be a fuzzy relation. A fuzzy relation R between two sets $A \subset X$ and $B \subset Y$ is defined on the Cartesian product of A and B: the set of all ordered pairs (a, b) such that $a \in A$ and $b \in B$. In mathematical notations, the Cartesian product is expressed by the following set:

$$A \times B = \{(a, b)\colon a \in A \text{ and } b \in B\} \tag{2.96}$$

For discrete A and B, relations can be expressed in a matrix form as follows:

$$R = A \times B = \begin{array}{c|ccccc} & b_1 & b_2 & \cdots & b_m \\ \hline a_1 & R(a_1,b_1) & R(a_1,b_2) & \cdots & R(a_1,b_m) \\ a_2 & R(a_2,b_1) & R(a_2,b_2) & \cdots & R(a_2,b_m) \\ \vdots & \vdots & \vdots & \vdots & \vdots \\ a_n & R(a_n,b_1) & R(a_n,b_2) & \cdots & R(a_n,b_m) \end{array} \tag{2.97}$$

in which $A = \{a_1, a_2, \ldots, a_n\}$, $B = \{b_1, b_2, \ldots, b_m\}$, and $R(a_i, b_j)$ = strength of relationship for the ordered pair (a_i, b_j). The membership value can be determined based on judgment.

The membership degree $R(x_1, x_2, \ldots, x_n)$ of a particular n-tuple $(x_1, x_2, \ldots, x_n)$, where $x_i \in X_i$ for all i, indicates the *strength of relation* among elements of the n-tuple. Relations that are two-dimensional have special significance; they are usually called *binary relations.*

All concepts and operations applicable to fuzzy sets are applicable to fuzzy relations as well. However, fuzzy relations involve additional concepts and operations due to their multidimensionality.

The union of two relations, say R and S, is denoted by $R \cup S$ and has the following membership function:

$$(R \cup S)(x, y) = \max[R(x, y), S(x, y)] \tag{2.98}$$

where both relations R and S are defined on the Cartesian product space $X \times Y$. On the other hand, the intersection of two fuzzy relations, $R \cap S$, has the following membership function:

$$(R \cap S)(x, y) = \min\,[R(x, y), S(x, y)] \tag{2.99}$$

These *soft* definitions of the union and intersection of two fuzzy relations can be generalized to perform the union and intersection of several relations using the max and min operators, respectively. The soft definitions of these operations can also be generalized using the Yager classes, similar to the union and intersection of fuzzy sets.

The complement $\bar{R}$ of fuzzy relation R has the following membership function:

$$\bar{R}\,(x, y) = 1 - R(x, y) \tag{2.100}$$

Among the additional operations, two of them are applicable to any n-dimensional fuzzy relations ($n \geq 2$). They are called *projections* and *cylindric extensions.* For the sake of simplicity, they are discussed here in terms of two- and three-dimensional relations; a generalization to higher dimensions is quite obvious.

Let R denote a three-dimensional (ternary) fuzzy relation on $X \times Y \times Z$. A *projection* of R is an operation that converts R into a lower-dimensional fuzzy relation, which in this case is either a two-dimensional or one-dimensional (degenerate) relation. In each projection, some dimensions are suppressed (not recognized), and the remaining dimensions are consistent with R in the sense that each α-cut of the projection is a projection of α-cut of R in the sense of classical set theory. Formally, the three two-dimensional projections of R on $X \times Y$, $X \times Z$, and $Y \times Z$, R_{XY}, R_{XZ}, and R_{YZ} are defined for all $x \in X$, $y \in Y$, and $z \in Z$ by the following formulas:

$$R_{XY}(x, y) = \max_{z \in Z} R(x, y, z) \tag{2.101a}$$

$$R_{XZ}(x,z) = \max_{y \in Y} R(x,y,z) \tag{2.101b}$$

$$R_{YZ}(y,z) = \max_{x \in X} R(x,y,z) \tag{2.101c}$$

Moreover, the three one-dimensional projections of R on X, Y, and Z, R_X, R_Y, and R_Z can then be obtained by similar formulas from the two-dimensional projections:

$$\begin{aligned} R_X(x) &= \max_{y \in Y} R_{XY}(x,y) \\ &= \max_{z \in Z} R_{XZ}(x,z) \end{aligned} \tag{2.102a}$$

$$\begin{aligned} R_Y(y) &= \max_{x \in X} R_{XY}(x,y) \\ &= \max_{z \in Z} R_{YZ}(y,z) \end{aligned} \tag{2.102b}$$

$$\begin{aligned} R_Z(z) &= \max_{x \in X} R_{XZ}(x,z) \\ &= \max_{y \in Y} R_{YZ}(y,z) \end{aligned} \tag{2.102c}$$

Any relation on $X \times Y \times Z$ that is consistent with a given projection of R is called an *extension* of R. The largest among the extensions is called a *cylindric extension*. Let R_{EXY} and R_{EX} denote the cylindric extensions of projections R_{XY} and R_X, respectively. Then R_{EXY} and R_{EX} are defined for all triples $(x, y, z) \in X \times Y \times Z$ by the formula

$$R_{EXY}(x,\ y,\ z) = R_{XY}(x,\ y) \tag{2.103a}$$

$$R_{EX}(x,\ y,\ z) = R_X(x) \tag{2.103b}$$

Cylindric extensions of the other two-dimensional and one-dimensional projections are defined in a similar way. This definition of cylindric extension for fuzzy relations is a cutworthy generalization of the classical concept of cylindric extension.

Given any set of projections of a given relation R, their standard fuzzy intersection (expressed by the minimum operator) is called a *cylindric closure* of the projections. This is again a cutworthy concept. Regardless of the given projections, it is guaranteed that their cylindric closure contains the fuzzy relation R.

Projections, cylindric extensions, and cylindric closures are the main operations for dealing with n-dimensional relations. For dealing with binary relations, an additional important operation is a *relational composition*.

Consider two binary fuzzy relations P and Q that are defined on sets $X \times Y$ and $Y \times Z$, respectively. Any such relations that are connected via the common set Y can be composed to yield a relation on $Y \times Z$. The *standard composition* of these relations, which is denoted by $P \circ Q$, produces a relation R on $X \times Z$ defined by the following formula for all pairs $(x, z) \in X \times Z$:

$$R(x, z) = (P \circ Q)(x, z) = \max_{y \in Y}[\min[P(x,y), Q(y,z)]] \tag{2.104}$$

Other definitions of a composition of fuzzy relations, in which the min and max operations are replaced with other unions (t-conorms) and intersections (t-norms), respectively, are possible and useful in some applications. All compositions are associative, i.e.,

$$(P \circ Q) \circ Q = P \circ (Q \circ Q) \tag{2.105}$$

However, the standard fuzzy composition is the only one that is cutworthy. Sets of Equation 2.104, which describe $R = P \circ Q$, are called *fuzzy relation equations*. Normally, it is assumed that P and Q are given and R is determined by Equation 2.104. However, two *inverse problems* play important roles in many applications. In one of them, R and P are given and Q is to be determined; in the other one, R and Q are given and P is to be determined. Various methods for solving these problems exactly as well as approximately have been developed, but are out of this book's scope. For further information, see Klir and Yuan (1995a), and Di Nola et al. (1989).

An interesting case of fuzzy composition is the composition of a fuzzy subset A defined on the universe X, with a relation R defined on the universe $X \times Y$. The result is a fuzzy subset B defined on the universe Y, with a membership function given by

$$(A \circ R)(y) = \max_{all\, x \in X}\{\min[A(x), R(x,y)]\} \tag{2.106}$$

A common application of the above operations of fuzzy relations is in constructing an approximate logic based on conditional propositions of the following type: if A_1, then B_1; else, if A_2, then B_2 … else, if A_n, then B_n. This statement can be modeled using the operations of fuzzy sets and relations in the following form:

$$(A_1 \times B_1) \cup (A_2 \times B_2) \ldots \cup (A_n \times B_n) \tag{2.107}$$

where $A_1, A_2, \ldots, A_n$ and $B_1, B_2, \ldots, B_n$ are fuzzy sets. Equation 2.106 is used for developing controllers based on fuzzy logic (Klir and Folger, 1988; Hassan et al., 1992; Hassan and Ayyub, 1993a, 1993b, 1994, 1997; Ayyub and Hassan, 1992a, 1992b, 1992c).

Additional information about the above operations and other operations with examples are provided by Kaufmann (1975), Kaufmann and Gupta (1985), and Klir and Folger (1988).

EXAMPLE 2.19 FUZZY RELATION FOR EXPERIENCE AND QUALITY

A fuzzy relation can be expressed in a conditional form. For example, the relation R can be defined as if experience of workers on a production line is short, then the quality of the product is medium. Defining short experience and a medium product quality, respectively, as

$$\text{Short experience, } A = \{40/0.1,\ 30/0.5,\ 20/0.7,\ 10/0.9,\ 0/1\} \tag{2.108}$$

$$\text{Medium quality} = \{70/0.2,\ 60/0.7,\ 50/1,\ 40/0.7,\ 30/0.2\} \tag{2.109}$$

the fuzzy relation R can be computed based on the minimum operator according to *cylindric closure* as follows:

R = (2.110)

		short experience				
		40	30	20	10	0
medium	70	0.1	0.2	0.2	0.2	0.2
product	60	0.1	0.5	0.7	0.7	0.7
quality	50	0.1	0.5	0.7	0.9	1.0
	40	0.1	0.5	0.7	0.7	0.7
	30	0.1	0.2	0.2	0.2	0.2

Note that the fuzzy sets "short experience" and "medium product quality" are from two different universes, namely, experience and quality, respectively. The membership values of the first row in Equation 2.110 were evaluated as follows:

$$\begin{aligned} R(70, 40) &= \min(0.2, 0.1) = 0.1 \\ R(70, 30) &= \min(0.2, 0.5) = 0.2 \\ R(70, 20) &= \min(0.2, 0.7) = 0.2 \\ R(70, 10) &= \min(0.2, 0.9) = 0.2 \\ R(70, 0) &= \min(0.2, 1.0) = 0.2 \end{aligned} \tag{2.111}$$

2.4.9 FUZZIFIED AND FUZZY FUNCTIONS

Engineers and scientists are commonly interested in developing relationships among underlying variables for a system. These relationships are based on the underlying physics of modeled problems or system behavior, such as economic forecasts, power consumption forecasting, extreme loads on a structure, etc. Approximate functions can be developed using fuzzified functions. In engineering and the sciences, we are interested in cause–effect relationships expressed as

$$f: X \rightarrow Y \tag{2.112a}$$

This function can be expressed as

$$y = f(x) \tag{2.112b}$$

where X and Y are crisp sets; y is the value of the criterion variable, also called dependent variable; x is the predictor variable, also called independent variable; and f is the functional relationship. The function is *fuzzified* when it is extended to act on fuzzy sets defined on X and Y. That is, the fuzzified function maps, in general, fuzzy sets defined on X to fuzzy sets defined on Y. Formally, the fuzzified function, F, has the form

$$F: F(X) \rightarrow F(Y) \tag{2.113}$$

where $F(X)$ and $F(Y)$ denote the fuzzy power sets (the sets of all fuzzy subsets) of X and Y, respectively. To qualify as a fuzzified version of f, function F must conform to f within the extended domain $F(X)$ and $F(Y)$ using the *extension principle* that employs the maximum operator.

Another functional form is the concept of fuzzy functions defined using α-cuts. A fuzzy function can be expressed using the following triplet functions:

$$Y = ({}^{\alpha}\underline{f}(x), {}^{\alpha}f_m(x), {}^{\alpha}\overline{f}(x)) \tag{2.114}$$

where ${}^{\alpha}\underline{f}(x)$ is the lower α-cut function at $\alpha = 0$, ${}^{\alpha}\overline{f}(x)$ is the upper α-cut function at $\alpha = 0$, and ${}^{\alpha}f_m(x)$ is the middle α-cut function at $\alpha = 1$. Other α-cut functions can be developed using linear interpolation as follows:

$${}^{\alpha}f(x) = \begin{cases} {}^{\alpha}\underline{f}(x) + \alpha({}^{\alpha}f_m(x) - {}^{\alpha}\underline{f}(x)) & \text{for } {}^{\alpha}\underline{f}(x) \le {}^{\alpha}f(x) \le {}^{\alpha}f_m(x) \\ {}^{\alpha}f_m(x) + \alpha({}^{\alpha}\overline{f}(x) - {}^{\alpha}f_m(x)) & \text{for } f_m(x) \le {}^{\alpha}f(x) \le {}^{\alpha}\overline{f}(x) \end{cases} \tag{2.115}$$

Fuzzy functions can be used to extrapolate empirical function to regions beyond data availability, for example, developing forecasting models. Functional operations such as derivatives, integrals, roots, maximums, and minimums can be defined using the α-cut concepts. These computations can be performed using numerical techniques performed on the function values at the α-cuts. Ayyub and McCuen (1996) describe commonly used numerical methods with practical examples.

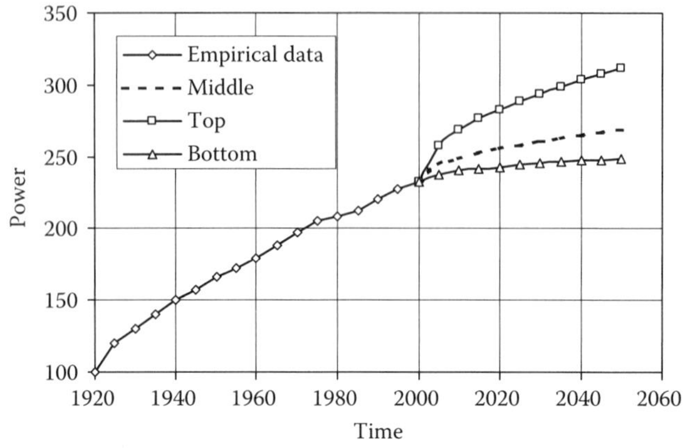

FIGURE 2.13 A fuzzy function for forecasting power needs.

Example 2.20 Forecasting Power Needs Using a Fuzzy Function

The power needs of a city can be forecasted to help city planners in making zoning and power plant construction decisions. Figure 2.13 shows an empirical power consumption trend over time and a subjectively developed forecast of the city's needs. The forecasted segment of the curve is provided for demonstration purposes. The following fuzzy function is used to develop Figure 2.13:

$$Y = \begin{cases} {}^{\alpha u}f(x) = 233 + 5\sqrt{5(\text{Year} - 2000)} \\ {}^{\alpha m}f(x) = 233 + 5\sqrt{5(\text{Year} - 2000)} \\ {}^{\alpha l}f(x) = 233 + 5\sqrt{5(\text{Year} - 2000)/5} \end{cases} \tag{2.116}$$

where Year is in a four-digit number format. The figure shows the empirical data and three functions that correspond to middle, and top and bottom at α-cuts of 0 and 1, respectively.

2.5 GENERALIZED MEASURES

A generalized measure is defined as a function μ from a family of subsets C of a universal set X to the interval [0, 1]. Commonly C is the power set of X as defined in Section 2.3.9, i.e., P_X. The family of subsets usually is constructed to meet some analytical objective and exhibits some algebraic structure. This mapping relation can be expressed as

$$\mu : C \rightarrow [0,1] \tag{2.117}$$

This function must have the following properties in addition to continuity from above and continuity from below:

$$\mu(\varnothing) = 1 \quad \text{if} \quad \varnothing \in C \tag{2.118a}$$

$$\mu(X) = 1 \quad \text{if} \quad X \in C \tag{2.118b}$$

$$\mu(A) \leq \mu(B) \quad \text{if} \quad A \,\&\, B \in C \text{ and } A \subseteq B \tag{2.118c}$$

For some historical reasons of little significance, generalized measures that satisfy the monotonicity requirement (Equation 2.118c) are often called fuzzy measures. This name is somewhat confusing since no fuzzy sets are involved in defining these generalized measures. It is more appropriate to refer to them as monotone measures. Wang and Klir (1992) provide additional information on generalized (or fuzzy) measures.

Example 2.21 Fuzzy Measures for Sets

For a universal set X of objects that are vague and numbered by integer values from 1 to n for identification purposes, the set can be expressed as

$$X = \{1, 2, 3, \ldots, n\} \tag{2.119}$$

The power set of $P(X)$ can be defined, and for any set $E \in P(X)$, the following fuzzy measure μ can be defined for illustration purposes for a notion related to its number of elements:

$$\mu(E) = \left(\frac{|E|}{n} \right)^2 \tag{2.120}$$

where $|E|$ is the cardinality of E, i.e., the number of elements in E. In this case the measure function of Equation 2.120 is called a *regular measure* since it meets all the conditions stated in Section 2.5.

2.6 ROUGH SETS AND OPERATIONS

2.6.1 Rough Set Definitions

Rough sets were introduced by Pawlak (1991) and are described with examples by Pal and Skowron (1999). Rough sets provide the means to represent crisp sets under restricted resolution capability. The representation is based on a partition of the universal space involved that should be constructed to facilitate this representation. Each rough set represents a given crisp set by two subsets of the partition, a *lower approximation* and an *upper approximation*. The lower approximation consists of all subsets of the partition that are included in the crisp set represented, whereas the

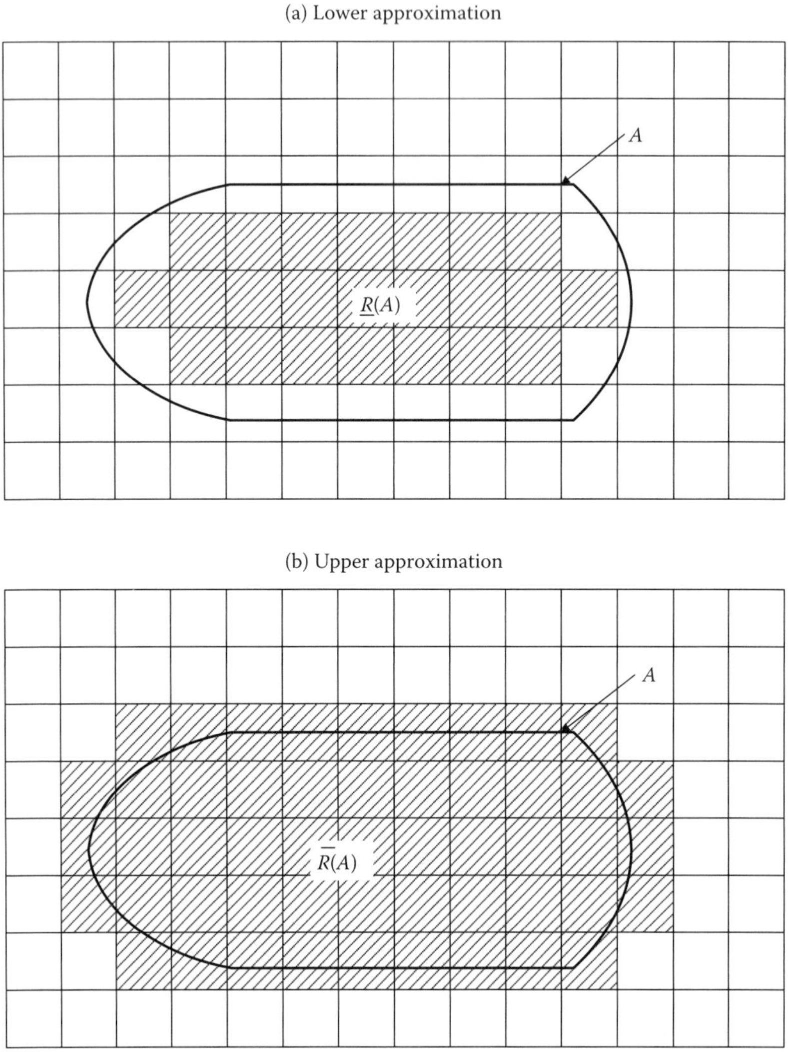

FIGURE 2.14 Rough set approximations.

upper approximation consists of all subsets that overlap with the set represented. Figure 2.14 shows a crisp set A that belongs to a universal space S. The universal space is partitioned using the grid shown in Figure 2.14. The lower and upper approximations of A, $\underline{R}(A)$ and $\overline{R}(A)$, respectively, are shown in the figure. These two subsets of approximations constitute the rough set approximation of A. The set difference $\underline{R}(A) - \overline{R}(A)$, is called a rouph boundary of set A.

Fuzzy sets are different from rough sets. The former represents vagueness of a quantity, such as obtaining linguistic quantities from experts, whereas the latter represents coarseness as an approximation of a crisp set. Since both types can be relevant in some applications, they can be combined as either *fuzzy rough sets* or *rough fuzzy sets*. Fuzzy rough sets are rough sets that are based on fuzzy partitions, whereas rough fuzzy sets are rough set approximations of fuzzy sets based on crisp partitions.

2.6.2 Rough Set Operations

Rough sets can be manipulated using operations of unions, intersections, and complements. Figure 2.14 shows a rough set approximation of the crisp set A. The lower and upper set approximations of A can be written as

$$\underline{R}(A) \subseteq A \subseteq \overline{R}(A) \tag{2.121}$$

where $\underline{R}(A)$ and $\overline{R}(A)$ are the lower and upper approximations of A. The subset $\underline{R}(A)$ includes all the sets of the universal space (S) that are contained in A. The subset $\overline{R}(A)$ includes all the sets of the universal space that contain any part of A. Based on this definition, the following operations can be provided for two crisp sets A and B (Pal and Skowron, 1999):

$$\underline{R}(\varnothing) = \varnothing = \overline{R}(\varnothing) \tag{2.122}$$

$$\underline{R}(S) = S = \overline{R}(S) \tag{2.123}$$

$$\overline{R}(A \cup B) = \overline{R}(A) \cup \overline{R}(B) \tag{2.124}$$

$$\underline{R}(A \cap B) = \underline{R}(A) \cap \underline{R}(B) \tag{2.125}$$

$$A \subseteq B \Rightarrow \underline{R}(A) \subseteq \underline{R}(B) \text{ and } \overline{R}(A) \subseteq \overline{R}(B) \tag{2.126}$$

$$\underline{R}(A \cup B) \supseteq \underline{R}(A) \cup \underline{R}(B) \tag{2.127}$$

$$\overline{R}(A \cap B) \subseteq \overline{R}(A) \cap \overline{R}(B) \tag{2.128}$$

$$\overline{R}(\overline{A}) = \overline{\underline{R}(A)} \tag{2.129}$$

$$\underline{R}(\overline{A}) = \overline{\overline{R}(A)} \tag{2.130}$$

$$\underline{R}(\underline{R}(A)) = \overline{R}(\underline{R}(A)) = \underline{R}(A) \tag{2.131}$$

$$\overline{R}(\overline{R}(A)) = \underline{R}(\overline{R}(A)) = \overline{R}(A) \tag{2.132}$$

A measure of the *accuracy of an approximation* (δ) can be expressed as

$$\delta = \left|\frac{\underline{R}(A)}{\overline{R}(A)}\right| \tag{2.133}$$

where $|\underline{R}(A)|$ is the cardinality of $\underline{R}(A)$, i.e., the number of elements in $\underline{R}(A)$. The range of δ is [0, 1].

2.6.3 Membership Functions for Rough Sets

In classical sets, the concept of a characteristic function was introduced as provided by Equations 2.17 and 2.18. For each element, the membership function in the case of crisp sets takes on either 1 or 0, which correspond to belonging and not belonging of the element to a set A of interest. The membership function for rough sets has a different meaning than in the case of crisp sets.

Example 2.22 Rough Sets to Represent Information Systems

This example deals with a quality assurance department in a factory that inspects all hand-made products produced by the factory. For each item produced, the inspectors of the department record who made it and the result of the inspection, either acceptable or unacceptable. Skilled workers are used to produce these items who have varying levels of experience in terms of the number of months at the job and number of items that each has produced. The data shown in Table 2.5 were collected based on inspecting 10 items. For each item, the table shows the experience level for the person who made the item, measured as the number of months at the job and number of items that have been produced by this person, and the outcome of the inspection. Representing each item by its pair of the number of months and number of items, Figure 2.15 can be

TABLE 2.5
Inspection Data for Quality

Item Number	Experience (Number of Months)	Experience (Number of Items Produced)	Inspection Outcome
x_1	5	10	Unacceptable
x_2	5	10	Unacceptable
x_3	12	5	Unacceptable
x_4	12	12	Acceptable
x_5	12	12	Unacceptable
x_6	20	10	Acceptable
x_7	24	12	Acceptable
x_8	6	8	Unacceptable
x_9	18	10	Acceptable
x_{10}	10	10	Unacceptable

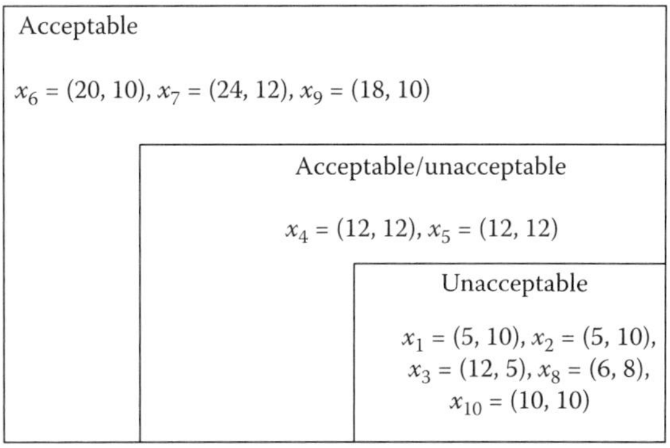

FIGURE 2.15 Rough sets for presenting product quality.

developed using rough sets. The figure shows x_1, x_2, x_3, x_8, and x_{10} as items that are unacceptable, x_4 and x_5 as borderline items of the rough set that are either acceptable or unacceptable, and x_6, x_7, and x_9 as items that are acceptable. Based on this figure, the following rough set of unacceptable items (A) can be defined:

$$\begin{aligned} \underline{R}(A) &= \{x_1, x_2, x_3, x_8, x_{10}\} \\ \overline{R}(A) &= \{x_1, x_2, x_3, x_4, x_5, x_8, x_{10}\} \end{aligned} \tag{2.134}$$

Similarly, the following rough set of acceptable items ($\bar{A}$) can be defined:

$$\begin{aligned} \underline{R}(\bar{A}) &= \{x_6, x_7, x_9\} \\ \overline{R}(\bar{A}) &= \{x_4, x_5, x_6, x_7, x_9\} \end{aligned} \tag{2.135}$$

These rough sets are shown in Figure 2.15.

2.6.4 Rough Functions

The concept of rough functions was introduced by Pawlak (1999) to present a coarse approximation of unknown functions. Rough functions can be an effective means to meet the needs of engineers and scientists, as an example, for the purpose of expressing expert opinions. In developing relationships among variables underlying an engineering system, these rough functions can be used to articulate and express the opinion of experts in cases such as economic forecasts, power consumption forecasting, and assessing extreme loads on a structure.

This section provides a format for presenting these subjective relationships. Although the mathematics needed to manipulate these functions are not provided, they can be obtained and developed based on the work of Pawlak (1999). In engineering and science, we are interested in cause–effect relationships expressed as

$$y = f(x) \tag{2.136}$$

where y is the value of the criterion variable, or dependent variable; x is the predictor variable, or independent variable; and f is the functional relationship. Using the concept of lower and upper approximations of f, a rough function can be expressed using the following pair:

$$y = (\underline{R}(f), \overline{R}(f)) \tag{2.137}$$

$\underline{R}(f)$ and $\overline{R}(f)$ are lower and upper approximations of f. Rough functions can be used to extrapolate empirical functions to regions beyond data availability for developing forecasting models. Functional operations such as derivatives, integrals, roots, maximum, and minimum can be defined on the two approximations. Also, they can be performed using numerical techniques performed on values at the lower and upper approximations. Ayyub and McCuen (1996) describe commonly used numerical methods with practical examples.

EXAMPLE 2.23 FORECASTING POWER NEEDS USING A ROUGH FUNCTION

The power needs of a city can be forecasted to help city planners in making zoning and power plant construction decisions, as discussed in Example 2.20. Figure 2.16 shows an empirical power consumption trend over time and a subjectively developed forecast of the city's needs using lower and upper approximations of the needs (f). The forecasted segment of the curve is provided for demonstration purposes. The rough function was developed by establishing convenient grid lines, every 10 units of power and every 5 years, as shown Figure 2.16. Then, lower and upper approximations of the unknown city's needs of power were made by identifying the coordinate points that would include the unknown function.

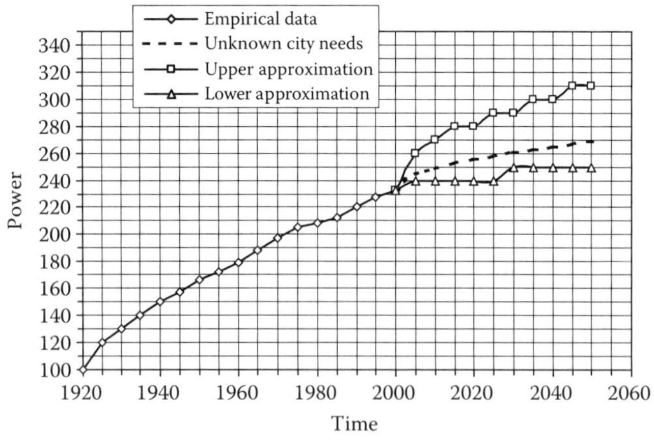

FIGURE 2.16 A rough function for forecasting power needs.

2.7 GRAY SYSTEMS AND OPERATIONS

Many systems in engineering and the sciences, such as social, economic, agricultural, industrial, ecological, and biological systems, are modeled with incomplete information about the system and interactions among variables within the system and with other systems. Commonly, not even all such variables are known with certainty to analysts performing the modeling. Systems that are not fully understood by engineers and scientists can be viewed as black boxes that receive an input and produce an output. In this case, what goes inside the system is unknown. Such systems are called black systems according to this theory. For a black system, internal structure and information processing is unknown, appearance is dark, and the process is unknown or new, and might have the property of chaotic behavior. For example, in control theory, the darkness of colors has been commonly used to indicate the degree of clarity of information. An accepted representation in control is the so-called black box method, which stands for an object with its internal relations or structure totally unknown to analysts. Contrarily, *white* can be used for cases involving completely known structure, and *gray* can be used for cases that are partially known and partially unknown. Accordingly, systems with complete knowledge on structure are called white systems, systems with completely unknown structure are called black systems, and systems with partially known and partially unknown structure are called gray systems. For example, prediction of yield of agricultural projects and consequent annual income is quite difficult, even with known information such as the quality of seeds, fertilizers, temperature, and irrigation distribution, due to various unknown or vague information related to labor quality, technology level employed, natural environment, infestation, weather conditions, etc. Another example is the effectiveness of pesticide to control insects. Economic and social systems display such uncertainty that might require the use of gray system modeling methods. A complete treatment of gray systems is beyond the scope of this book; however, Deng (1993) and Forrest (1997) provide additional information and examples.

EXERCISE PROBLEMS

2.1. A construction manager needs to procure building materials for the construction of an office building. The following sources are identified:

Material Type	Sources
Concrete	Sources A and B
Reinforcing steel	Sources C and D
Timber	Sources D and E
Structural steel	Sources C and F
Hardware	Sources F, G, and H

Define the sample space of all possible combinations of sources supplying the construction project, assuming that each material type can be procured from one source only, but a source may supply more than one material type at the same time.

2.2. A construction tower crane can operate up to a height H of 300 feet, a range (radius) R of 50 feet, and an angle ϕ of $\pm 90^\circ$ in a horizontal plane. Sketch the sample space of operation of the crane. Sketch the following events:

Event	Definition
A	$30 < H < 80$ and $R < 30$
B	$H > 50$ and $0^\circ < \phi < 50^\circ$
C	$H < 40$ and $R > 60$
D	$H > 80$ and $-30^\circ < \phi < 50^\circ$

2.3. Construct Venn diagrams for each of the following:
a. A deck of playing cards
b. The roll of a die
c. Letter grades on a test assuming equal probabilities for each grade
d. Letter grades on a test assuming the following probabilities:
A: 15%
B: 25%
C: 30%
D: 20%
E: 10%
e. Options at an intersection with the following probabilities:
Left turn: 20%
Straight ahead: 40%
Right turn: 25%
U-turn: 10%
Remain stopped: 5%

2.4. For the data and events of Problem 2.2, sketch the following events:

$$A \cup B,\ A \cap B,\ C \cup D,\ C \cap D,\ A \cup C,\ A \cup (B \cap C),\ \bar{A}, \text{ and } \bar{A} \cap B$$

2.5. The traffic that makes a left turn at an intersection consists of two types of vehicles, types A and B. A type A vehicle is twice the length of type B. The left-turn lane can accommodate eight vehicles of type B, four of type A, or combinations of A and B. Define the sample space of all possible combinations of vehicles waiting for a left turn at the intersection. Also define the following events: (1) at least one vehicle of type A waiting for left turn, (2) two vehicles of type B waiting for left turn, and (3) exactly one of type A and one of type B waiting for left turn.

2.6. Construct a Venn diagram for a deck of playing cards (4 suits, 13 cards per suit). Show the following events:
a. A = All diamonds and all aces
b. B = All face cards

c. C = The intersection of red cards and face cards
d. D = The union of black cards and cards with values of 4 or smaller
What are the cardinalities of these events?

2.7. Using the α-cut concept, compute the intersection, union, and complements of the following fuzzy sets:
a. Triangular sets defined as A = [10, 15, 20] and B = [15, 18, 22]
b. Trapezoidal sets defined as C = [10, 12, 18, 20] and D = [15, 18, 20, 22]
Plot your results.

2.8. Using the α-cut concept, compute the intersection, union, and complements of the triangular set defined as A = [10, 15, 20] and trapezoidal set defined as B = [10, 12, 18, 20]. Plot your results.

2.9. Using the α-cut concept, evaluate the following operations on the triangular set defined as A = [10, 15, 20] and trapezoidal set defined as B = [10, 12, 18, 20], and plot your results:
a. $A + B$
b. $A - B$
c. $A \times B$
d. A/B
e. $A + A$, with the equality constraint
f. $A - A$, with the equality constraint
g. $A \times A$, with the equality constraint
h. A/A, with the equality constraint

2.10. Using the α-cut concept, evaluate the following operations on the triangular number defined as A = [–10, 15, 20] and trapezoidal number defined as B = [10, 12, 18, 20], and plot your results:
a. $A + B$
b. $A - B$
c. $A \times B$
d. A/B
e. $A + A$, with the equality constraint
f. $A - A$, with the equality constraint
g. $A \times A$, with the equality constraint
h. A/A, with the equality constraint

2.11. Compute the cardinality of the following sets:

$$A = \{100/0.1, 80/0.2, 60/0.4, 40/0.6, 20/0.8, 0/1\}$$

$$B = \{50/0.1, 60/0.2, 70/0.4, 80/0.6, 90/0.8, 100/1\}$$

2.12. Compute the cardinality of the following sets:

$$A = \{\text{I}/0.1, \text{II}/0.2, \text{III}/0.4, \text{IV}/0.6, \text{V}/0.8, \text{VI}/1\}$$

$$B = \{\text{c}/0.1, \text{d}/0.2, \text{e}/0.4, \text{f}/0.6, \text{g}/0.8, \text{h}/1\}$$

2.13. Compute the cardinality of the following sets:

$$A = \{1/0.1,\ 2/0.2,\ 3/0.4,\ 4/0.6,\ 5/0.8,\ 6/1\}$$

$$B = \{6/0.1,\ 5/0.2,\ 4/0.4,\ 3/0.6,\ 2/0.8,\ 1/1\}$$

2.14. Develop two illustrative computational examples from engineering or the sciences that requires the use of constrained fuzzy arithmetic.

2.15. Use two illustrative computational examples to compare the t-conorms (unions) provided in Table 2.3. Discuss your results. Provide recommendations on when each type should be used by stating limitations, advantages, and disadvantages.

2.16. Use two illustrative computational examples to compare the t-norms (intersections) provided in Table 2.4. Discuss your results. Provide recommendations on when each type should be used by stating limitations, advantages, and disadvantages.

2.17. Using the α-cut concept, compute the complements of the following fuzzy sets using the Yager and Sugeno classes:
a. Triangular sets defined as A = [10, 15, 20]
b. Trapezoidal sets defined as B = [10, 12, 18, 20]
Discuss your results. Provide recommendations on when each type should be used by stating limitations, advantages, and disadvantages.

2.18. Using the α-cut concept, evaluate the generalized fuzzy operations of Figure 2.6 and display the results in a figure similar to Figure 2.6 using the Yager classes for the following two triangular sets:

$$A = [10,\ 15,\ 18]$$

$$B = [12,\ 18,\ 24]$$

Discuss your results.

2.19. Using the α-cut concept, evaluate the generalized fuzzy operations of Figure 2.6 and display the results in a figure similar to Figure 2.6 using the Yager classes for the following two triangular sets:

$$A = [5,\ 10,\ 15]$$

$$B = [10,\ 15,\ 25]$$

Discuss your results.

2.20. Construct a fuzzy relation $A \times B$ based on the triangular set defined as A = [10, 15, 20] and the trapezoidal set defined as B = [10, 12, 18, 20], and plot your results.

2.21. Develop an equation to compute an average of n fuzzy data points that are of:
a. Triangular fuzzy membership functions

b. Trapezoidal fuzzy membership functions
c. Mixed triangular and trapezoidal fuzzy membership functions

2.22. Building on Problem 2.21, develop equations needed to compute variance of n fuzzy data points that are of:
a. Triangular fuzzy membership functions
b. Trapezoidal fuzzy membership functions
c. Mixed triangular and trapezoidal fuzzy membership functions
What are the limitations of your equations?

2.23. Provide two examples of unknown set and function that can be represented by a rough set and a rough function, respectively, with sample computations.

2.24. Provide two examples of systems that can be modeled using generalized measures with sample computations.

2.25. Provide two examples of gray systems.

2.26. Develop a spreadsheet for performing fuzzy arithmetic without constraints for the following types of fuzzy numbers:
a. Two triangular fuzzy numbers
b. Two trapezoidal fuzzy numbers
Demonstrate and validate your spreadsheet.

2.27. Develop a spreadsheet for performing fuzzy arithmetic with constraints for the following types of fuzzy numbers:
a. Two triangular fuzzy numbers
b. Two trapezoidal fuzzy numbers
Use the following constraints:
a. $x = y$
b. $x \leq y$
Demonstrate and validate your spreadsheet.

2.28. Develop a spreadsheet for solving two simultaneous linear equations with fuzzy coefficients using fuzzy arithmetic with constraints. Demonstrate and validate your spreadsheet.

2.29. Develop a spreadsheet for solving three simultaneous linear equations with fuzzy coefficients using fuzzy arithmetic with constraints. Demonstrate and validate your spreadsheet.

3 Uncertainty and Information Synthesis

3.1 SYNTHESIS FOR A GOAL

Data and information collected about a system require synthesis for the purpose of achieving an analytical goal or a mission. Engineers and scientists are always interested in understanding and predicting system behavior (or performance) for the purpose of making appropriate decisions. The behavior or performance of the system can only be assessed according to available information, thereby involving some uncertainty, and to the best of their knowledge, that involves some aspects of ignorance. This synthesis requires the employment of fundamental measures to assess these uncertainties and ignorance types. Various measures, including probability measures as special and commonly used measures, are introduced in this chapter, building on data-encoding and information expression methods, i.e., formalized languages, presented in Chapter 2. The identification (or selection) of appropriate measures for uncertainty-based synthesis of information requires the definition of an analytical goal that can be used as a basis for recognizing ignorance types most relevant to the system under consideration, developing universal spaces and appropriate structures, and selecting and using appropriate measures.

3.2 KNOWLEDGE, SYSTEMS, UNCERTAINTY, AND INFORMATION

The recognition that *scientific knowledge* is organized, by and large, in terms of *systems* of various types is an important outcome of systems science, as discussed in Chapter 1. Systems are constructed for various *purposes*, such as predicting, retrodicting, prescribing, diagnosing, controlling, etc. Every system involves a *relation* among some variables, which is utilized in a given purposeful way for determining unknown states of some variables on the basis of known states of other variables. When the unknown states are determined uniquely, the systems are called *deterministic*. Otherwise, they are called *nondeterministic*. It is a common observation that nondeterministic systems are far more prevalent than deterministic systems in contemporary science and technology.

Each nondeterministic system inevitably involves some *uncertainty*, which is associated with the purpose for which the system has been constructed. In each nondeterministic system, the relevant uncertainty must be properly incorporated into the formal description of the system in set theoretic terms or in terms of some other formalized language, as was discussed in Chapters 1 and 2.

Although nondeterministic systems have been accepted in science and engineering since the early 20th century (primarily under the influence of statistical mechanics), it was tacitly assumed for a long time that uncertainty in these systems can be adequately formalized and dealt with by probability theory. This assumption was challenged shortly after World War II, when the emerging computer technology opened new methodological possibilities. It was increasingly realized, as most eloquently described by Weaver (1948), that the established methods of science were not applicable to a broad class of important problems for which Weaver coined the suggestive term *organized complexity*. These are problems that involve considerable numbers of entities that are interrelated in complex ways. They are typical in life, cognitive, social, and environmental sciences, as well as in applied fields such as modern technology, medicine, and management. They almost invariably involve uncertainties of various types, but rarely uncertainties resulting from randomness, which can yield meaningful statistical averages. That is, they cannot be adequately represented by probability theory alone.

Uncertainty liberated from its probabilistic confines is a phenomenon of the second half of the 20th century. It is closely connected with two important generalizations in mathematics. One of them is the generalization of classical measure theory (Halmos, 1950) to the theory of generalized measures, which was first suggested by Gustave Choquet (1953–1954) in his theory of capacities. The second one is the generalization of classical set theory to fuzzy set theory, introduced by Lotfi Zadeh (1965) and overviewed in Chapter 2. Generalized measures are obtained by abandoning the requirement of additivity of classical measures. Fuzzy sets are obtained by abandoning the requirement of sharp boundaries of classical sets. These generalizations enlarged substantially the framework for formalizing uncertainty. As a consequence, they made it possible to conceive of new theories of uncertainty.

In general, uncertainty is an expression of some information deficiency. This suggests that information could be measured in terms of uncertainty reduction. To reduce relevant uncertainty in a situation formalized within a mathematical theory requires that some relevant action be taken by a cognitive agent, such as performing a relevant experiment, searching for a relevant fact, or accepting and interpreting a relevant message. If results of the action taken (an experimental outcome, a discovered fact, etc.) reduce uncertainty involved in the situation, then the amount of information obtained by the action is measured by the amount of uncertainty reduced — the difference between *a priori* and *a posteriori* uncertainty. Measuring information in this way is clearly contingent upon our capability to measure uncertainty within the various mathematical frameworks. Information measured solely by uncertainty reduction is an important, even though restricted, notion of information. To distinguish it from the various other conceptions of information, it is common to refer to it as *uncertainty-based information* (Klir and Wierman, 1999, Klir, 2006).

A research program whose objective is to develop a broader treatment of uncertainty-based information, not restricted to probabilistic formalization of uncertainty, was introduced in the early 1990s under the name generalized information theory (GIT) (Klir, 1991a). The ultimate goal of GIT is to develop the capability to deal with any type of uncertainty-based information that is recognized on intuitive grounds. To be able to deal with each recognized type of uncertainty (and the associated type of information) in an operational way, relevant issues must be addressed at each of the following four levels:

- *Level 1*: Find an appropriate mathematical representation of the conceived type of uncertainty.
- *Level 2*: Develop a calculus by which this type of uncertainty attributes can be properly quantified and manipulated.
- *Level 3*: Find a meaningful way of measuring relevant uncertainty in any situation formalized in the theory.
- *Level 4*: Develop methodological aspects of the theory, including procedures for making the various uncertainty principles operational within the theory.

In this book, our aim is to view uncertainty and uncertainty-based information from the broad perspective of GIT and to illustrate the utility of results emerging from GIT in science and engineering. In this chapter, GIT is introduced in more specific terms and surveys some of the uncertainty theories that are subsumed under it. After introducing the concept of a monotone measure in Section 3.4.1 and describing a broad framework for formalizing uncertainty in Section 3.4.2, this chapter examines several well-developed uncertainty theories, including the classical theories based on the notions of possibility and probability. The fundamentals of measure theory and various measures that are appropriate for uncertainty modeling are presented. The chapter introduces monotone measures, evidence theory and random sets, possibility theory, probability theory, Bayesian methods, interval probabilities, interval cumulative distribution functions, and probability bounds. Examples are used in this chapter to demonstrate the various methods and concepts.

3.3 MEASURE THEORY AND CLASSICAL MEASURES

In mathematics, a *measure* is a function that assigns a number to quantify a *notion* of a *metric* representing a subset of a given set, e.g., a size, volume, or probability. The concept is important in mathematical analysis, including uncertainty analysis. *Classical measures* that are used in *probability theory* meet axiomatic constraints as discussed in this section.

Measures, in general, build on the concepts of a universal set (X), a nonempty family C of subsets of X with an appropriate algebraic structure, sets (such as A), and the power set (P_A), in order to establish a logical measure that can be used to characterize some system attributes of interest, i.e., uncertainty, likelihood, probability, possibility, belief, etc. Classical measures are formulated for a universal set X and a family of subsets C such that if $A_i \in C$, it leads to $A_i \subset X$. The family C is called an *algebra* if the following conditions are met:

$$C \text{ contains the empty set, i.e., } \emptyset \in C \tag{3.1}$$

$$C \text{ contains the entire set } X\text{, i.e., } X \in C \tag{3.2}$$

$$\text{For any } A_i \in C, \text{ the complementary set } \overline{A_i} \in C \tag{3.3}$$

The family is called a σ-*algebra* if it has the following additional property:

$$\text{For } A_i \in C,\ i = 1, 2, \ldots, \bigcup_{all\, i} A_i \in C \tag{3.4}$$

In other words, Equation 3.4 states that the countable union of any family of subsets in C belongs to C (Halmos, 1950; Royden, 1988).

A measure μ can be defined in its broadest form as a function that maps C to the real line (R). This function can be defined mathematically as follows:

$$\mu : C \rightarrow R \tag{3.5}$$

Of special interest for the purposes of this book is a function that is limited to nonnegative real values (R_+) as follows:

$$\mu : C \rightarrow R_+ = [0, \infty] \tag{3.6}$$

In probability theory, the probability measure that is introduced in a later section imposes additional requirements on μ consisting of the following:

$$\mu : C \rightarrow [0, 1] \tag{3.7}$$

$$\mu(\emptyset) = 0 \tag{3.8}$$

$$\text{For disjoint } A_i \in C,\ i = 1, 2, \ldots, \mu\left(\bigcup_{all\, i} A_i\right) = \sum_{all\, i} \mu(A_i) \tag{3.9}$$

where any events A_i and A_j meet the following condition:

$$A_i \cap A_j = \emptyset \tag{3.10}$$

Equation 3.7 limits the mapping to the closed interval of [0, 1], with the measure for the null set being zero according to Equation 3.8. Equation 3.9 states that the function μ for the union of several disjoint (i.e., with null intersections) subsets is the sum of the measures (i.e., μ values) of these subsets. This *additive property* is unique to this *classical measure* of *probability*. Although the development and evolution of probability theory was based more on intuition than on mathematical axioms during its early development, an axiomatic basis for probability theory was established, and it is now universally accepted.

3.4 MONOTONE MEASURES AND THEIR CLASSIFICATION

3.4.1 Definition of Monotone Measures

When generalized measures are employed for representing uncertainty, it makes sense to require that the additivity property of classical measures be replaced with a weaker property of monotonicity with respect to the subsethood relationship. Such measures are called *monotone measures*. Their range is usually the unit interval [0, 1], as in probability measures, and it is required that the measure of the universal set be 1. Such measures are called *regular monotone measures*. They are formally defined in this section.

A *regular monotone measure* (the kind suitable for formalizing uncertainty) can be defined based on a nonempty family C of subsets from P_X (i.e., the power set of X) for a given universal set X, that contains ϕ and X, with an appropriate algebraic structure as a mapping from C to [0, 1], as follows:

$$\mu: C \rightarrow [0, 1] \tag{3.11}$$

A monotone measure must satisfy the following conditions:

1. *Boundary conditions*: The monotone measure must meet the following

$$\mu(\varnothing) = 0 \quad \text{and} \quad \mu(X) = 1 \tag{3.12a}$$

2. *Monotonicity:*

$$\text{For all } A_i \text{ and } A_j \in C, \text{ if } A_i \subseteq A_j \text{ then } \mu(A_i) \leq \mu(A_j) \tag{3.12b}$$

3. *Continuity from below*:

For any increasing sequence $A_1 \subseteq A_2 \subseteq \ldots$ of sets in C,

$$\text{if } \bigcup_{all\, i} A_i \in C, \text{ then } \lim_{i \to \infty} \mu(A_i) = \mu\left(\bigcup_{all\, i} A_i\right) \tag{3.12c}$$

4. *Continuity from above*:

For any decreasing sequence $A_1 \supseteq A_2 \supseteq \ldots$ of sets in C,

$$\text{if } \bigcap_{all\, i} A_i \in C, \text{ then } \lim_{i \to \infty} \mu(A_i) = \mu\left(\bigcap_{all\, i} A_i\right) \tag{3.12d}$$

Functions μ that satisfy requirements of Equation 3.12a, Equation 3.12b, and either Equation 3.12c or Equation 3.12d, which are called *semicontinuous from below and from above*, respectively, are functions that allow us to formalize *upper and lower probabilities* of various types.

For any pair A_1 and $A_2 \in C$ such that $A_1 \cap A_2 = \emptyset$, a monotone measure μ is capable of capturing any of the following situations (Wang and Klir, 1992; Klir and Wierman, 1999):

$$\mu(A_1 \cup A_2) > \mu(A_1) + \mu(A_2) \tag{3.13}$$

called *superadditivity*, which expresses a cooperative action or synergy between A_1 and A_2 in terms of the measured property;

$$\mu(A_1 \cup A_2) = \mu(A_1) + \mu(A_2) \tag{3.14}$$

called *additivity*, which expresses the fact that A_1 and A_2 are *noninteractive* with respect to the measured property; and

$$\mu(A_1 \cup A_2) < \mu(A_1) + \mu(A_2) \tag{3.15}$$

called *subadditivity*, which expresses some sort of inhibitory effect or incompatibility between A_1 and A_2 as far as the measured property is concerned.

Probability theory, which is based on the classical measure theory, is capable of capturing only situations of Equation 3.14. This demonstrates that the theory of monotone measures provides us with a considerably broader framework than probability theory for formalizing uncertainty. As a consequence, it allows us to capture types of uncertainty that are beyond the scope of probability theory. In general, *lower probabilities* are superadditive and *upper probabilities* are subadditive.

For some historical reasons of little significance, monotone measures are often referred to in the literature as *fuzzy measures* (Wang and Klir, 1992). This name is somewhat confusing since no fuzzy sets are involved in the definition of monotone measures. To avoid this confusion, the term *fuzzy measures* should be reserved for measures (additive or nonadditive) that are defined on families of fuzzy sets. That is, the term *fuzzy measures* should refer to monotone measures that are fuzzified (Wang and Klir, 1992).

Monotone measures are needed to model *inconsistency* and *incompleteness* ignorance types, as shown in Figure 1.19. Fuzzy sets generalize classical sets by abandoning the requirement that sets have sharp boundaries. As a consequence, objects may belong to fuzzy sets with *various degrees of membership*, which in standard fuzzy sets are expressed by numbers in the unit interval of real numbers [0, 1]. Monotone measures allows for superadditivity and subadditivity in its analytical computations.

3.4.2 Classifying Monotone Measures

Based on Section 3.4.1, probability theory builds on two requirements: (1) families of classical sets (or classical propositions), each with the underlying structure of a

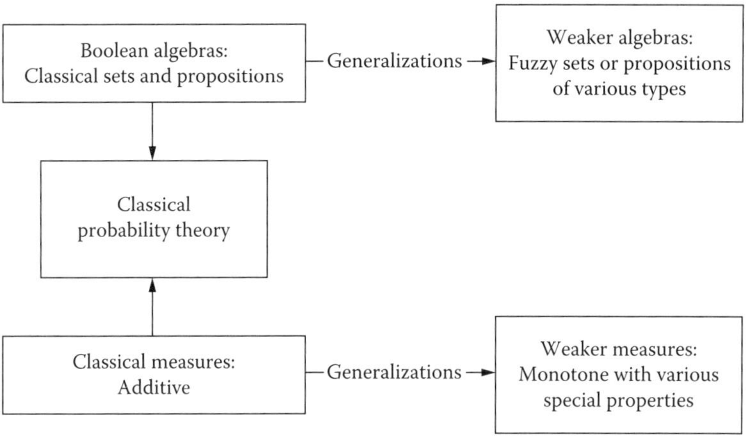

FIGURE 3.1 Probability theory and its generalizations.

Boolean algebra, and (2) classical (additive) measures defined on these families of sets, as shown in Figure 3.1. These two components define the framework within which probability theory operates.

The emergence of fuzzy set theory and the theory of monotone measures made it possible to expand this classical framework considerably. This expansion is two-dimensional. In one dimension, the formalized language of classical set theory is expanded to the more expressive language of fuzzy set theory, where further distinctions (and the underlying algebraic structures) are based on special types of fuzzy sets. In the other dimension, the classical (additive) measure theory is expanded to the less restrictive theory of monotone measures, within which various branches can be distinguished by monotone measures with different special properties, as shown in Figure 3.1.

The two-dimensional expansion of possible uncertainty theories is illustrated by the matrix in Figure 3.2, where the rows represent various types of monotone measures, while the columns represent various types of formalized languages (Klir, 2006). Under the entry of nonadditive measures in Figure 3.2, only a few representative types are listed. Some of them are presented as pairs of dual measures employed jointly in some uncertainty theories. Under formalized languages in Figure 3.2, not only theories of classical sets and standard fuzzy sets are listed, but also theories based on some of the nonstandard fuzzy sets. Nonstandard fuzzy sets are still less common and are not covered in this book.

An uncertainty theory of a particular type is formed by choosing a formalized language of a particular type and expressing relevant uncertainty (predictive, prescriptive, diagnostic, etc.) involved in situations described in this language in terms of a measure (or a pair of measures) of a certain type. This means that each entry in the matrix in Figure 3.2 represents an uncertainty theory of a particular type.

Figure 3.2 classifies uncertainty theories, including some fairly well developed ones. Among them are two classical uncertainty theories: classical probability theory and classical (crisp) possibility theory. Figure 3.2 shows a cross-classification of

Uncertainty theories			Formalized languages		
			Classical sets	Non-classical sets: Standard fuzzy sets	Non-classical sets: Non-standard fuzzy sets
Monotone measures	Additive	Classical numerical probability	Crisp probability theory	Probability of fuzzy events	NA
Monotone measures	Non-additive	Possibility and necessity	Crisp possibility theory	Graded possibility theory	NA
Monotone measures	Non-additive	Sugeno λ-measures	Sugeno λ-measures	Fuzzified λ-measures	NA
Monotone measures	Non-additive	Belief and plausibility (capacities of order ∞)	Dempster-Shafer theory (DST) (capacities of order ∞)	Fuzzified DST	NA
Monotone measures	Non-additive	Capacities of various finite orders	Capacities of finite orders, such as 1, 2, ...,k	NA	NA
Monotone measures	Non-additive	Interval-valued probability distributions	Feasible interval-valued probability distribution	Feasible fuzzy probability distributions	NA
Monotone measures	Non-additive	⋮	⋮	⋮	⋮
Monotone measures	Non-additive	General lower and upper probabilities	General lower and upper probabilities	NA	NA

NA = Not available

FIGURE 3.2 Classification of uncertainty theories.

available uncertainty theories according to the classification of monotone measures provided in Equations 3.13 to 3.15 and an underlying formalized language describing the structure of a problem. Not all the theories are fully mature and developed for practical use by engineers and scientists. As a result, only the mature and fully developed ones are discussed in subsequent sections of this chapter. Figure 3.3 provides a conceptual relationship structure among the theories; i.e., it shows what assumptions or generalizations are needed to obtain one theory from another. The following example theories are presently used in practice to model uncertainty:

1. *Classical probability theory*: Classical probability (additive) functions defined on classical (crisp) sets.

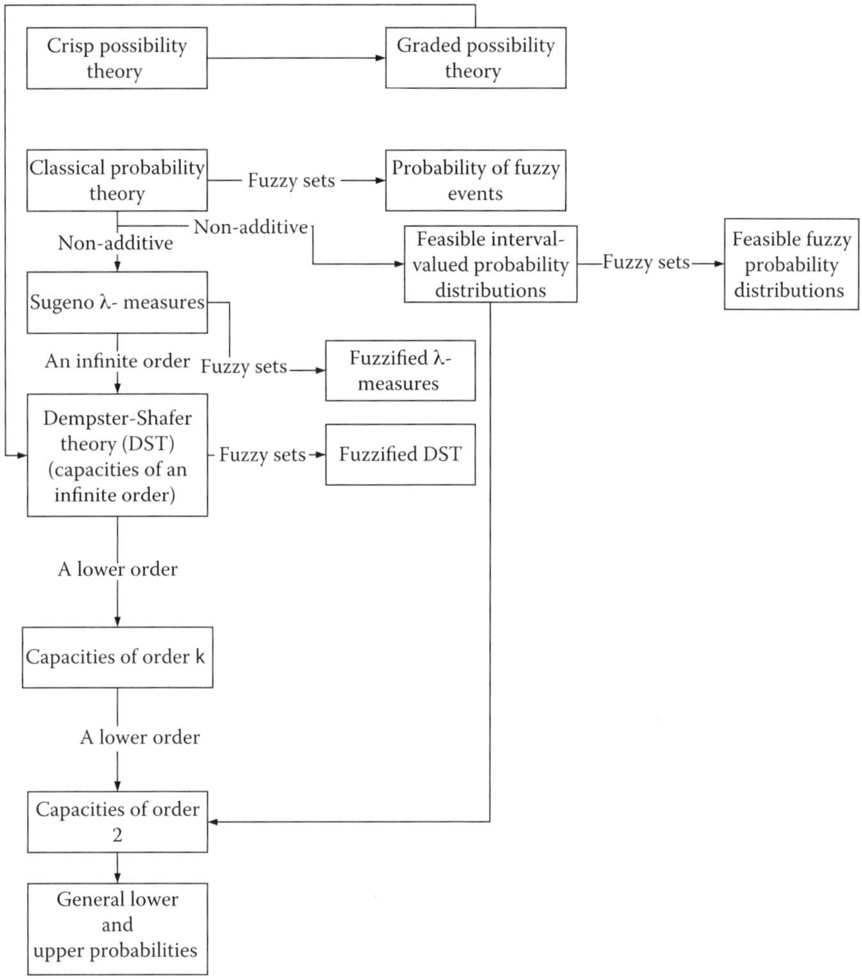

FIGURE 3.3 Ordering uncertainty theories by levels of their generality.

2. *Probability theory based on fuzzy events*: Classical probability (additive) functions defined on fuzzy sets.
3. *Classical possibility theory*: A pair of classical (crisp) *possibility and necessity functions* defined on classical sets.
4. *Theory of graded possibilities*: A pair of possibility and necessity functions defined on fuzzy sets.
5. *Dempster–Shafer theory (DST) of evidence*: A pair of special semicontinuous monotone measures, called *belief and plausibility measures*, which are defined on classical sets and which conveniently represent lower and upper probabilities, respectively.
6. *Fuzzified Dempster–Shafer theory of evidence*: Belief and plausibility functions of DST defined on fuzzy sets.

7. *Theory based on feasible interval-valued probability distributions (FIPDs)*: According to the FIPD, lower and upper probabilities $\underline{\mu}(A)$ and $\bar{\mu}(A)$ are determined for all sets $A \in P_X$ by intervals $[\underline{\mu}(\{x\}), \bar{\mu}(\{x\})]$ of probabilities on singletons ($x \in X$).
8. *Fuzzified FIPD*: Feasible interval-valued probability distributions defined on fuzzy sets.
9. *Other uncertainty theories*: Including a theory based on λ-measures, a theory based on probability boxes, theories based on various decomposable fuzzy measures, and theories based on *p*-additive measures (Klir, 2006).

The evidence, probability, possibility, and imprecise probability theories are the ones primarily featured in this chapter for their potential use in synthesizing information based on modeling and measuring uncertainties within the framework of ignorance types. The theory of evidence is introduced in the next section to form the basis for introducing the other theories.

3.5 DEMPSTER–SHAFER EVIDENCE THEORY

The theory of evidence, also called the Dempster–Shafer theory (DST), was developed by Shafer (1976) and Dempster (1976a, 1976b). This theory is based on *belief measures* and *plausibility measures*.

3.5.1 Belief Measures

A *belief measure* (*Bel*) is defined on a universal set *X* as a function that maps the power set of *X* to the range [0, 1]. Formally,

$$Bel: P_X \rightarrow [0, 1] \tag{3.16}$$

where P_X is the *power set* of *X*. Belief functions have to meet the following three conditions:

$$Bel(\varnothing) = 0 \tag{3.17}$$

$$Bel(X) = 1 \tag{3.18}$$

$$\begin{aligned} Bel(A_1 \cup A_2 \cup \ldots \cup A_N) \geq \sum_{j=1}^{N} Bel(A_j) - \sum_{k=1}^{N}\left[\sum_{j=k+1}^{N} Bel(A_j \cap A_k)\right] + \\ \sum_{k=1}^{N}\left\{\sum_{j=k+1}^{N}\left[\sum_{l=j+1}^{N} Bel(A_j \cap A_k \cap A_l)\right]\right\} - \ldots + (-1)^{N+1} Bel(A_1 \cap A_2 \cap \ldots \cap A_N) \end{aligned} \tag{3.19}$$

where ϕ is the null set and $A_1, A_2, \ldots, A_N$ is any possible family of subsets of X. The inequality provided by Equation 3.19 shows that belief measures have the property of being superadditive, i.e., $Bel(A \cup B) \geq Bel(A) + Bel(B)$ where $A \cap B = \phi$.

3.5.2 Plausibility Measure

Each belief measure (Bel) has a *dual measure*, called a *plausibility measure* (Pl), which is defined by the following equations:

$$Pl(A) = 1 - Bel(\bar{A}) \tag{3.20a}$$

$$Pl(\bar{A}) = 1 - Bel(A) \tag{3.20b}$$

$$Bel(A) = 1 - Pl(\bar{A}) \tag{3.20c}$$

$$Bel(\bar{A}) = 1 - Pl(A) \tag{3.20d}$$

where A is a subset that belongs to the power set P_X. Plausibility measures must satisfy the following conditions:

$$Pl(\emptyset) = 0 \tag{3.21}$$

$$Pl(X) = 1 \tag{3.22}$$

$$Pl(A_1 \cap A_2 \cap \ldots \cap A_N) \leq \sum_{j=1}^{N} Pl(A_j) - \sum_{k=1}^{N}\left[\sum_{j=k}^{N} Pl(A_j \cup A_k)\right] + \ldots \tag{3.23}$$
$$+(-1)^{N+1} Pl(A_1 \cup A_2 \cup \ldots \cup A_N)$$

where $A_1, A_2, \ldots, A_N$ is any family of subsets of X. The pair Bel and Pl forms a duality. The inequality provided by Equation 3.23 shows that the plausibility measure has the property of being subadditive, i.e., $Pl(A \cup B) \leq Pl(A) + Pl(B)$. It can be shown that the belief and plausibility functions satisfy the following condition:

$$Pl(A) \geq Bel(A) \tag{3.24}$$

for each A in the power set.

3.5.3 Interpretation of Belief and Plausibility Measures

Belief and plausibility measures characterized by Equations 3.16 to 3.24 can be interpreted in some applications as a lower limit (Bel) and upper limit (Pl) on the strength of evidence at hand. For example, Dong and Wong (1986a, 1986b, 1986c) and Bogler (1987) use this interpretation to develop lower and upper limits on probabilities in earthquake and target identification applications, as described in Examples 3.1 and 3.2, respectively. This probability interpretation of evidential measures can be justified based on the work of Dempster (1976a, 1976b).

3.5.4 Möbius Representation as a Basic Assignment

In engineering and science, a body of evidence represented by a family of sets (A_1, A_2, …, A_N) can be characterized by a *Möbius representation* called a *basic assignment* constructed for convenience and for facilitating the synthesis of data and information. A basic assignment is given to the sets (i.e., not the elements) and is termed a *Möbius* (m) representation. A basic assignment can be related to the belief and plausibility measures; however, its creation is commonly easier and more relevant to problems than directly developing the belief and plausibility measures. A basic assignment provides an assessment of the likelihood of each set in a family of sets identified by the analyst. The family of sets are crisp sets, whereas the element of interest x of X can be imprecise in its boundaries, and hence uncertain in its belonging to the sets in the family of sets. Nguyen (1978) interpreted the basic assignment as probability masses assigned to subsets of the power set, calling this interpretation and representation random sets.

A basic assignment (m) can be conveniently characterized by the following function that maps the power set (P_X) to the range [0, 1]:

$$m\colon P_X \to [0, 1] \tag{3.25}$$

A basic assignment must satisfy the following two conditions:

$$m(\varnothing) = 0 \tag{3.26}$$

$$\sum_{\text{all } A \in P_X} m(A) = 1 \tag{3.27}$$

If $m(A_i) > 0$ for an i, A_i is also called a *focal element*. The belief measure and plausibility measure can be computed based on a particular basic assignment m, for any set $A_i \in P_X$, as follows:

$$Bel(A_i) = \sum_{\text{all } A_j \subseteq A_i} m(A_j) \tag{3.28}$$

$$Pl(A_i) = \sum_{\text{all } A_j \cap A_i \neq \phi} m(A_j) \tag{3.29}$$

The summation in Equation 3.28 should be performed over all sets A_j that are contained or equal to A_i, whereas the summation in Equation 3.29 should be performed over all sets A_j that belong to or intersect with the set A_i. The three functions *Bel*, *Pl*, and m can be viewed as alternative representations of the same information or evidence regarding the element x. These functions express the likelihood that x belongs to each A_i as a belief measure (strongest), plausibility measure (weakest), and a basic assignment (collected evidence). Once one of the three functions is defined, the other two functions can be uniquely computed. For example, the basic assignment m for $A_i \in P_X$ can be computed based on the belief function as follows:

$$m(A_i) = \sum_{all\ A_j \subseteq A_i} (-1)^{|A_i - A_j|} Bel(A_j) \tag{3.30}$$

where $|A_i - A_j|$ is the cardinality of the difference between the two sets. Equation 3.20 can be used to compute the *Bel* from the *Pl* for cases where *Pl* values are given, then Equation 3.30 can be used to compute m.

3.5.5 Combination of Evidence

Several evidence combination methods are available without having a universally accepted single method. In this section four methods are presented.

3.5.5.1 Dempster's Rule of Combination

Basic assignments (m_1 and m_2) produced by two experts on the same element and a family of sets of interest can be combined using *Dempster's rule of combination* to obtain a combined opinion ($m_{1,2}$) as follows:

$$m_{1,2}(A_i) = \frac{\sum_{all\ A_j \cap A_k = A_i} m_1(A_j) m_2(A_k)}{1 - \sum_{all\ A_j \cap A_k = \varnothing} m_1(A_j) m_2(A_k)} \tag{3.31}$$

where A_i must be a nonempty set and $m_{1,2}(\varnothing) = 0$. The term $1 - \sum_{all\ A_j \cap A_k = \varnothing} m_1(A_j) m_2(A_k)$ of Equation 3.31 is a normalization factor that can be interpreted as a normalization factor for the contradiction or conflict among the evidential information (Shafer, 1976; Wang and Klir, 1992; Klir and Folger, 1988). Equation 3.31 provides an example rule to combine expert opinions that does not account for the reliability of the source and other relevant considerations.

The Dempster rule of combination provides a unique solution that can be proven under the axiomatic conditions of the Dempter–Shafer theory and under the assumption that the two sources of information are independent of each other. Independence in this case means that observations made by one source do not constrain observations made by the other source (Dubois and Prade, 1986a). On the other hand, the requirement that $m_{1,2}(\emptyset) = 0$ in Equation 3.31 leads to the normalization factor in this equation. This requirement of $m_{1,2}(\emptyset) = 0$ is considered unnecessarily restrictive and could lead, in some cases, to counterintuitive results, as was demonstrated by Zadeh (1979b, 1986) and Smets (1990). Allowing $m_{1,2}(\emptyset) \neq 0$ can be interpreted as the nonzero value of $m_{1,2}(\emptyset)$ supporting a combined evidence of the hypothesis of having a value outside the universal set under consideration. Contrarily, requiring $m_{1,2}(\emptyset) = 0$ implies that all relevant hypotheses in a given context are included in the accepted universal set (i.e., a closed-world assumption or position); on the other hand, allowing $m_{1,2}(\emptyset) \neq 0$ recognizes that the universal set might be incomplete in a given context (i.e., an open-world assumption or position). The assumption of an open world for the universal set is discussed further in Chapter 5.

3.5.5.2 Yager's Rule of Combination

The primary difference between Dempster's rule and Yager's rule of combination is in the handling of the contradiction provided by $1 - \sum_{\text{all } A_j \cap A_k = \emptyset} m_1(A_j)m_2(A_k)$ in Equation 3.31. Yager (1987c) suggested allocating the contradiction to the universe X by introducing what is called the ground probability mass function ($q_{1,2}$) for combining evidence from two sources that can be computed as follows with the property $q_{1,2}(\emptyset) \geq 0$:

$$q_{1,2}(A_i) = \sum_{\text{all } A_j \cap A_k = A_i} m_1(A_j)m_2(A_k) \tag{3.32a}$$

$$m_{1,2}(A_i) = q_{1,2}(A_i) \quad \text{for } A_i \neq \emptyset \text{ and } A_i \neq X \tag{3.32b}$$

$$m_{1,2}(X) = q_{1,2}(X) + q_{1,2}(\emptyset) \tag{3.32c}$$

3.5.5.3 Inagaki's Rule of Combination

Inagaki (1991) introduced a unified combination rule with Dempster's rule and Yager's rule of combination as special cases based on a combination parameter k 0. By setting this parameter to zero, Yager's rule of combination is obtained, whereas by setting this parameter to $1/(1 - q(\emptyset))$, Dempster's rule of combination is obtained. Other combinations can be obtained for values of this parameter in the following range:

$$0 \leq k \leq \frac{1}{1 - q_{1,2}(X) - q_{1,2}(\emptyset)} \tag{3.33a}$$

This unified rule of combination is given by

$$m_{1,2}(A_i) = [1 + kq_{1,2}(\varnothing)]q_{1,2}(A_i) \quad \text{for } A_i \neq \varnothing \text{ and } A_i \neq X \tag{3.33b}$$

$$m_{1,2}(X) = [1 + kq_{1,2}(\varnothing)]q_{1,2}(X) + [1 + kq_{1,2}(\varnothing) - k]q_{1,2}(\varnothing) \tag{3.33c}$$

$$m_{1,2}(\varnothing) = 0 \tag{3.33d}$$

The procedure to assign a k value is not well justified, and Inagaki's rule is not associative except at the k values that correspond to Dempster's rule.

3.5.5.4 Mixed or Averaging Rule of Combination

According to this method, the combination is performed based on the assumption that all sources are not equally credible and any contradiction or conflict among them is not taken into account. The combination rule is expressed as (Bae et al., 2004):

$$m_{1,2,3,\ldots,n}(A_i) = \frac{1}{n}\sum_{k=1}^{n} w_k m_k(A_i) \tag{3.34}$$

The weight factors are assigned based on the credibility of the evidence and its source.

EXAMPLE 3.1 ESTIMATING THE LOCATION OF AN EARTHQUAKE EPICENTER

In some applications, expert opinions in the form of subjective information, such as the location of an epicenter of an earthquake or the probability of an event, need to be combined into a single value, and perhaps confidence intervals for their use in probabilistic and risk analyses. Cooke (1991) and Rowe (1992) provided a summary of methods for combining expert opinions. The methods can be classified into consensus methods and mathematical methods (Clemen, 1989; Ferrell, 1985). The mathematical methods can be based on assigning equal weights to the experts or different weights. Sometimes it might be desirable to elicit probabilities or consequences using linguistic terms, such as linguistic probabilities discussed in subsequent sections. Linguistic terms of this type can be translated into intervals or fuzzy numbers. Intervals are considered a special case of fuzzy numbers that are in turn a special case of fuzzy sets.

Assessing the potential consequences of an earthquake requires knowing the location of its epicenter. Dong and Wong (1986a) developed an example for estimating the location of the epicenter based on the opinions of experts that was summarized by Wang and Klir (1992). In this example, a group of 15 experts provided their estimates of possible location of the epicenter of the earthquake in the form of zones, as provided in Figure 3.4, called E_1 to E_{15}. The location estimates are both nonspecific (i.e., provided as regions) and conflicting (i.e., not in agreement) with each other. The expert

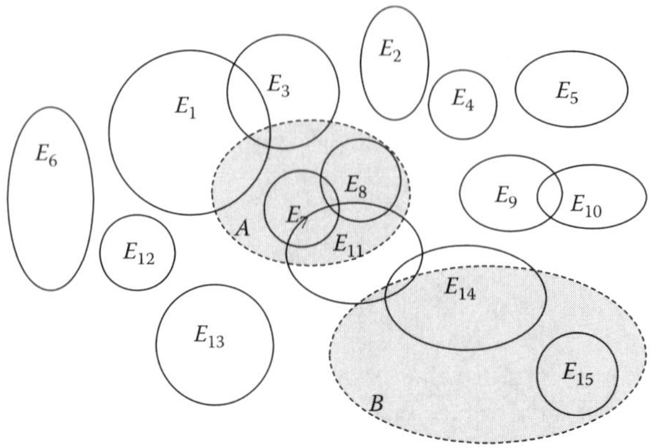

FIGURE 3.4 Estimates of the locations of the epicenter of an earthquake.

opinions can be treated as a body of evidence in this case. This evidence body can be used to estimate the likelihood that the epicenter is inside particular areas of interest representing two cities or densely populated areas *A* and *B*, as shown in Figure 3.4. The experts can be assumed for the purposes of this example to have an equal level of credibility and reliability, therefore assigning their respective regions equal weights of evidence of 1/15 — in other words, equally credible and reliable (i.e., equally equivalent in their opinions).

The degrees of belief and plausibility can be calculated based on this assignment of evidence and the areas provided in Figure 3.4. The belief that the epicenter is in region *A* is given by

$$Bel(A) = 2/15 = 0.13 \tag{3.35a}$$

The plausibility that the epicenter is in region *A* is given by

$$Pl(A) = 5/15 = 0.33 \tag{3.35b}$$

Similarly for region *B*, the belief and plausibility are

$$Bel(B) = 1/15 = 0.07 \tag{3.35c}$$

$$Pl(B) = 3/15 = 0.2 \tag{3.35d}$$

These belief and plausibility values can be used to construct the following interval-valued estimates of respective probabilities of *A* and *B*:

$$0.13 \le P(A) \le 0.33 \tag{3.36}$$

$$0.07 \le P(B) \le 0.2 \tag{3.37}$$

TABLE 3.1
Belief Computations for Classifying Bridge Failures

Subset[a] (Failure Cause)	Expert 1		Expert 2		Combined Judgment	
	m_1	Bel_1	m_2	Bel_2	$m_{1,2}$	$Bel_{1,2}$
Design error (D)	0.10	0.10	0.05	0.05	0.167147	0.167147
Construction error (C)	0.05	0.05	0.10	0.10	0.152738	0.152738
Human error (H)	0.10	0.10	0.15	0.15	0.181556	0.181556
$D \cup C$	0.20	0.35	0.25	0.40	0.230548	0.550433
$D \cup H$	0.10	0.30	0.10	0.30	0.086455	0.435158
$C \cup H$	0.05	0.20	0.10	0.35	0.066282	0.400576
$D \cup C \cup H$	0.40	1.	0.25	1.	0.115274	1.

[a] The subsets could also be written as $\{D\}$, $\{C\}$, $\{H\}$, $\{D, C\}$, $\{D, H\}$, $\{C, H\}$, and $\{D, C, H\}$.

Example 3.2 Causes of a Bridge Failure during Construction

Bridges can collapse during construction due to many causes (Eldukair and Ayyub, 1991). Consider three common causes: (1) design error (D), (2) construction error (C), and (3) human error (H). A database of bridges failed during construction can be created. For each bridge failure case, the case needs to be classified in terms of its causes and entered in the database. The sets D, C, and H belong to the universal set X of failure causes. Two experts were asked to review a bridge failure case and subjectively provide basic assignments for this case in terms of the sets D, C, and H. The experts provided the assignments in Table 3.1 for D, C, H, $D \cup C$, $D \cup H$, and $C \cup H$. The assignment for $D \cup C \cup H$ was computed based on Equation 3.27 to obtain a total of 1 for the assignments provided by each expert. The Bel values for each expert in Table 3.1 were computed using Equation 3.28.

In order to combine the opinions of the experts according to Equation 3.31, the normalizing factor needs to be computed as follows:

$$1 - \sum_{all\, A_j \cap A_k = \varnothing} m_1(A_j)m_2(A_k) =$$

$$1 - \begin{bmatrix} m_1(D)m_2(C) + m_1(D)m_2(H) + m_1(D)m_2(C \cup H) + \\ m_1(C)m_2(D) + m_1(C)m_2(H) + m_1(C)m_2(D \cup H) + \\ m_1(H)m_2(D) + m_1(H)m_2(C) + m_1(H)m_2(C \cup D) + \\ m_1(D \cup C)m_2(H) + m_1(D \cup H)m_2(C) + m_1(C \cup H)m_2(D) \end{bmatrix} \quad (3.38)$$

Substituting the values of m produces the following normalizing factor:

$$1 - \sum_{all\,A_j \cap A_k = \varnothing} m_1(A_j) m_2(A_k) =$$

$$1 - \begin{bmatrix} 0.1(0.1) + 0.1(0.15) + 0.1(0.1) + \\ (0.05)(0.05) + (0.05)(0.15) + (0.05)(0.1) + \\ (0.1)(0.05) + (0.1)(0.1) + (0.1)(0.25) + \\ (0.2)(0.15) + (0.1)(0.1) + (0.05)(0.05) \end{bmatrix} = 0.8675 \tag{3.39}$$

The combined opinions can then be computed using Equation 3.31 as follows:

$$m_{1,2}(D) =$$

$$\frac{\begin{bmatrix} m_1(D)m_2(D) + m_1(D)m_2(D \cup C) + m_1(D)m_2(D \cup H) + \\ m_1(D)m_2(D \cup C \cup H) + m_1(D \cup C)m_2(D) + m_1(D \cup C)m_2(D \cup H) + \\ m_1(D \cup H)m_2(D) + m_1(D \cup H)m_2(D \cup C) + m_1(D \cup C \cup H)m_2(D) \end{bmatrix}}{0.8675} \tag{3.40}$$

or

$$m_{1,2}(D) = \frac{\begin{bmatrix} 0.1(0.05) + 0.1(0.25) + 0.1(0.1) + \\ 0.1(0.25) + 0.2(0.05) + 0.2(0.1) + \\ 0.1(0.05) + 0.1(0.25) + 0.4(0.05) \end{bmatrix}}{0.8675} = 0.167147 \tag{3.41}$$

$$m_{1,2}(C) = \frac{\begin{bmatrix} 0.05(0.1) + 0.05(0.25) + 0.05(0.1) + \\ 0.05(0.25) + 0.2(0.1) + 0.2(0.1) + \\ 0.05(0.1) + 0.05(0.25) + 0.4(0.1) \end{bmatrix}}{0.8675} = 0.152738 \tag{3.42}$$

$$m_{1,2}(H) = \frac{\begin{bmatrix} 0.1(0.15) + 0.1(0.1) + 0.1(0.1) + \\ 0.1(0.25) + 0.1(0.15) + 0.1(0.1) + \\ 0.05(0.15) + 0.05(0.1) + 0.4(0.15) \end{bmatrix}}{0.8675} = 0.181556 \tag{3.43}$$

$$m_{1,2}(D \cup C) = \frac{[0.2(0.25) + 0.2(0.25) + 0.4(0.25)]}{0.8675} = 0.230548 \tag{3.44}$$

$$m_{1,2}(D \cup H) = \frac{[0.1(0.1) + 0.1(0.25) + 0.4(0.1)]}{0.8675} = 0.086455 \tag{3.45}$$

TABLE 3.2
Plausibility Computations for Classifying Bridge Failures

Subset[a] (Failure Cause)	Expert 1 m_1	Expert 1 Pl_1	Expert 2 m_2	Expert 2 Pl_2	Combined Opinion $m_{1,2}$	Combined Opinion $Pl_{1,2}$
Design error (D)	0.10	0.80	0.05	0.65	0.167147	0.599424
Construction error (C)	0.05	0.70	0.10	0.70	0.152738	0.564842
Human error (H)	0.10	0.65	0.15	0.60	0.181556	0.449567
$D \cup C$	0.20	0.90	0.25	0.85	0.230548	0.818444
$D \cup H$	0.10	0.95	0.10	0.90	0.086455	0.847262
$C \cup H$	0.05	0.90	0.10	0.95	0.066282	0.832853
$D \cup C \cup H$	0.40	1.	0.25	1.	0.115274	1.

[a] The subsets could also be written as $\{D\}$, $\{C\}$, $\{H\}$, $\{D, C\}$, $\{D, H\}$, $\{C, H\}$, and $\{D, C, H\}$.

$$m_{1,2}(C \cup H) = \frac{[0.05(0.1) + 0.05(0.25) + 0.4(0.1)]}{0.8675} = 0.066282 \tag{3.46}$$

$$m_{1,2}(C \cup D \cup H) = \frac{[0.4(0.25)]}{0.8675} = 0.115274 \tag{3.47}$$

The values provided by Equations 3.41 to 3.47 must add up to 1. The $Bel_{1,2}$ values in Table 3.1 were computed using Equation 3.28. The plausibility computations for classifying bridge failures are shown in Table 3.2 and were based on Equation 3.29.

Example 3.3 Target Identification

The identification of targets can be based on data from sensors. Bogler (1987) provides an example involving multiple-sensor target identification in which intelligence reports are also employed as a source of information. For the purposes of the example, a list of 100 possible target types are considered and denoted as $X = \{x_1, x_2, \ldots, x_{100}\}$, i.e., X = the universal set. A primary intelligence source (called source 1), which can identify only 40% of these 100 target types, i.e., $\{x_1, x_2, \ldots, x_{40}\}$, indicates that a target type belonging to $\{x_1, x_2, \ldots, x_{40}\}$ has entered a relevant tactical area. The information from this source can be formulated according to the Dempster–Shafer theory by defining $A = \{x_1, x_2, \ldots, x_{40}\}$. Based on the evidence of this source, the assignment $m_1(A) = 0.4$ and $m_1(X) = 0.6$ can be made. This assignment can be used to construct the following belief and plausibility values for this primary intelligence source:

$$Bel_1(A) = 0.4 \tag{3.48a}$$

$$Pl_1(A) = 1 \tag{3.48b}$$

TABLE 3.3
Evidence for Target Identification

Subset (Evidence)	Measures			
	Assignment (m_i)	Belief (Bel_i)	Plausibility (Pl_i)	Probability (P_i)
(1) A primary intelligence source				
Event (A)	0.4	0.4	1	[0.4, 1]
$\bar{A}$	0*	0	0.6	[0, 0.6]
Universal set (X)	0.6	1	1	[1, 1]
(2) A secondary intelligence source				
Event (B)	0.5	0.5	1.0	[0.5, 1]
$\bar{B}$	0[a]	0.0	0.5	[0, 0.5]
Universal set (X)	0.5	1	1	[1, 1]

[a] Not provided in the assignment.

$$Bel_1(\bar{A}) = 0 \tag{3.49a}$$

$$Pl_1(\bar{A}) = 0.6 \tag{3.49b}$$

Interpreting these values of belief and plausibility as lower and upper limits on probabilities, the following probability ranges can be constructed:

$$0.4 \le P_1(A) \le 1 \tag{3.50}$$

$$0 \le P_1(\bar{A}) \le 0.6 \tag{3.51}$$

The model can be expanded by introducing a secondary intelligence source (called source 2) that indicates not only target type x_1 being in the population of incoming targets, but also 10 other target types, defining $B = \{x_1, x_2, \ldots, x_{50}\}$ with $m_2(B) = 0.5$, and subsequently $Bel_2(B) = 0.5$, $Pl_2(B) = 1$, $Bel_2(\bar{B}) = 0$, and $Pl_2(\bar{B}) = 0.5$. These values are summarized in Table 3.3. Using the rule of combination as provided by Equation 3.31, a combined body of evidence can be computed as shown in Table 3.4.

A variant to this example was also introduced by considering only two types of targets of the 100 possible targets, say a fighter (F) or a bomber (B). In this case, a short-range sensor and a radar warning system identified an incoming and fast-moving target. The short-range sensor provides a support of 0.6 that the target is a fighter, and the radar warning provides a support of 0.95 that it is a bomber. The degrees of support for the target types based on this combined evidence can be computed using Equation 3.31 as summarized in Table 3.4.

TABLE 3.4
Combined Evidence for Identification of a Fighter and a Bomber

	Measures					
Subset (Evidence)	**Short-Range Sensor Assignment (m_1)**	**Radar Warning System Assignment (m_2)**	**Assignment (m_{12})**	**Belief (Bel_{12})**	**Plausibility (Pl_{12})**	**Probability (P_{12})**
Fighter (F)	0.6	0	0.07	0.07	0.12	[0.07, 0.12]
Bomber (B)	0	0.95	0.88	0.88	0.93	[0.88, 0.93]
Universal set (X)	0.4	0.05	0.05	1	1	[1, 1]

3.6 POSSIBILITY THEORY

3.6.1 Classical Possibility Theory

Classical possibility theory and its monotone measures of necessity and possibility are based on crisp sets and the nonadditive properties of Equations 3.12 and 3.15 as described by Klir (2006). Assume that, according to given evidence, we know that alternatives in a particular set E (such that $E \subseteq X$) are possible, while those outside E are not possible. This means that, according to the evidence, the true alternative is in set E. This simple evidence can be formalized by defining a *possibility measure*, Pos_E, based on evidence focusing on E:

$$Pos_E(\{x\}) = \begin{cases} 1 & \text{when } x \in E \\ 0 & \text{when } x \in \bar{E} \end{cases} \tag{3.52}$$

for all $x \in X$.

The outcome is called a *possibility distribution* function, $r(x)$, defined as a mapping from the universal set X to the values $\{0, 1\}$ according to the following equation:

$$r : X \rightarrow \{0,1\} \tag{3.53}$$

The function $r(x)$ assigns either 0 or 1 to each element, describing its occurrence as either possible or impossible, respectively. The possibility measure (Pos) for a subset $A_i \subseteq X$ can be uniquely determined based on $r(x)$ as follows:

$$Pos(A_i) = \max_{x \in A_i}(r(x)) \tag{3.54}$$

The possibility measure is therefore a mapping from the power set of X to the values $\{0, 1\}$. The corresponding necessity (*Nec*) measure for a crisp subset $A_i \subseteq X$ can be defined as

$$Nec(A_i) = 1 - Pos(\bar{A}_i) \tag{3.55}$$

where $\bar{A}_i$ is the complement of A. Therefore, the necessity measure is also a mapping from the power set of X to the values $\{0, 1\}$.

3.6.2 Theory of Graded Possibilities

The evidence E used in Equation 3.52a is a crisp evidence. Representing the evidence using a standard, normal fuzzy set E requires defining the possibility measure based on the α-cut of E, Pos_E, as follows:

$$^{\alpha}Pos_E(\{x\}) = \begin{cases} 1 & \text{when } x \in {}^{\alpha}E \\ 0 & \text{when } x \in {}^{\alpha}\bar{E} \end{cases} \tag{3.56}$$

for all $x \in X$. For a set $A \in P_X$, the *graded possibility measure* for A based on the evidence $^{\alpha}E$ is

$$^{\alpha}Pos_E(A) = \sup_{x \in A} {}^{\alpha}Pos_E(\{x\}) \tag{3.57}$$

where sup is the *supremum* of A defined as the least upper bound of the set. Using the α-cut representation of E, the membership function of E is given by

$$E(x) = \sup_{\alpha \in (0,1]} \alpha\, {}^{\alpha}E(x) \tag{3.58}$$

for all $x \in X$, where $^{\alpha}E$ denotes the characteristic function of the crisp α-cut of E. Equation 3.56 can be rewritten as

$$^{\alpha}Pos_E(\{x\}) = {}^{\alpha}E(x) \tag{3.59}$$

for $\alpha \in (0, 1]$ and $x \in X$. Hence, the possibility measure, Pos_E, can be defined in terms of $^{\alpha}Pos_E$, for $\alpha \in (0, 1]$, as follows:

$$Pos_E(\{x\}) = \sup_{\alpha \in (0,1]} \alpha\, {}^{\alpha}Pos_E(\{x\}) \tag{3.60}$$

or

$$Pos_E(\{x\}) = E(x) \tag{3.61}$$

This definition of possibility (Zadeh, 1987) is referred to as the standard fuzzy set interpretation of possibility theory. Given the $Pos_E(\{x\})$ for all $x \in X$, $Pos_E(A)$ can be computed by

$$Pos_E(A) = \sup_{x \in X} Pos_E(\{x\}) \tag{3.62}$$

for all $A \in P_X$. Where A is fuzzy, the following general equation should be used:

$$Pos_E(A) = \sup_{x \in X} \min[A(x), Pos_E(\{x\})] \tag{3.63}$$

The associated necessity measure, Nec_E, is defined by

$$Nec_E(A) = 1 - Pos_E(\bar{A}) \tag{3.64}$$

for each set A, which may be crisp or fuzzy.

The possibility theory can also be viewed as a special case of the Dempster–Shafer theory of evidence and its monotone measures of belief and plausibility by requiring the underlying subsets of a universe X to be nested, i.e., $A_1 \subset A_2 \subset \ldots \subset X$. Nested subsets on X are called *chains*. Nested subsets for an evidence body result in minimal conflicts with each other; therefore, their belief and plausibility measures, called necessity and possibility measures, respectively, in this case are described to be *consonant*. An example of five nested sets (A_i) with 10 discrete elements (x_j) is shown in Figure 3.5. For nested subsets, the associated belief and plausibility measures, i.e., necessity and possibility measures, respectively, satisfy the following conditions:

$$Bel(A_1 \cap A_2) = \min[Bel(A_1), Bel(A_2)] \quad \text{for any } A_1 \text{ and } A_2 \in P_X \tag{3.65}$$

and

$$Pl(A_1 \cup A_2) = \max[Pl(A_1), Pl(A_2)] \quad \text{for any } A_1 \text{ and } A_2 \in P_X \tag{3.66}$$

The following properties of possibility and necessity measures are provided for any pairs of subsets $A_i \subseteq X$ and $A_j \subseteq X$:

$$Pos(A_i \cup A_j) = \max[Pos(A_i), Pos(A_j)] \tag{3.67a}$$

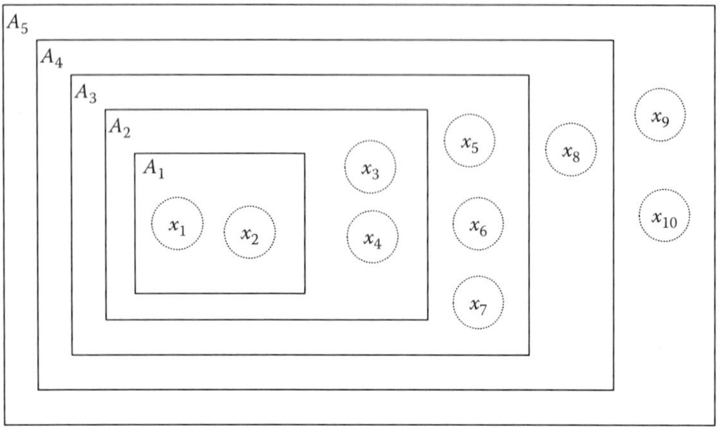

FIGURE 3.5 Nested sets and singletons for a possibility distribution.

$$Nec(A_i \cap A_j) = \min[Nec(A_i), Nec(A_j)] \tag{3.67b}$$

$$Pos(A_i \cap A_j) \leq \min[Pos(A_i), Pos(A_j)] \tag{3.67c}$$

$$Nec(A_i \cup A_i) \geq \max[Nec(A_i), Nec(A_j)] \tag{3.67d}$$

$$Nec(A_i) \leq Pos(A_i) \quad \text{for all } A_i \tag{3.67e}$$

$$Pos(A_i) + Pos(\bar{A}_i) \geq 1 \quad \text{for all } A_i \tag{3.67f}$$

$$Nec(A_i) + Nec(\bar{A}_i) \leq 1 \quad \text{for all } A_i \tag{3.67g}$$

$$\max[Pos(A_i), Pos(\bar{A}_i)] = 1 \quad \text{for all } A_i \tag{3.67h}$$

$$\min[Nec(A_i), Nec(\bar{A}_i)] = 0 \quad \text{for all } A_i \tag{3.67i}$$

$$Pos(A_i) < 1 \Rightarrow Nec(A_i) = 0 \quad \text{for all } A_i \tag{3.67j}$$

$$Nec(A_i) > 0 \Rightarrow Pos(A_i) = 1 \quad \text{for all } A_i \tag{3.67k}$$

The nested structure of a family of sets, i.e., $A_1 \subset A_2 \subset \ldots \subset X$, is compatible with the α-cuts of convex fuzzy sets, making fuzzy set interpretation of possibility theory logical. Klir and Folger (1988), Klir and Wierman (1999), Dubois and Prade

(1988), and De Cooman (1997) provide additional details on possibility theory and its applications.

3.7 PROBABILITY THEORY

Probability theory and related concepts are briefly introduced in this section for the purpose of comparing and discussing probabilistic methods with nonprobabilistic methods. The coverage of probability theory is not intended to be complete, and readers should consult other sources for a more complete coverage, such as Ayyub and McCuen (2003).

3.7.1 Relationship between Evidence Theory and Probability Theory

Probability theory can be treated as a special case of the theory of evidence. For the cases where all focal elements for a given basic assignment, m, i.e., body of evidence, are *singletons*, the associated belief measure and plausibility measure collapse into a single measure, a classical probability measure. The term *singleton* means that each subset A_i of the family A of subsets, presenting an evidence body, contains only one element. The resulting probability measure is additive in this case; i.e., it follows Equation 3.14. The following differences between evidence theory and probability theory can be noted based on this reduction of evidence theory to probability theory:

1. A basic assignment in evidence theory can be used to compute the belief and plausibility measures that map the power set of X to the range [0, 1].
2. A probability assignment, such as a probability mass function, in probability theory is a mapping function from the universal set X to the range [0, 1].

Dempster (1976a, 1976b) examined the use of evidence theory to construct a probability distribution for singletons based on a basic assignment for some subsets of a universal set. The solution can be expressed in the form of minimum and maximum probabilities for the singletons for cases where the evidence body, i.e., the basic assignment, is not contradictory.

3.7.2 Classical Definitions of Probability

The concept of probability has its origin in games of chance. In these games, probabilities are determined based on many repetitions of an experiment and counting the number of outcomes of an event of interest. Then, *the probability of the outcome of interest can be measured by dividing the number of occurrences of an event of interest by the total number of repetitions*. Quite often, probability is specified as a percentage; for example, when the weather bureau indicates that there is a 30% chance of rain, experience indicates that under similar meteorological conditions it has rained 3 of 10 times. In this example, the probability was estimated empirically using the concept of relative frequency, expressed as

$$P(X = x) = \frac{n}{N} \tag{3.68}$$

in which n = number of observations on the random variable X that results in an outcome of interest x, and N = total number of observations of x. The probability of a value associated with an event x in this equation was defined as the *relative frequency* of its occurrence. Also, probability can be defined as a *subjective probability* (also called *judgmental probability*) of the occurrence of the event. The type of definition depends on the nature of the underlying event. For example, in an experiment that can be repeated N times with n occurrences of the underlying event, the relative frequency of occurrence can be considered the probability of occurrence. In this case, the probability of occurrence is n/N. However, there are many engineering problems that do not involve large numbers of repetitions, and still we are interested in estimating the probability of occurrence of some event. For example, during the service life of an engineering product, the product either fails or does not fail in performing a set of performance criteria. The events of unsatisfactory performance and satisfactory performance are mutually exclusive and collectively exhaustive of the universal set (that is, the space of all possible outcomes). The probability of unsatisfactory performance (or satisfactory performance) can be considered a *subjective probability.*

Another example is the failure probability of a dam due to an extreme flooding condition. An estimate of such probabilities can be achieved by modeling the underlying system, its uncertainties and performances. The resulting subjective probability is expected to reflect the *status of our knowledge* about the system regarding occurrence of the events of interest. Therefore, subjective probabilities can be associated with degrees of belief and can form a basis for Bayesian methods (Ayyub and McCuen, 2003). It is important to keep in mind both definitions, so that results are not interpreted beyond the range of their validity.

An axiomatic definition of probability is commonly provided in the literature, such as Ayyub and McCuen (2003). For an event A, the notation $P(A)$ means the probability of occurrence of the event A. The probability P should satisfy the following axioms, i.e., Equations 3.7 to 3.9:

1. $0 \leq P(A) \leq 1$, for any A that belongs to the *set of all possible outcomes* (i.e., universal set X) for the system.
2. The probability of having X, $P(X) = 1$.
3. The occurrence probability of the union of mutually exclusive events is the sum of their individual occurrence probabilities.

The first axiom states that the probability of any event is inclusively between 0 and 1. Therefore, negative probabilities or probabilities larger than 1 are not allowed. The second axiom comes from the definition of the universal set. Since the universal set is the set of all possible outcomes, one or more of these outcomes must occur, resulting in the occurrence of S. If the probability of the universal set does not equal 1, this means that the universal set was incorrectly defined. The third axiom sets a basis for the mathematics of probability. These axioms as a single entity can be viewed as a definition of probability; i.e., any numerical structure that adheres to these axioms will provide a probability structure. Therefore, the relative frequency and subjective probability meet this definition of probability.

The relative frequency and subjective probability concepts are tools that help engineers and planners to deal with and model uncertainty, and should be used appropriately as engineering systems and models demand. In the case of relative frequency, increasing the number of repetitions according to Equation 3.68 would produce an improved estimate with a diminishing return on invested computational and experimental resources until a limiting (i.e., long-run or long-term) frequency value is obtained. This limiting value can be viewed as the *true probability*, although the absolute connotation in this terminology might not be realistic and cannot be validated. Philosophically, a true probability might not exist, especially when dealing with subjective probabilities. This, however, does not diminish the value of probabilistic analysis and methods since they provide a consistent, systematic, rigorous, and robust framework for dealing with uncertainty and decision making.

3.7.3 Linguistic Probabilities

Probability as described in the previous section provides a measure of the likelihood of occurrence of an event. It is a numerical expression of uncertainty; however, it is common for subjects (such as experts) to express uncertainty verbally using linguistic terms, such as *likely*, *probable*, *improbable*, etc. Although, these linguistic terms are somewhat fuzzy, they are meaningful. Lichtenstein and Newman (1967) developed a table that translates commonly used linguistic terms into probability values using responses from subjects. The Lichtenstein and Newman (1967) summary is shown in Table 3.5. The responses of the subjects show encouraging consistency in defining each term; however, the ranges of responses are large. Moreover, mirror-image pairs sometimes produce asymmetric results. The term "rather unlikely" is repeated in the table, as it was used twice in the questionnaire to the subjects, at almost the start and at the end of the questionnaire, to check consistency. It can be concluded from this table that verbal descriptions of uncertainty can be useful as an initial assessment, but other analytical techniques should be used to assess uncertainty; for example, the linguistic terms in Table 3.5 can be modeled using fuzzy sets (Haldar et al., 1997; Ayyub et al., 1997; Ayyub and Gupta, 1997; Ayyub, 1998).

3.7.4 Failure Rates

A failure rate can be defined as the number of failures per unit time or a unit of operation, such as cycle, revolution, rotation, start-up, etc. For example, a constant failure rate for an electronic device of 0.1 per year means that on *average*, the device fails once per 10 years. Another example that does not involve time is an engine with a failure rate of 10^{-5} per cycle of operation (or it can be in terms of mission length). In this case, the failure rate means that on average, the engine fails once per 100,000 cycles. Due to manufacturing, assembly, and aging effects, failure rates

TABLE 3.5
Linguistic Probabilities and Translations

Rank	Phrase	No. of Responses	Mean	Median	Standard Deviation	Range
1	Highly probable	187	0.89	0.90	0.04	0.60–0.99
2	Very likely	185	0.87	0.90	0.06	0.60–0.99
3	Very probable	187	0.87	0.89	0.07	0.60–0.99
4	Quite likely	188	0.79	0.80	0.10	0.30–0.99
5	Usually	187	0.77	0.75	0.13	0.15–0.99
6	Good chance	188	0.74	0.75	0.12	0.25–0.95
7	Predictable	146	0.74	0.75	0.20	0.25–0.95
8	Likely	188	0.72	0.75	0.11	0.25–0.99
9	Probable	188	0.71	0.75	0.17	0.01–0.99
10	Rather likely	188	0.69	0.70	0.09	0.15–0.99
11	Pretty good chance	188	0.67	0.70	0.12	0.25–0.95
12	Fairly likely	188	0.66	0.70	0.12	0.15–0.95
13	Somewhat likely	187	0.59	0.60	0.18	0.20–0.92
14	Better than even	187	0.58	0.60	0.06	0.45–0.89
15	Rather	124	0.58	0.60	0.11	0.10–0.80
16	Slightly more than half the time	188	0.55	0.55	0.06	0.45–0.80
17	Slight odds in favor	187	0.55	0.55	0.08	0.05–0.75
18	Fair chance	188	0.51	0.50	0.13	0.20–0.85
19	Toss-up	188	0.50	0.50	0.00	0.45–0.52
20	Fighting chance	186	0.47	0.50	0.17	0.05–0.90
21	Slightly less than half the time	188	0.45	0.45	0.04	0.05–0.50
22	Slight odds against	185	0.45	0.45	0.11	0.10–0.99
23	Not quite even	180	0.44	0.45	0.07	0.05–0.60
24	Inconclusive	153	0.43	0.50	0.14	0.01–0.75
25	Uncertain	173	0.40	0.50	0.14	0.08–0.90
26	Possible	178	0.37	0.49	0.23	0.01–0.99
27	Somewhat unlikely	186	0.31	0.33	0.12	0.03–0.80
28	Fairly unlikely	187	0.25	0.25	0.11	0.02–0.75
29	Rather unlikely	187	0.24	0.25	0.12	0.01–0.75
30	Rather unlikely	187	0.21	0.20	0.10	0.01–0.75
31	Not very probable	187	0.20	0.20	0.12	0.01–0.60
32	Unlikely	188	0.18	0.16	0.10	0.01–0.45
33	Not much chance	186	0.16	0.15	0.09	0.01–0.45
34	Seldom	188	0.16	0.15	0.08	0.01–0.47
35	Barely possible	180	0.13	0.05	0.17	0.01–0.60
36	Faintly possible	184	0.13	0.05	0.16	0.01–0.50
37	Improbable	187	0.12	0.10	0.09	0.01–0.40
38	Quite unlikely	187	0.11	0.10	0.08	0.01–0.50
39	Very unlikely	186	0.09	0.10	0.07	0.01–0.50
40	Rare	187	0.07	0.05	0.07	0.01–0.30
41	Highly improbable	181	0.06	0.05	0.05	0.01–0.30

Source: Adapted from Lichtenstein, S. and Newman, J.R., *Psychonometric Sci.*, 9, 563–564, 1967.

can generally be variant with time (or other units of operation), therefore requiring sometimes a statement of limitation on their applicability. Failure rates can be used in probabilistic analysis. There are analytical methods to convert failure rates into probabilities of some events of interest.

3.7.5 Central Tendency Measures

A very important descriptor of data is central tendency measures. The central tendency can be measured using, for example, (1) the mean (or average) value or (2) the median value.

The average value is the most commonly used central tendency descriptor. The definition of the sample mean (or average) value herein is based on a sample of size n. The sample consists of n values of a random variable X. For n observations, if all observations are given equal weights, then the average value is given by

$$\bar{X} = \frac{1}{n}\sum_{i=1}^{n} x_i \tag{3.69a}$$

where x_i = a sample point, $i = 1, 2, \ldots, n$, and

$$\sum_{i=1}^{n} x_i = x_1 + x_2 + x_3 + \ldots + x_n \tag{3.69b}$$

Since the average value ($\bar{X}$) is based on a sample, it has statistical error due to two reasons: (1) it is sample dependent, i.e., a different sample might produce a different average, and (2) it is sample-size dependent, i.e., as the sample size is increased, the error is expected to reduce. The mean value has another mathematical definition that is based on probability distributions according to probability theory, which is not described herein.

The average time between failures can be computed as the average ($\bar{X}$), where x_i = a sample point indicating the age at failure of a failed component and $i = 1, 2, \ldots, n$. The failed components are assumed to be replaced by new identical ones or repaired to a state "as good as new." The average time between failures is related to the failure rate as its reciprocal. For example, a component with a failure rate of 0.1 per year has an average time between failures of 1/0.1 = 10 years. Similar to failure rates, the average time between failures can be constant or time dependent.

According to probability theory, the mean (μ) can be computed from a probability mass function ($P(x)$) or a probability density function ($f(x)$) for a random variable (X) according to the respective equations:

$$\mu = \sum_{all\ x} x_i P(x_i) \text{ for discrete } X \tag{3.70a}$$

$$\mu = \int_{-\infty}^{\infty} xf(x)dx \text{ for continuous } X \tag{3.70b}$$

The median value x_m is another measure of central tendency. It is defined as the point that divides the data into two equal parts, i.e., 50% of the data are above x_m and 50% are below x_m. The median value can be determined by ranking the n values in the sample in decreasing order, 1 to n. If n is an odd number, then the median is the value with a rank of $(n + 1)/2$. If n is an even number, then the median equals the average of the two middle values, i.e., those with ranks $n/2$ and $(n/2) + 1$.

The advantage of using the median value as a measure of central tendency over the average value is its insensitivity to extreme values such as outliers. Consequently, this measure of central tendency is commonly used in combining expert judgments in an expert opinion elicitation process.

3.7.6 Dispersion (or Variability)

Although the central tendency measures convey certain information about the underlying sample, they do not completely characterize the sample. Two random variables can have the same mean value, but different levels of data scatter around the computed mean. Thus, measures of central tendency cannot fully characterize the data. Other characteristics are also important and necessary. The dispersion measures describe the level of scatter in the data about the central tendency location.

The most commonly used measure of dispersion is the variance and other quantities that are derived from it, such as the standard deviation and coefficient of variation. For n observations in a sample that are given equal weight, the variance (S^2) is given by

$$S^2 = \frac{1}{n-1}\sum_{i=1}^{n}(x_i - \bar{X})^2 \tag{3.71a}$$

The units of the variance are the square of the units of the variable x; for example, if the variable is measured in pounds per square inch (psi), the variance has units of $(\text{psi})^2$. Computationally, the variance of a sample can be determined using the following alternative equation:

$$S^2 = \frac{1}{n-1}\left[\sum_{i=1}^{n} x_i^2 - \frac{1}{n}\left(\sum_{i=1}^{n} x_i\right)^2\right] \tag{3.71b}$$

By definition the standard deviation (S) is the square root of the variance as follows:

$$S = \sqrt{\frac{1}{n-1}\left[\sum_{i=1}^{n} x_i^2 - \frac{1}{n}\left(\sum_{i=1}^{n} x_i\right)^2\right]} \tag{3.72}$$

It has the same units as both the underlying variable and the central tendency measures. Therefore, it is a useful descriptor of the dispersion or spread of a sample of data. The coefficient of variation (*COV*) is a normalized quantity based on the standard deviation and the mean and is different from the covariance (discussed in Section 3.8.7). Therefore, the *COV* is dimensionless and is defined as

$$COV = \frac{S}{\bar{X}} \tag{3.73}$$

It is also used as an expression of the standard deviation in the form of a percent of the average value. For example, consider $\bar{X}$ and S to be 50 and 20, respectively; therefore, $COV(X) = 0.4$, or 40%. In this case, the standard deviation is 40% of the average value.

According to probability theory, the variance (σ^2) can be computed from a probability mass function ($P(x)$) or a probability density function ($f(x)$) for a random variable (X) according to the respective equations:

$$\sigma^2 = \sum_{all\,x} (x_i - \mu)^2 P(x_i) \text{ for discrete } X \tag{3.74a}$$

$$\sigma^2 = \int_{-\infty}^{\infty} (x - \mu)^2 f(x)dx \text{ for continuous } X \tag{3.74b}$$

3.7.7 Percentile Values

A p-percentile value (x_p) for a random variable based on a sample is the value of the parameter such that p% of the data is less or equal to x_p, where $0 \le p \le 1$. On the basis of this definition, the median value is considered to be the 50th percentile value.

Aggregating information or data, such as the opinions of experts, sometimes requires the computation of the 25th, 50th, and 75th percentile values. The computation of these values depends on the number of experts providing opinions. Table 3.6 provides a summary of the needed equations for 4 to 20 experts. In the table, x_i means the opinion of an expert with the ith smallest value; i.e., $X_1 \ge X_2 \ge X_3 \ge \ldots \ge x_n$, where n = number of experts. In the table, the arithmetic average was used to compute the percentiles of the powers of the values. In some cases, where the values of x_i differ by power order of magnitude, the geometric average can be used.

TABLE 3.6
Computations of Percentiles

Number of Experts (n)	25th Percentile Arithmetic Value	25th Percentile Geometric Value	50th Percentile Arithmetic Value	50th Percentile Geometric Value	75th Percentile Arithmetic Value	75th Percentile Geometric Value
4	$(X_1 + X_2)/2$	$\sqrt{X_1X_2}$	$(X_2 + X_3)/2$	$\sqrt{X_2X_3}$	$(X_3 + X_4)/2$	$\sqrt{X_3X_4}$
5	X_2	X_2	X_3	X_3	X_4	X_4
6	X_2	X_2	$(X_3 + X_4)/2$	$\sqrt{X_3X_4}$	X_5	X_5
7	$(X_2 + X_3)/2$	$\sqrt{X_2X_3}$	X_4	X_4	$(X_5 + X_6)/2$	$\sqrt{X_5X_6}$
8	$(X_2 + X_3)/2$	$\sqrt{X_2X_3}$	$(X_4 + X_5)/2$	$\sqrt{X_5X_6}$	$(X_6 + X_7)/2$	$\sqrt{X_6X_7}$
9	$(X_2 + X_3)/2$	$\sqrt{X_2X_3}$	X_5	X_5	$(X_7 + X_8)/2$	$\sqrt{X_7X_8}$
10	$(X_2 + X_3)/2$	$\sqrt{X_2X_3}$	$(X_5 + X_6)/2$	$\sqrt{X_5X_6}$	$(X_8 + X_9)/2$	$\sqrt{X_8X_9}$
11	X_3	X_3	X_6	X_6	X_9	X_9
12	X_3	X_3	$(X_6 + X_7)/2$	$\sqrt{X_6X_7}$	X_{10}	X_{10}
13	$(X_3 + X_4)/2$	$\sqrt{X_3X_4}$	X_7	X_7	$(X_{10} + X_{11})/2$	$\sqrt{X_{10}X_{11}}$
14	$(X_3 + X_4)/2$	$\sqrt{X_3X_4}$	$(X_7 + X_8)/2$	$\sqrt{X_7X_8}$	$(X_{11} + X_{12})/2$	$\sqrt{X_{11}X_{12}}$
15	X_4	X_4	X_8	X_8	X_{12}	X_{12}
16	X_4	X_4	$(X_8 + X_9)/2$	$\sqrt{X_8X_9}$	X_{13}	X_{13}
17	$(X_4 + X_5)/2$	$\sqrt{X_5X_6}$	X_9	X_9	$(X_{13} + X_{14})/2$	$\sqrt{X_{13}X_{14}}$
18	$(X_4 + X_5)/2$	$\sqrt{X_5X_6}$	$(X_9+X_{10})/2$	$\sqrt{X_9X_{10}}$	$(X_{14} + X_{15})/2$	$\sqrt{X_{14}X_{15}}$
19	X_5	X_5	X_{10}	X_{10}	X_{15}	X_{15}
20	X_5	X_5	$(X_{10} + X_{11})/2$	$\sqrt{X_{10}X_{11}}$	X_{15}	X_{15}

3.7.8 Statistical Uncertainty

Values of random variables obtained from sample measurements are commonly used in making important decisions. For example, samples of river water are collected to estimate the average level of a pollutant in the entire river at that location. Samples of stopping distances are used to develop a relationship between the speed of a car at the time the brakes are applied and the distance traveled before the car comes to a complete halt. The average of sample measurements of the compressive strength of concrete collected during the pouring of a large concrete slab, such as the deck of a parking garage, is used to help decide whether the deck has the strength specified in the design specifications. It is important to recognize the random variables involved in these cases. In each case, the individual measurements or samples are values of a random variable, and the computed mean is also the value of a random variable. For example, the transportation engineer measures the stopping distance; each measurement is a sample value of the random variable. If 10 measurements are made for a car stopping from a speed of 50 mph, then the sample consists of 10 values of the random variable. Thus, there are two random variables in this example, the stopping distance and the estimated mean of the stopping distance; this is also true for the water quality pollutant and compressive strength examples.

The estimated mean for a random variable is considered by itself to be a random variable, because different samples about the random variable can produce different estimated mean values, hence the randomness in the estimated mean. When a sample of n measurements of a random variable is collected, the n values are not necessarily identical. The sample is characterized by variation. For example, let us assume that five independent estimates of the compressive strength of the concrete in a parking garage deck are obtained from samples of the concrete obtained when the concrete was poured. For illustration purposes, let us assume that the five compressive strength measurements are 3250, 3610, 3460, 3380, and 3510 psi. This produces a mean of 3442 psi and a standard deviation of 135.9 psi. Assume that another sample of five measurements of concrete strength was obtained from the same concrete pour; however, the values were 3650, 3360, 3328, 3420, and 3260 psi. In this case, the estimated mean and standard deviation are 3404 and 149.3 psi, respectively. Therefore, the individual measurement and the mean are values of two different random variables, i.e., X and $\bar{X}$.

It would greatly simplify decision making if the sample measurements were identical; i.e., there was no sampling variation, so the standard deviation was zero. Unfortunately, that is never the case, so decisions must be made in the presence of uncertainty. For example, let us assume in the parking garage example that the building code requires a mean compressive strength of 3500 psi. Since the mean of 3442 psi based on the first sample is less than the required 3500 psi, should we conclude that the garage deck does not meet the design specifications? Unfortunately, decision making is not that simple. If a third sample of five measurements had been randomly collected from other locations on the garage deck, the following values are just as likely to have been obtained: 3720, 3440, 3590, 3270, and 3610 psi. This sample of five produces a mean of 3526 psi and a standard deviation of 174.4 psi. In this case, the mean exceeds the design standard of 3500 psi. Since the sample mean is greater than the specified value of 3500 psi, can we conclude that the concrete is of adequate strength? Unfortunately, we cannot conclude with certainty that the strength is adequate any more than we could conclude with the first sample that the strength was inadequate. The fact that different samples lead to different means is an indication that we cannot conclude that the design specification is not met just because the sample mean is less than the design standard. We need to have more assurance.

The data that are collected on some variable or parameter represent sample information, but it is not complete by itself, and predictions are not made directly from the sample. The intermediate step between sampling and prediction is the identification of the underlying population. The sample is used to identify the population, and then the population is used to make predictions or decisions. This sample-to-population-to-prediction sequence is true for the univariate methods of this chapter.

The need then is for a systematic decision process that takes into account the variation that can be expected from one sample to another. The decision process must also be able to reflect the risk of making an incorrect decision. This decision can be made using, for example, hypothesis testing as described by Ayyub and McCuen (2003).

3.7.9 Bayesian Probabilities

Engineers commonly need to solve a problem and make decisions based on limited information about one or more of the parameters of the problem. The types of information available to them can be classified using the common terminology in the Bayesian literature as follows:

1. *Objective* or empirical information based on experimental results or observations
2. *Subjective* information based on experience, intuition, other previous problems that are similar to the one under consideration, or the physics of the problem

The first type of information can be dealt with using the theories of probability and statistics, as described in the previous chapters. In this type, probability is interpreted as the frequency of occurrence assuming sufficient repetitions of the problem, its outcomes, and parameters, as a basis of the information. The second type of information is subjective and can depend on the engineer or analyst studying the problem. In this type, uncertainty that exists needs to be dealt with using probabilities. However, the definition of probability is not the same as the first type because it is viewed herein as a subjective probability that reflects the state of knowledge of the engineer or the analyst.

It is common in engineering to encounter problems with both objective and subjective types of information. In these cases, it is desirable to utilize both types of information to obtain solutions or make decisions. The subjective probabilities are assumed to constitute a prior knowledge about a parameter, with gained objective information (or probabilities). Combining the two types produces posterior knowledge. The combination is performed based on Bayes' theorem, as described by Ayyub and McCuen (2003). If $A_1, A_2, \ldots, A_n$ represents the prior (subjective) information, or a partition of a universal set X, and $E \subset X$ represents the objective information (or arbitrary event) as shown in Figure 3.6, the theorem of total probability states that

$$P(E) = P(A_1)\, P(E|A_1) + P(A_2)\, P(E|A_2) + \ldots + P(A_n)\, P(E|A_n) \qquad (3.75)$$

where $P(A_i)$ = the probability of the event A_i and $E|A$ = the occurrence of E given A_i, where $i = 1, 2, \ldots, n$. This theorem is very important in computing the probability of the event E, especially in practical cases where the probability cannot be computed directly, but the probabilities of the partitioning events and the conditional probabilities can be computed.

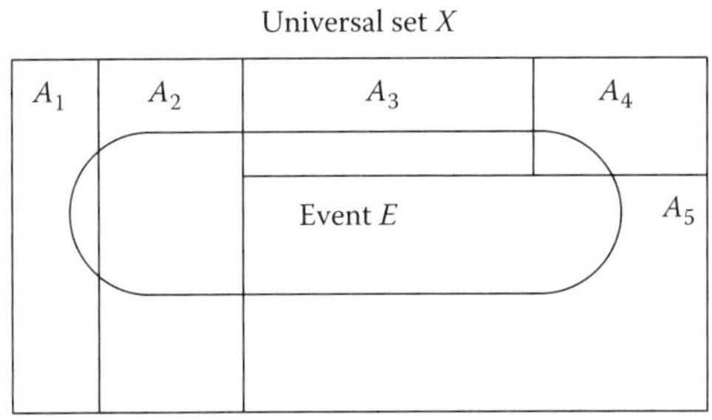

FIGURE 3.6 Bayes' theorem.

Bayes' theorem is based on the same conditions of partitioning and events as the theorem of total probability and is very useful in computing the posterior (or reverse) probability of the type $P(A_i|E)$, for $i = 1, 2, \ldots, n$. The posterior probability can be computed as follows:

$$P(A_i|E) = \frac{P(A_i)P(E|A_i)}{P(A_1)P(E|A_1) + P(A_2)P(E|A_2) + \ldots + P(A_n)P(E|A_n)} \quad (3.76)$$

The denominator of this equation is $P(E)$, which is based on the theorem of total probability. According to Equation 3.75, the prior knowledge, $P(A_i)$, is updated using the objective information, $P(E)$, to obtain the posterior knowledge, $P(A_i|E)$.

Example 3.4 Defective Products in Manufacturing Lines

A factory has three production lines. The three lines manufacture 20, 30, and 50% of the components produced by the factory, respectively. The quality assurance department of the factory determined that the probability of having defective products from lines 1, 2, and 3 are 0.1, 0.1, and 0.2, respectively. The following events were defined:

$$L_1 = \text{Component produced by line 1} \quad (3.77a)$$

$$L_2 = \text{Component produced by line 2} \quad (3.77b)$$

$$L_3 = \text{Component produced by line 3} \quad (3.77c)$$

$$D = \text{Defective component} \quad (3.77d)$$

Therefore, the following probabilities are given:

$$P(D|L_1) = 0.1 \quad (3.78a)$$

$$P(D|L_2) = 0.1 \quad (3.78b)$$

$$P(D|L_3) = 0.2 \quad (3.78c)$$

table. The last row in the table is the average nondefective component probability ($\overline{p}(ND)$) for cases where a nondefective component results from a test. This value $\overline{p}(ND)$ can be computed similar to Equation 3.81 or 3.83. For example, the $\overline{p}(ND)$ in case of a nondefective test according to the prior distribution is

$$\begin{aligned}\overline{p}(ND) &= (1-0.1)0.225 + (1-0.2)0.430 + (1-0.4)0.100 + \\ &\quad (1-0.6)0.120 + (1-0.8)0.080 + (1-0.9)0.045 \\ &= 0.675\end{aligned} \tag{3.84}$$

The computations for other cases are similarly performed as shown in Table 3.7. It should be noted that

$$\overline{p}(D) + \overline{p}(ND) = 1.0 \tag{3.85}$$

Now assume that a third component from line 3 was tested and found to be nondefective; the posterior distribution in column "Post. 2 D" of Table 3.7 needs to be revised to reflect the new (objective) information. The revised distribution is the posterior distribution ($P_p'(p)$), and can be computed using Equation 3.76 as follows:

$$P_p'(0.1) = \frac{0.0692(1-0.1)}{0.4883} = 0.1275 \tag{3.86a}$$

Similarly, the following posterior probabilities can be computed:

$$P_p'(0.2) = \frac{0.2646(1-0.2)}{0.4883} = 0.4335 \tag{3.86b}$$

$$P_p'(0.4) = \frac{0.1231(1-0.4)}{0.4883} = 0.1512 \tag{3.86c}$$

$$P_p'(0.6) = \frac{0.2215(1-0.6)}{0.4883} = 0.1815 \tag{3.86d}$$

$$P_p'(0.8) = \frac{0.1969(1-0.8)}{0.4883} = 0.0807 \tag{3.86e}$$

$$P_p'(0.9) = \frac{0.1246(1-0.9)}{0.4883} = 0.0255 \tag{3.86f}$$

The resulting probabilities in Equation 3.86a to f add up to 1. The probability $\overline{p}(ND)$ of 0.4883 was used in these calculations. The results of these calculations and the mean

TABLE 3.7
Prior and Posterior Distributions for Line 3

Probability, p	$P(p)$	Post. 1 D	Post. 2 D	Post. 3 ND	Post. 4 D	Post. 5 D	Post. 6 D	Post. 7 D	Post. 8 D	Post. 9 D	Post. 10 D
0.1	0.45	0.225	0.0692308	0.127599244	0.035809	0.0070718	0.0011355	0.0001638	2.22693E-05	2.912E-06	3.703E-07
0.2	0.43	0.43	0.2646154	0.433522369	0.2433245	0.0961062	0.0308633	0.0089068	0.002421135	0.0006332	0.0001611
0.4	0.05	0.1	0.1230769	0.151228733	0.1697613	0.1341016	0.08613	0.0497125	0.027026626	0.0141359	0.0071914
0.6	0.04	0.12	0.2215385	0.18147448	0.3055703	0.3620744	0.3488266	0.3020033	0.246280127	0.1932203	0.1474458
0.8	0.02	0.08	0.1969231	0.080655325	0.1810787	0.2860835	0.3674882	0.4242131	0.461254413	0.482506	0.4909318
0.9	0.01	0.045	0.1246154	0.025519849	0.0644562	0.1145626	0.1655564	0.2150004	0.262995429	0.3095016	0.3542696
$\overline{p}(D)$	0.2	0.325	0.5116923	0.356332703	0.506366	0.6227868	0.6930255	0.7357556	0.764764597	0.7862698	0.8029643
$\overline{p}(ND)$	0.8	0.675	0.4883077	0.643667297	0.493634	0.3772132	0.3069745	0.2642444	0.235235403	0.2137302	0.1970357

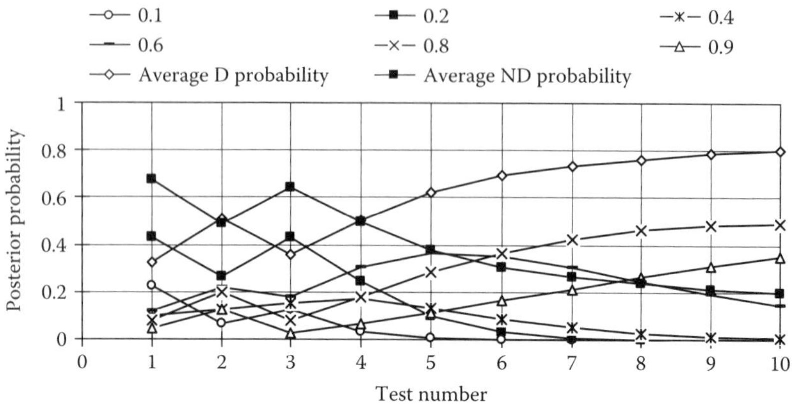

FIGURE 3.8 Posterior distributions for line 3.

probability $\overline{p}(ND)$ are shown in Table 3.7. It can be noted from the table that the mean defective component probability decreases as nondefective components are obtained through testing.

If the next seven tests result in defective components, the resulting posterior distributions are shown in Table 3.7. The results are also shown in Figure 3.8. It can be observed from the figure that the average probability is approaching 1 as more and more defective tests are obtained. Also, the effect of a nondefective component on the posterior probabilities can be seen in this figure.

3.8 IMPRECISE PROBABILITIES

The theory of imprecise probabilities was developed by Walley (1991) as a generalization of probability theory for cases where probability assignments to elements (x) of a set (A) are not available in a precise manner. For a finite set (A) from a universe X, the probability measure for its elements is defined as

$$P : X \rightarrow [0, 1] \tag{3.87}$$

The function P assigns a probability value to each element $x \in X$ according to the axioms of probability theory. In cases where P is uncertain and can only be assessed in imprecise terms, lower and upper probability values can be used, i.e., $\underline{P}$ and $\overline{P}$, respectively. These two functions are defined respectively as follows:

$$\underline{P} : X \rightarrow [0,1] \text{ such that } \sum_{all\, x \in X} \underline{P}(x) \leq 1 \tag{3.88a}$$

$$\overline{P} : X \rightarrow [0,1] \text{ such that } \sum_{all\, x \in X} \overline{P}(x) \geq 1 \tag{3.88b}$$

It can be shown that for a given lower probability $\underline{P}$, a unique dual (i.e., conjugate) upper probability $\overline{P}$ exists for all $A \in P_X$ according to the following relationship:

$$\overline{P}(A) = 1 - \underline{P}(\overline{A}) \tag{3.89}$$

Imprecise probabilities can be used to construct a *Möbius representation m* for all $A \in P_X$ as follows:

$$m(A) = \sum_{\text{all } B \text{ such that } B \subseteq A} (-1)^{|A-B|} \underline{P}(B) \tag{3.90}$$

where $|A - B|$ is the cardinality of the set of elements of A that do not belong to B. Equation 3.90 results in meeting the following conditions:

$$m(\phi) = 0 \tag{3.91a}$$

$$\sum_{\text{all } A \subseteq P_X} m(A) = 1 \tag{3.91b}$$

The inverse of Equation 3.90 is

$$\underline{P}(B) = \sum_{\text{all } B \text{ such that } B \subseteq A} m(A) \tag{3.92}$$

In subsequent subsections, modeling frameworks that utilize these concepts are provided and discussed.

3.8.1 Interval Probabilities

The term *interval probabilities* has more than one usage, as reported by Dempster (1976a, 1976b), Cui and Blockley (1990), and Ferson et al. (1999), and within a broader framework of interval arithmetic as provided by Moore (1979). Various models for dealing with interval probabilities are provided in subsequent sections, although this section is devoted to summarize the model suggested by Cui and Blockley (1990), who introduced interval probabilities based on the axioms of probability and by maintaining the additive condition of Equation 3.14. For an event A that represents a proposition on a universal set X, the probability measure for A is given by

$$P(A) \in [\underline{P}(A), \overline{P}(A)] \tag{3.93}$$

where $\underline{P}(A)$ and $\overline{P}(A)$ are the lower (left) and upper (right) estimates of the probability of A, $P(A)$, respectively. According to Equation 3.93, the probability of A falls in this range as follows:

$$\underline{P}(A) \leq P(A) \leq \overline{P}(A) \tag{3.94a}$$

The probability of the complement of A can be computed as follows:

$$1 - \overline{P}(A) \leq P(\overline{A}) \leq 1 - \underline{P}(A) \tag{3.94b}$$

The interval probability can be interpreted as a measure of belief in having a true proposition A as follows:

$P(A) \in [0, 0]$ represents a belief that A is certainly false or not dependable (3.95a)

$P(A) \in [1, 1]$ represents a belief that A is certainly true or dependable (3.95b)

$P(A) \in [0, 1]$ represents a belief that A is known (3.95c)

The use of the term *belief* in Equation 3.95a to c should not be confused with the belief measure provided in the theory of evidence. Interval probabilities are related to probability theory. Hall et al. (1998) provided an example application of interval probabilities, discussed at the end of the section.

The theory presented herein can be used to propagate uncertainty in the form of interval probabilities into a logical structure using inference. The underlying mathematics for inference in this case requires the use of the theorem of total probability for an event A for a partition that involves a set E and its complement $\overline{E}$ as follows:

$$P(A) = P(A|E)P(E) + P(A|\overline{E})P(\overline{E}) \tag{3.96a}$$

Equation 3.96a can be extended for multiple sets similar to E. Dubois and Prade (1991) developed the following expressions for computing an interval $P(A)$ based on intervals for $P(A|E)$ and $P(A|\overline{E})$:

$$\underline{P}(A) = \begin{cases} \underline{P}(A|E)\underline{P}(E) + \underline{P}(A|\overline{E})(1 - \underline{P}(E)) & \text{if } \underline{P}(A|E) \leq \underline{P}(A|\overline{E}) \\ \underline{P}(A|E)\overline{P}(E) + \underline{P}(A|\overline{E})(1 - \overline{P}(E)) & \text{if } \underline{P}(A|E) > \underline{P}(A|\overline{E}) \end{cases} \tag{3.96b}$$

$$\overline{P}(A) = \begin{cases} \overline{P}(A|E)\overline{P}(E) + \overline{P}(A|\overline{E})(1 - \overline{P}(E)) & \text{if } \overline{P}(A|E) \leq \overline{P}(A|\overline{E}) \\ \overline{P}(A|E)\underline{P}(E) + \overline{P}(A|\overline{E})(1 - \underline{P}(E)) & \text{if } \overline{P}(A|E) > \overline{P}(A|\overline{E}) \end{cases} \tag{3.96c}$$

TABLE 3.8
Situation for Logical Inference

Logical Relationship between E and A	$P(A\|E)$	$P(A\|\bar{E})$
E may be a *necessary* condition for A	$P(A\|E) \le 1$	$P(A\|\bar{E}) = 0$
E may be a *sufficient* condition for A	$P(A\|E) = 1$	$P(A\|\bar{E}) \ge 0$
E is a *necessary and sufficient* condition for A	$P(A\|E) = 1$	$P(A\|\bar{E}) \ge 0$
E is a *relevant or partially sufficient* condition for A	$0 < P(A\|E) \le 1$	$0 \le P(A\|\bar{E}) \le 1$

The conditional probabilities in Equation 3.96b and c can be interpreted in logical inference as provided in Table 3.8, depending on the application or situation. In this table, necessity is defined as a measure of the extent to which E will cause A, whereas sufficiency is defined as a measure of influence amount of E on A. These definitions might require further refinement in a context-dependent manner.

Interval probably computations can account for dependency by introducing a parameter ρ that represents the degree of dependency between two events A and B. Cui and Blockley (1990) defined this parameter as follows:

$$\rho = \frac{P(A \cap B)}{\min(P(A), P(B))} \tag{3.97}$$

The parameter ρ has the following properties:

$$0 \le \rho \le 1 \tag{3.98}$$

$$\rho = 1 \quad \text{for } A \subseteq B \text{ (i.e., nested propositions)} \tag{3.99a}$$

$$\rho = 1 \quad \text{for } B \subseteq A \text{ (i.e., nested propositions)} \tag{3.99b}$$

$$\rho = 0 \quad \text{for } A \cap B = \varnothing \tag{3.100}$$

$$\rho = \max(P(A), P(B)) \quad \text{for fully dependent } A \text{ and } B \tag{3.101}$$

$$P(A \cap B) = P(A)P(B) \quad \text{for independent } A \text{ and } B \tag{3.102}$$

$$\rho_{min} = \max\left[\frac{P(A)+P(B)-1}{\min(P(A),P(B))},0\right] \quad (3.103)$$

The concept of dependency is discussed and fully developed in Section 3.7.3. In cases where the parameter ρ is defined as an interval $[\underline{\rho},\overline{\rho}]$, the following relations can be used:

$$\underline{P}(A \cap B) = \underline{\rho}(\min(\underline{P}(A),\underline{P}(B))) \quad (3.104)$$

$$\overline{P}(A \cap B) = \overline{\rho}(\min(\overline{P}(A),\overline{P}(B))) \quad (3.105)$$

$$\underline{P}(A \cup B) = \underline{P}(A) + P(B) - \underline{\rho}(\min(\underline{P}(A),\underline{P}(B))) \quad (3.106)$$

$$\overline{P}(A \cup B) = \overline{P}(A) + \overline{P}(B) - \overline{\rho}(\min(\overline{P}(A),\overline{P}(B))) \quad (3.107)$$

Dependency in inference can be related to common sources used to obtain data or information, or the two sets being influenced by common factors.

Example 3.5 Fault Tree Analysis

The fault tree provided in Figure 3.9 is used in this example to illustrate the use of Equations 3.104 to 3.107. The basic events A, B, C, D, and E have occurrence probabilities that are provided in the form of the following intervals:

Prob(Event A) = [0.25, 1.00]

Prob(Event B) = [0.50, 0.75]

Prob(Event C) = [0.30, 0.50]

Prob(Event D) = [0.60, 0.70]

Prob(Event E) = [0.50, 0.50]

The following assumptions are made regarding dependencies among the events:

- Events A and B have possible dependency.
- Events C and D have possible dependency.
- Events A and C have no dependency.
- Events A and D have no dependency.
- Events B and C have no dependency.
- Events B and D have no dependency.
- Event E is independent of all other events (A, B, C, and D).

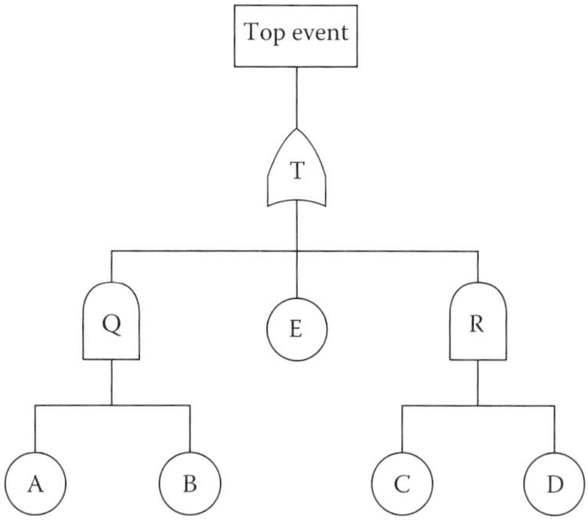

FIGURE 3.9 Fault tree model.

The following intervals specify the dependency between A and B, and C and D:

$$\rho(\text{A, B}) = [0.50, 0.75]$$

$$\rho(\text{C, D}) = [0.25, 0.65]$$

Using Equations 3.104 to 3.107 and the logic of the fault tree in Figure 3.9, the interval probabilities of the top event can be computed as follows:

$$Q = A \text{ and } B = [\underline{\rho}(\min(\underline{P}(A), \underline{P}(B))), \bar{\rho}(\min(\bar{P}(A), \bar{P}(B)))]$$

$$= [0.5(\min(0.25, 0.5)), 0.75(\min(1, 0.75))] = [0.13, 0.56]$$

Similarly,

$$\text{R} = \text{C and D} = [0.08, 0.33]$$

and

$$\text{T} = \text{Q or R or E} = [0.60, 0.85]$$

3.8.2 Interval Cumulative Distribution Functions

Probabilistic models are effective in expressing uncertainties in various variables that appear in engineering and science problems. Such models can be viewed as *certain* representations of uncertainty that demand knowledge of underlying distributions, parameters, or a lot of data. Systems that are represented by these models might not be known fully to the levels demanded by the models, hence the need of methods to deal with limited or incomplete information. Analysts commonly encounter situations where data are not available, limited, or available in intervals only.

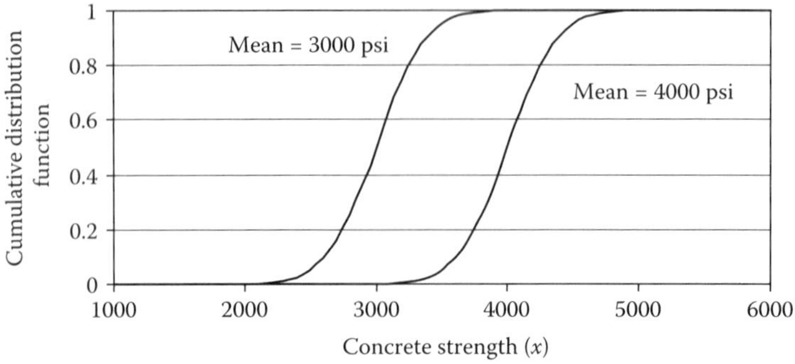

FIGURE 3.10 Normal cumulative distribution function using an interval mean.

This section provides methods that were selected or developed to deal with such situations. It covers three cases as follows: (1) uncertain parameters of a known probability distribution, (2) an uncertain probability distribution for known parameters, and (3) uncertain parameters and probability distribution due to limited data. These three cases are discussed with illustrative examples.

For some random variables, the distribution type might be known from historical information; however, the parameters relevant to a problem under consideration might not be known and can only be subjectively assessed using intervals or fuzzy numbers. The presentation herein is provided for interval parameters, but can be easily extended to fuzzy parameters expressed as fuzzy numbers using the α-cut concept. If we consider a concrete structural member with an unknown strength, the following state of knowledge can be used to demonstrate the construction of an interval-based distribution:

Normal probability distribution:
Mean value = [3000, 4000] psi
Standard deviation = 300 psi

The bounds of the cumulative distribution function ($F_X(x)$) are shown in Figure 3.10 based on evaluating the following integral:

$$F_X(x) = \int_{-\infty}^{x} \frac{1}{\sigma\sqrt{2\pi}} \exp\left[-\frac{1}{2}\left(\frac{x-\mu}{\sigma}\right)^2\right] \sigma dx \tag{3.108}$$

where μ = mean and σ = standard deviation. Another case is shown in Figure 3.11 using the following assumptions:

Normal probability distribution:
Mean value = [3000, 4000] psi
Standard deviation = [300, 400] psi, i.e., coefficient of variation = 0.10

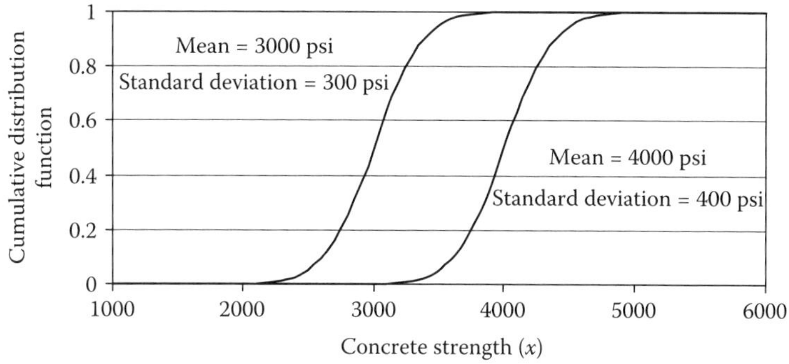

FIGURE 3.11 Normal cumulative distribution function using interval mean and standard deviation.

Some random variables might have known moments with unknown or uncertain distribution types. If data are available, someone could use hypothesis testing to select a distribution that best fits the data (Ayyub and McCuen, 2003). However, data might not be available, requiring the use of a bounding method. In this case, a short list of distributions can be subjectively identified. The cumulative distribution functions based on the known parameters can be determined, and a range on possible values of the cumulative distribution function can be assessed.

Some random variables might have limited data that are not sufficient to construct a histogram and select a probability distribution. In this case, the Kolmogorov–Smirnov (KS) one-sample method can be used to construct a confidence interval on the cumulative distribution function, also called bounds. The KS method, as described by Ayyub and McCuen (2003), constructs a sample cumulative distribution function as follows:

$$F_S(x) = \begin{cases} 0 & \text{for } x < x_1 \\ \dfrac{i}{n} & \text{for } x_i \leq x < x_{i+1} \\ 1 & \text{for } x \geq x_n \end{cases} \tag{3.109}$$

where $x_i = i^{\text{th}}$ largest value based on rank-ordering the sample values from the smallest (x_1) to the largest (x_n) for a sample of size n, and $F_S(x)$ = the sample cumulative distribution function. The KS method provides tabulated limits on the maximum deviation between the sample cumulative distribution function and an acceptable model for the cumulative distribution function. These tabulated limits correspond to various sample sizes and significance levels, i.e., 1 minus the confidence level defined as the conditional probability of accepting a model given it is an incorrect model. Table 3.9 shows critical values for the KS method as a function of sample sizes and the significance levels.

TABLE 3.9
Critical Values for the Kolmogorov–Smirnov Test

	Level of Significance			
Sample Size, *n*	**0.20**	**0.10**	**0.05**	**0.01**
1	0.900	0.950	0.975	0.995
2	0.684	0.776	0.842	0.929
3	0.565	0.642	0.708	0.828
4	0.494	0.564	0.624	0.733
5	0.446	0.510	0.565	0.669
6	0.410	0.470	0.521	0.618
7	0.381	0.438	0.486	0.577
8	0.358	0.411	0.457	0.543
9	0.339	0.388	0.432	0.514
10	0.322	0.368	0.410	0.490
11	0.307	0.352	0.391	0.468
12	0.295	0.338	0.375	0.450
13	0.284	0.325	0.361	0.433
14	0.274	0.314	0.349	0.418
15	0.266	0.304	0.338	0.404
20	0.231	0.264	0.294	0.356
30	0.195	0.222	0.248	0.298
40	0.169	0.193	0.215	0.258
50	0.151	0.173	0.192	0.231
60	0.138	0.158	0.176	0.210
70	0.125	0.146	0.163	0.195
80	0.120	0.136	0.152	0.182
90	0.113	0.129	0.143	0.172
100	0.107	0.122	0.136	0.163
>50	$1.07/\sqrt{n}$	$1.22/\sqrt{n}$	$1.36/\sqrt{n}$	$1.63/\sqrt{n}$

EXAMPLE 3.6 WATER QUALITY

The following set of five measurements of a water quality parameter in ppm can be used to construct KS bounds on a cumulative distribution function: {47, 53, 61, 57, 65}. If a level of significance of 5% is used, a sample cumulative distribution function and bounds can be computed using Equation 3.109 and Table 3.9. Table 3.10 shows the calculations for sample cumulative distribution function and the two bounds. For a 5% level of significance, the critical value is 0.565. The sample, left, and right cumulative distribution functions are shown in Figure 3.12.

3.8.3 DEPENDENCE MODELING AND MEASURES

The two events *A* and *B* (subsets of a universe *X*) are used in this section to develop models and measures for various dependency levels between them, starting from the extreme cases of independence and perfect dependence, followed by other cases of opposite dependence and partial dependence.

TABLE 3.10
Left and Right Bounds Using the Kolmogorov–Smirnov Limits

Sorted Data Point Rank i	Sample Value x	Sample CDF	Right CDF	Left CDF
0	20	0	0	0.565
1	47	0	0	0.565
1	47	0.2	0	0.765
2	53	0.2	0	0.765
2	53	0.4	0	0.965
3	57	0.4	0	0.965
3	57	0.6	0.035	1
4	61	0.6	0.035	1
4	61	0.8	0.235	1
5	65	0.8	0.235	1
5	65	1	0.435	1

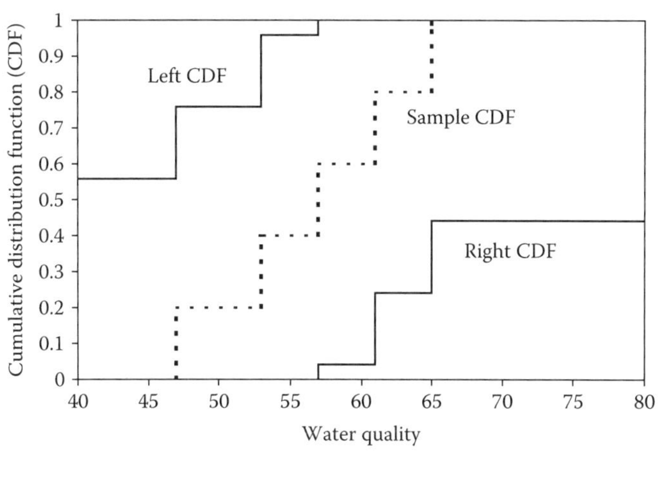

FIGURE 3.12 The Kolmogorov–Smirnov bounds on a cumulative distribution function.

3.8.3.1 Perfect Independence

For the two events A and B with perfect independence, the probabilities of these events can be used to compute the probabilities of their intersection and union, i.e., conjunction and disjunction, respectively, as follows:

$$P(A \cap B) = P(A)P(B) \tag{3.110a}$$

$$P(A \cup B) = 1 - ((1 - P(A))(1 - P(B))) \tag{3.110b}$$

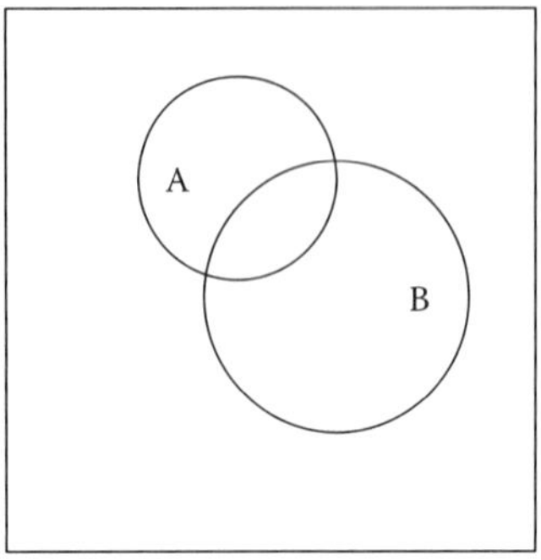

FIGURE 3.13 Perfect independencies of two events.

Figure 3.13 shows this case of independence using a Venn diagram. The overlap between the two events represents the intersection, and the total area of the two events represents the union.

3.8.3.2 Mutual Exclusion

The two events A and B do not have an overlap if they are mutually exclusive; i.e., the occurrence of one event precludes the other from occurring. In order for the events to have no overlap, the sum of their probabilities must be less than 1. The probabilities of these events can be used to compute the probabilities of their intersection and union, i.e., conjunction and disjunction, respectively, as follows:

$$P(A \cap B) = 0 \tag{3.111a}$$

$$P(A \cup B) = P(A) + P(B) \tag{3.111b}$$

Figure 3.14 shows this case using a Venn diagram. The overlap between the two events represents the intersection, and the total area of the two events represents the union.

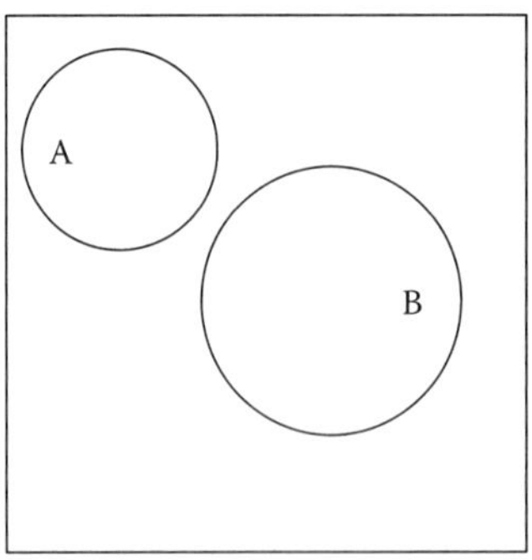

FIGURE 3.14 Mutually exclusive events.

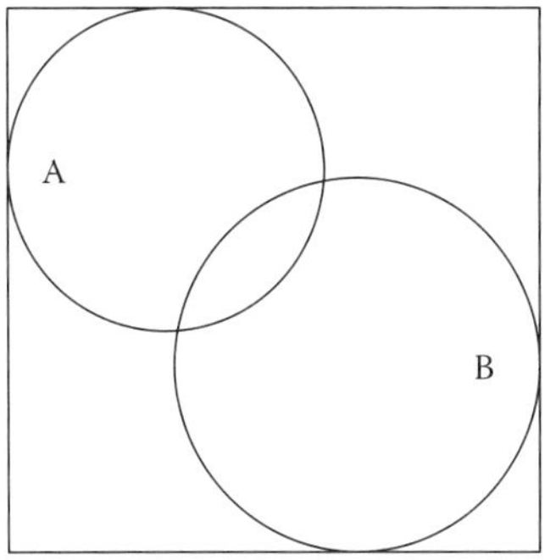

FIGURE 3.15 Opposite dependence for two events.

3.8.4 Opposite Dependence

In this case, the two events A and B have the minimum possible overlap. The case of mutually exclusive events is a special case of opposite dependence if the sum of their probabilities is less than 1. The probabilities of these events can be used to compute the probabilities of their intersection and union, i.e., conjunction and disjunction, respectively, as follows:

$$P(A \cap B) = \max(P(A) + P(B) - 1, 0) \tag{3.112a}$$

$$P(A \cup B) = \min(1, P(A) + P(B)) \tag{3.112b}$$

Figure 3.15 shows this case using a Venn diagram. The overlap between the two events represents the intersection, and the total area of the two events represents the union.

3.8.5 Perfect Dependence

For the two events A and B with perfect dependence, the Venn diagram is shown in Figure 3.16. The probabilities of these events can be used to compute the

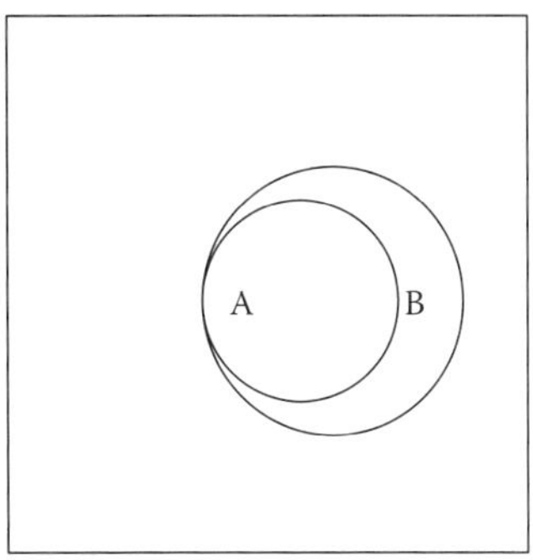

FIGURE 3.16 Perfect dependence of two events.

probabilities of their intersection and union, i.e., conjunction and disjunction, respectively, as follows:

$$P(A \cap B) = \min(P(A), P(B)) \tag{3.113a}$$

$$P(A \cup B) = \max(P(A), P(B)) \tag{3.113b}$$

The overlap between the two events represents the intersection, and the total area of the two events represents the union.

3.8.6 Partial Dependence

The previous cases of perfect dependence, perfect independence, and opposite dependence, including its special case of mutual exclusion, can be used to bound cases involving partial dependence. These cases are used in subsequent sections to measure correlations between events and between random variables.

3.8.7 Correlation between Random Variables

3.8.7.1 Correlation Based on Probability Theory

The covariance (Cov) of two random variables X_1 and X_2 is defined in terms of mathematical expectation as

$$\text{Cov}(X_1, X_2) = E[(X_1 - \mu_{X_1})(X_2 - \mu_{X_2})] \tag{3.114}$$

It is common to use the notation $\sigma_{X_1X_2}$, σ_{12}, or Cov(X_1, X_2) for the covariance of X_1 and X_2. The covariance for two random variables can also be determined using the following equation that results from Equation 3.114:

$$\text{Cov}(X_1, X_2) = \text{E}(X_1X_2) - \mu_{X_1}\mu_{X_2} \tag{3.115}$$

where the expected value of the product (X_1X_2) is given by

$$E(X_1X_2) = \int_{-\infty}^{+\infty}\int_{-\infty}^{+\infty} x_1x_2 f_{X_1X_2}(x_1,x_2)dx_1dx_2 \tag{3.116}$$

If X_1 and X_2 are statistically uncorrelated, then

$$\text{Cov}(X_1, X_2) = 0 \tag{3.117a}$$

and

$$\text{E}(X_1X_2) = \mu_{X_1}\mu_{X_2} \tag{3.117b}$$

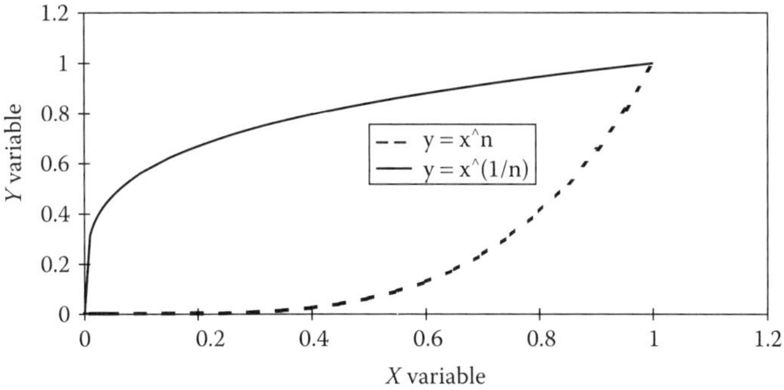

FIGURE 3.17 The two curves in Example 3.7.

The correlation coefficient is defined as a normalized covariance with respect to the standard deviations of X_1 and X_2 and is given by

$$\rho_{X_1X_2} = \frac{\text{Cov}(X_1, X_2)}{\sigma_{X_1}\sigma_{X_2}} \tag{3.118}$$

The correlation coefficient ranges inclusively between –1 and +1, i.e.,

$$-1 \le \rho_{X_1X_2} \le +1 \tag{3.119}$$

If the correlation coefficient is zero, then the two random variables are described to be uncorrelated. From the definition of correlation, in order for $\rho_{X_1X_2}$ to be zero, the Cov(X_1, X_2) must be zero. Therefore, X_1 and X_2 are statistically uncorrelated. However, the converse of this finding does not hold. The correlation coefficient can also be viewed as a measure of the degree of linear association between X_1 and X_2. The sign (– or +) indicates the slope for the linear association. It is important to note that the correlation coefficient does not give any indications about the presence of a nonlinear relationship between X_1 and X_2 (or the lack of it).

Example 3.7 Covariance and Correlation

The universal set for two random variables, X and Y, is defined by the region between the two curves shown in Figure 3.17. Assume that any pair (x, y) in this region is equally likely to occur; i.e., a uniform distribution is assumed over the region between the two curves. The two curves are given by the following equations:

$$y = x^{1/n} \quad \text{for } 0 \le x \le 1 \tag{3.120}$$

and

$$y = x^{n} \quad \text{for } 0 \le x \le 1 \tag{3.121}$$

Therefore, the joint density value for these random variables can be viewed on a third axis that is perpendicular to the plane of the curves in Figure 3.17. The joint density function in this case takes on a constant value over this region. The range of y for both curves is [0, 1]. Therefore, the value of the density function can be determined based on the following condition:

$$\int_0^\infty \int_0^\infty f_{XY}(x,y)dx\,dy = 1 \tag{3.122a}$$

where $f_{XY}(x, y)$ = constant (c). Therefore,

$$1 = \left[\int_0^1 x^{1/n}dx - \int_0^1 x^n dx \right] c \tag{3.122b}$$

Solving for c, the following expression can be obtained:

$$c = \frac{n+1}{n-1} \tag{3.123}$$

Therefore, the density function is given by

$$f_{XY}(x,y) = \begin{cases} \dfrac{n+1}{n-1} & \text{for } 0 \le x \le 1 \text{ and } 0 \le y \le 1 \\ 0 & \text{otherwise} \end{cases} \tag{3.124}$$

The marginal density function of X is given by

$$f_X(x) = c \int_{x^n}^{x^{1/n}} dy = \frac{n+1}{n-1}(x^{1/n} - x^n) \quad \text{for } 0 \le x \le 1 \tag{3.125}$$

Similarly, the marginal density function of Y is given by

$$f_Y(y) = \frac{n+1}{n-1}(y^{1/n} - y^n) \quad \text{for } 0 \le y \le 1 \tag{3.126}$$

Therefore, $f_Y(y)$ is similar to $f_X(x)$. Thus, the expected value of X is equal to the expected value of Y and is given by

$$E(X) = E(Y) = \int_0^1 x f_X(x)dx \tag{3.127a}$$

or

$$E(X) = E(Y) = \frac{n+1}{n-1}\left[\frac{n}{1+2n} - \frac{1}{n+2}\right] \tag{3.127b}$$

Also, the second moments of X and Y are equal and given by

$$E(X^2) = E(Y^2) = \int_0^1 x^2 f_X(x)dx \tag{3.128a}$$

or

$$E(X^2) = E(Y^2) = \frac{n+1}{n-1}\left[\frac{n}{1+3n} - \frac{1}{n+3}\right] \tag{3.128b}$$

Therefore, the variances of X and Y are

$$\text{Var}(X) = \text{E}(X^2) - [\text{E}(X)]^2 \tag{3.129a}$$

$$\text{Var}(Y) = \text{E}(Y^2) - [\text{E}(Y)]^2 \tag{3.129b}$$

The expected value of the product XY is

$$E(XY) = c\int_0^1 \int_{x^n}^{x^{1/n}} xy\,dy\,dx \tag{3.130a}$$

or

$$E(XY) = \frac{1}{2}\frac{n+1}{n-1}\left[\frac{n}{2+2n} - \frac{1}{2n+2}\right] = \frac{1}{4} \tag{3.130b}$$

Therefore, the covariance of X and Y is

$$\text{Cov}(X, Y) = \text{E}(XY) - \text{E}(X)\,\text{E}(Y) \tag{3.131}$$

For $n = 2$, these moments take the following values:

$$\text{E}(X) = \text{E}(Y) = \frac{9}{20} \tag{3.132}$$

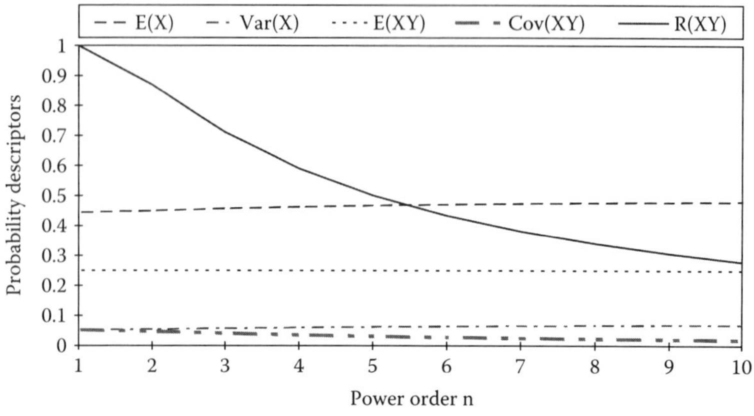

FIGURE 3.18 Probability descriptors as functions of n.

$$E(X^2) = E(Y^2) = \frac{9}{35} \tag{3.133}$$

$$Var(X) = Var(Y) = 0.0546 \tag{3.134}$$

$$E(XY) = \frac{1}{4} \tag{3.135}$$

$$Cov(X, Y) = 0.0475 \tag{3.136}$$

Therefore, the correlation coefficient is

$$\rho_{XY} = \frac{Cov(X,Y)}{\sqrt{Var(X)Var(Y)}} = \frac{0.0475}{0.0546} = 0.87 \tag{3.137}$$

For $n = 3$, the correlation coefficient ρ_{XY} is 0.71. Figure 3.18 shows selected probability descriptors, including ρ_{XY}, as functions of n. It is interesting to note that as the power order n approaches 1, the area between the two curves diminishes and the correlation coefficient approaches 1. Also, by increasing n, the area between the two curves increases, approaching a limiting case where it covers the entire area of Figure 3.18, and the correlation coefficient approaches zero.

3.8.7.2 Statistical Correlation

A set of observations on a random variable Y has a certain amount of variation, which may be characterized by the variance of the sample. The variance equals the sum of squares of the deviations of the observations from the mean of the observations divided by the degrees of freedom. Ignoring the degrees of freedom, the variation in the numerator can be separated into two parts: (1) variation associated with a second variable X and (2) variation not associated with X. That is, the total variation (TV), which equals the sum of the squares of the sample data points about the mean of the data points, is separated into the variation that is explained by variation in the second variable (EV) and the variation that is not explained, that is, the unexplained variation (UV). Thus, TV can be expressed as

$$\text{TV} = \text{EV} + \text{UV} \tag{3.138}$$

Using the general form of the variation of a random variable, each of the three terms in this equation can be represented by a sum of squares as follows:

$$\sum_{i=1}^{n}(y_i - \bar{Y})^2 = \sum_{i=1}^{n}(\hat{y}_i - \bar{Y})^2 + \sum_{i=1}^{n}(y_i - \hat{y}_i)^2 \tag{3.139}$$

where y_i = an observation on the random variable, $\hat{y}_i$ = the value of Y estimated from the best linear relationship with the second variable X, and $\bar{Y}$ = the mean of the observations on Y.

The separation of variation concept is useful for quantifying the Pearson correlation coefficient. Specifically, dividing both sides of Equation 3.138 by the total variation TV gives

$$1 = \frac{\text{EV}}{\text{TV}} + \frac{\text{UV}}{\text{TV}} \tag{3.140}$$

The ratio $\frac{\text{EV}}{\text{TV}}$ represents the fraction of the total variation that is explained by the linear relationship between Y and X; this is called the *coefficient of determination* and is given by

$$R^2 = \frac{\text{EV}}{\text{TV}} = \frac{\sum_{i=1}^{n}(\hat{y}_i - \bar{Y})^2}{\sum_{i=1}^{n}(y_i - \bar{Y})^2} \tag{3.141}$$

The square root of the ratio is the correlation coefficient, R. If the explained variation equals the total variation, the correlation coefficient will equal 1. If the relationship between X and Y is inverse, and the explained variation equals the total variation in magnitude, R will equal –1. These represent the extremes, but both values indicate a perfect association, with the sign only indicating the direction of the relationship. If the explained variation equals zero, R equals zero. Thus, a correlation coefficient of zero, which is sometimes called the null correlation, indicates no linear association between the two variables X and Y.

While Equation 3.141 provides the means to compute a value of the correlation coefficient, it can be shown that Equation 3.141 can be rearranged to the following form using a linear model for $\hat{Y}$ that minimizes the unexplained variation (UV):

$$R = \frac{\sum_{i=1}^{n} x_i y_i - \frac{1}{n}\left(\sum_{i=1}^{n} x_i\right)\left(\sum_{i=1}^{n} y_i\right)}{\sqrt{\sum_{i=1}^{n} x_i^2 - \frac{1}{n}\left(\sum_{i=1}^{n} x_i\right)^2}\sqrt{\sum_{i=1}^{n} y_i^2 - \frac{1}{n}\left(\sum_{i=1}^{n} y_i\right)^2}} \tag{3.142}$$

The linear model $\hat{Y}$ that minimizes UV is based on the principle of least squares.

3.8.8 Correlation between Events

The correlation between two events was introduced in Equation 3.97 (Cui and Blockley, 1990; Ferson et al., 2004; Davis and Hall, 2003; Lucas, 1995). The correlation can be measured using statistical methods based on data constructed by imagining the throwing of darts on one of the Venn diagrams provided in Figure 3.13 to Figure 3.16, and keeping scores of 0 or 1 for both events for the cases of a dart hitting or missing the events, respectively (Ferson et al., 2004). The resulting data of zeros and ones for the two events can be used to compute the correlation coefficient, such as the Pearson correlation coefficient provide in Equation 3.142. The resulting correlation coefficient (ρ) can be alternately computed using the following expression for two events A and B, modeled after Equation 3.118:

$$\rho = \frac{P(A \cap B) - P(A)P(B)}{\sqrt{P(A)(1-P(A))}\sqrt{P(B)(1-P(B))}} \tag{3.143a}$$

where $\sqrt{P(A)(1-P(A))}$ and $\sqrt{P(B)(1-P(B))}$ measure the standard deviations associated with the generated streams of zeros and ones for the two events. The term $P(A \cap B) - P(A)P(B)$ measures the covariance associated with the generated streams of zeros and ones for the two events. Equation 3.143a can be used to compute $P(A \cap B)$ as follows (Lucas, 1995):

$$P(A \cap B) = P(A)P(B) + \rho\sqrt{P(A)(1-P(A))}\sqrt{P(B)(1-P(B))} \tag{3.143b}$$

Equation 3.143a and b does offer an approximate solution and could violate the axioms in probability by producing negative probabilities in cases when $\rho = -1$. These negative values result from assigning values to ρ beyond a feasible range for ρ. Therefore, ρ must be limited to a range that is smaller than [1, –1] (Feller, 1968; Nelson, 1999). These limits on ρ can be demonstrated statistically using the dart example previously discussed by sorting the values of zeros and ones in the two columns for the two events in two extreme cases: (1) all zeros at the top of the columns and (2) all zeros at the bottom of the columns. These extreme cases will not produce the end of the range [1, –1], but a smaller nested range. The smaller nested range defines the limits on ρ to meet the axioms of probability theory. Values outside this range are problematic and not realistic. These limits on ρ can be computed as follows:

$$\underline{\rho} = \frac{\max(P(A)+P(B)-1,0)}{\sqrt{P(A)(1-P(A))}\sqrt{P(B)(1-P(B))}} \tag{3.144a}$$

$$\bar{\rho} = \frac{\min(P(A),P(B))-P(A)P(B)}{\sqrt{P(A)(1-P(A))}\sqrt{P(B)(1-P(B))}} \tag{3.144b}$$

Frank (1979) offered an alternate model to compute the probability of the intersection of two correlated events (A and B) based on the concept of copulas, which is used to characterize the dependence or association among random variables. The alternate model is expressed as (Frank, 1979):

$$P(A \cap B) = \begin{cases} \min((PA),P(B)) & \text{if } \rho = +1 \\ P(A)P(B) & \text{if } \rho = 0 \\ \min((PA)+P(B)-1,0) & \text{if } \rho = -1 \\ \log_s(1+(s^{P(A)}-1)(s^{P(B)}-1)/(s-1) & \text{otherwise} \end{cases} \tag{3.145}$$

where $s = \tan(\pi(1-\rho)/4)$, with the last expression approaching the values for the extreme cases of ρ. The probability of the union of two correlated events (A and B) based on the concept of copulas can be computed as follows:

$$P(A \cup B) = \begin{cases} \min((PA),P(B)) & \text{if } \rho = +1 \\ 1-(1-P(A))(1-P(B)) & \text{if } \rho = 0 \\ \min((PA)+P(B),1) & \text{if } \rho = -1 \\ 1-\log_s(1+(s^{1-P(A)}-1)(s^{1-P(B)}-1)/(s-1) & \text{otherwise} \end{cases} \tag{3.146}$$

3.8.9 Unknown Dependence between Events

In cases where the dependence is unknown, the conjunction (i.e., intersection) and disjunction (i.e., union) probabilities for two events A and B can be bounded, respectively, as follows based on the classical Fréchet inequalities (Fréchet, 1935, 1951), as provided by Ferson et al. (2004):

$$P(A \cap B) = [\max(0,P(A)+P(B)-1),\min(P(A),P(B))] \tag{3.147a}$$

$$P(A \cup B) = [\max(P(A), P(B)), \min(1, P(A) + P(B))] \quad (3.147b)$$

3.8.10 Unknown Positive Dependence between Events

In cases where the dependence is known to be only positive, the conjunction (i.e., intersection) and disjunction (i.e., union) probabilities for two events A and B can be bounded, respectively, as follows based on the work of Williamson (1989) and Wise and Henrion (1986), as provided by Ferson et al. (2004):

$$P(A \cap B) = [P(A)P(B), \min(P(A), P(B))] \quad (3.148a)$$

$$P(A \cup B) = [\max(P(A), P(B)), 1 - (1 - P(A))((1 - P(B))] \quad (3.148b)$$

3.8.11 Unknown Negative Dependence between Events

In cases where the dependence is known to be negative, the conjunction (i.e., intersection) and disjunction (i.e., union) probabilities for two events A and B can be bounded, respectively, as follows based on the work of Williamson (1989) and Wise and Henrion (1986), as provided by Ferson et al. (2004):

$$P(A \cap B) = [\max(P(A) + P(B) - 1), P(A)P(B)] \quad (3.149a)$$

$$P(A \cup B) = [1 - (1 - P(A))((1 - P(B)), \min(P(A), P(B))] \quad (3.149b)$$

3.8.12 Probability Bounds

Probability bounds can be viewed as a mix of probability theory and interval analysis (Ferson et al., 1999). They have similar bases as interval probabilities and concepts covered in probabilistic analysis using limited or incomplete information. Probabilities in this case are uncertain, and hence represented by probability bounds. Where random variables are used, cumulative distribution functions (CDFs) offer a complete description of their probabilistic characteristics. Uncertainty in underlying parameters or limited knowledge about these variables results in the need to construct bounds on them.

For example, a random variable X might be known only to the extent of the minimum (e.g., $x = 50$) and maximum (e.g., $x = 70$) values that the variable could possibly take. The probability bounds for this random variable can be expressed in the form of CDF bounds, as shown in Figure 3.19. The CDF bounds can be interpreted as the left and right limits on any possible CDF that meet the constraint given by the minimum and maximum values of X. These CDF bounds can be denoted $\underline{F}_X(x)$ and $\overline{F}_X(x)$ for left (or called lower on x) and right (or called upper on x) approximations of the CDF (i.e., F) of X. Increasing the level of information in this constraint results in reducing the gap between these bounds. For example, adding a median value at $x = 60$ to the minimum–maximum constraint produces the CDF bounds of Figure 3.20. Figure 3.10 to Figure 3.12 offer additional examples of CDF

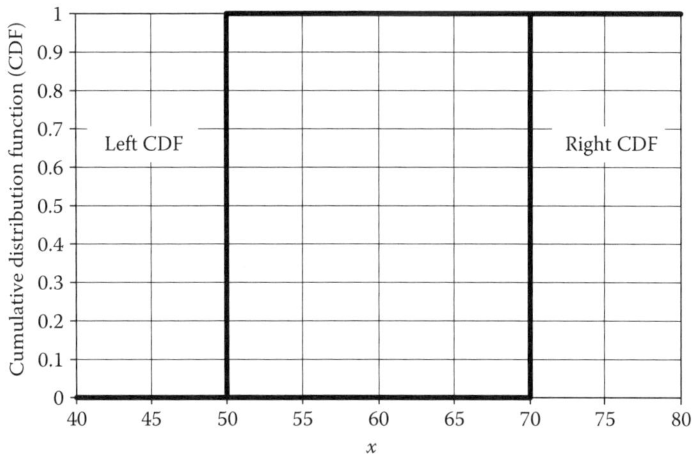

FIGURE 3.19 Bounds on a cumulative distribution function based on minimum and maximum values.

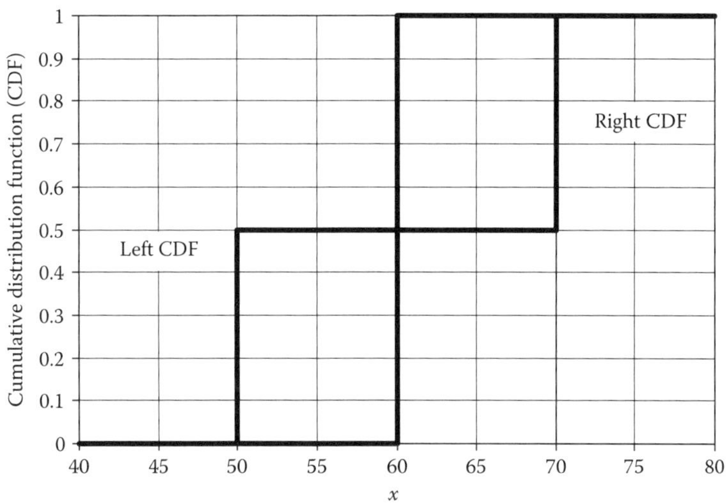

FIGURE 3.20 Bounds on a cumulative distribution function based on minimum, median, and maximum values.

bounds. Figure 3.10 and Figure 3.11 can be approximated using interval values on the underlying random variable X so that the resulting CDF bounds have the general step function shapes provided in Figure 3.12.

To facilitate the probability calculus for these probability bounds, left and right approximations of the CDF of a random variable can be represented as step functions with an added restriction, for computational convenience, that the steps for both left and right functions occur at the same CDF values. Using an underlying independence assumption for, say, two variables X and Y allows for computing, for example, X +

Y, by portioning the spaces of X and Y to convenient intervals, performing interval arithmetic on all the combinations of the Cartesian space of X and Y, and computing the corresponding probabilities of the resulting intervals as the product of the respective pairs.

The case of underlying, but unknown, dependencies between two random variables such as X and Y requires arithmetic operations on X and Y to be conducted using a probability-bound convolution with the constraint that the sum of probabilities must be 1. Frank et al. (1987), Nelson (1999), and Williamson and Downs (1990) provided the following probability bounds on $Z = X * Y$, where $* \in [+, -, \times, \div]$, for arithmetic operations without a dependence assumption between two random variables, such as X and Y:

$$\underline{F}_{X+Y}(z) = \sup_{\text{such that } z=u+v} \{\max[\underline{F}_X(u) + \underline{F}_Y(v) - 1, 0]\} \tag{3.150}$$

$$\overline{F}_{X+Y}(z) = \inf_{\text{such that } z=u+v} \{\min[\overline{F}_X(u) + \overline{F}_Y(v), 1]\} \tag{3.151}$$

$$\underline{F}_{X-Y}(z) = \sup_{\text{such that } z=u-v} \{\max[\underline{F}_X(u) - \overline{F}_Y(v), 0]\} \tag{3.152}$$

$$\overline{F}_{X-Y}(z) = 1 + \inf_{\text{such that } z=u-v} \{\min[\overline{F}_X(u) - \underline{F}_Y(v), 0]\} \tag{3.153}$$

$$\underline{F}_{X\times Y}(z) = \sup_{\text{such that } z=u\times v} \{\max[\underline{F}_X(u) + \underline{F}_Y(v) - 1, 0]\} \tag{3.154}$$

$$\overline{F}_{X\times Y}(z) = \inf_{\text{such that } z=u\times v} \{\min[\overline{F}_X(u) + \overline{F}_Y(v), 1]\} \tag{3.155}$$

$$\underline{F}_{X\div Y}(z) = \sup_{\text{such that } z=u\div v} \{\max[\underline{F}_X(u) - \overline{F}_Y(v), 0]\} \tag{3.156}$$

$$\overline{F}_{X\div Y}(z) = 1 + \inf_{\text{such that } z=u\div v} \{\min[\overline{F}_X(u) - \underline{F}_Y(v), 0]\} \tag{3.157}$$

where sup = *supremum*, defined as the least upper bound of a set, and inf = *infimum*, defined as the greatest lower bound of a set. Williamson and Downs (1990) and Regan et al. (2000) showed that the above bounds hold for the arithmetic operation of addition and multiplication for both positive and negative X and Y, and the arithmetic operation of subtraction and division regardless of the sign of X and Y, by using the interval mathematics of Equations 2.37 to 2.40 that, in the case of subtraction and division, combine the lower bound of one variable with the upper bound of another, and vice versa, as required by these equations. For two events A

and B with given probabilities $P(A)$ and $P(B)$, the limits provided by Equations 3.150 to 3.157 are partially based on the conjunction and disjunction (Fréchet) inequalities, introduced as Equation 3.147a and b, as follows, respectively:

$$\text{Conjunction: } \max(0, P(A)+P(B)-1) \leq P(A \cap B) \leq \min(P(A), P(B)) \quad (3.158)$$

$$\text{Disjunction: } \max(P(A), P(B)) \leq P(A \cup B) \leq \min(1, P(A)+P(B)) \quad (3.159)$$

Equations 3.158 and 3.159 usually result in wide limits, and their use for CDFs can violate the constraint that the sum of probabilities must be 1, whereas Equations 3.150 to 3.157 do not violate this constraint. Regan et al. (2000) showed the equivalency of Equations 3.150 to 3.157 in propagating uncertainty to methods offered by Walley (1991) for imprecise probabilities, and Dempster–Shafer belief functions as provided by Yager (1986).

Ferson et al. (2004) provide methods to address partial knowledge of dependence between two variables X and Y, and the use of this knowledge to tighten the bounds computed by Equations 3.150 to 3.157. The methods result in decreasing the range between the bounds.

Example 3.8 Computations of Probability Bounds for Independent Variables

Figure 3.21 provides examples of CDF bounds for two random variables X and Y. These figures express uncertainty in the CDF. For example, Figure 3.21 provides the CDF bounds at $x = 3.5$ of [0.2, 0.6]. Also, the same figure expresses the uncertainty in the value of X at a given CDF value. For a CDF value, i.e., percentile value, of 0.90, the value of x belongs to the interval [5, 6].

Random variables defined by CDF bounds can be combined using arithmetic operations such as addition, subtraction, multiplication, and division; however, information on underlying dependencies between the two random variables is needed in order to assess the combined result. Two cases are considered in this section as follows: (1) the case of an underlying independence between two random variables such as X and Y, and (2) the case of underlying, but unknown, dependencies between two random variables such as X and Y. The first case is covered below, whereas the second case is provided as an exercise at the end of the chapter.

The underlying independence assumption for X and Y allows for computing, for example, $X + Y$, by portioning the spaces of X and Y to convenient intervals, performing interval arithmetic on all the combinations of the Cartesian space of X and Y, and computing the corresponding probabilities of the resulting intervals as the product of the respective pairs. The computational procedure is demonstrated using the random variables X and Y of Figure 3.21 to evaluate their addition, i.e., $X + Y$, as shown in Table 3.11. The left and right probability bounds of the CDF of the addition result Z can be evaluated as shown in Table 3.12. Table 3.12 was constructed by identifying the range of Z from Table 3.11, from 3 to 18, generally in increments of 1. Then for

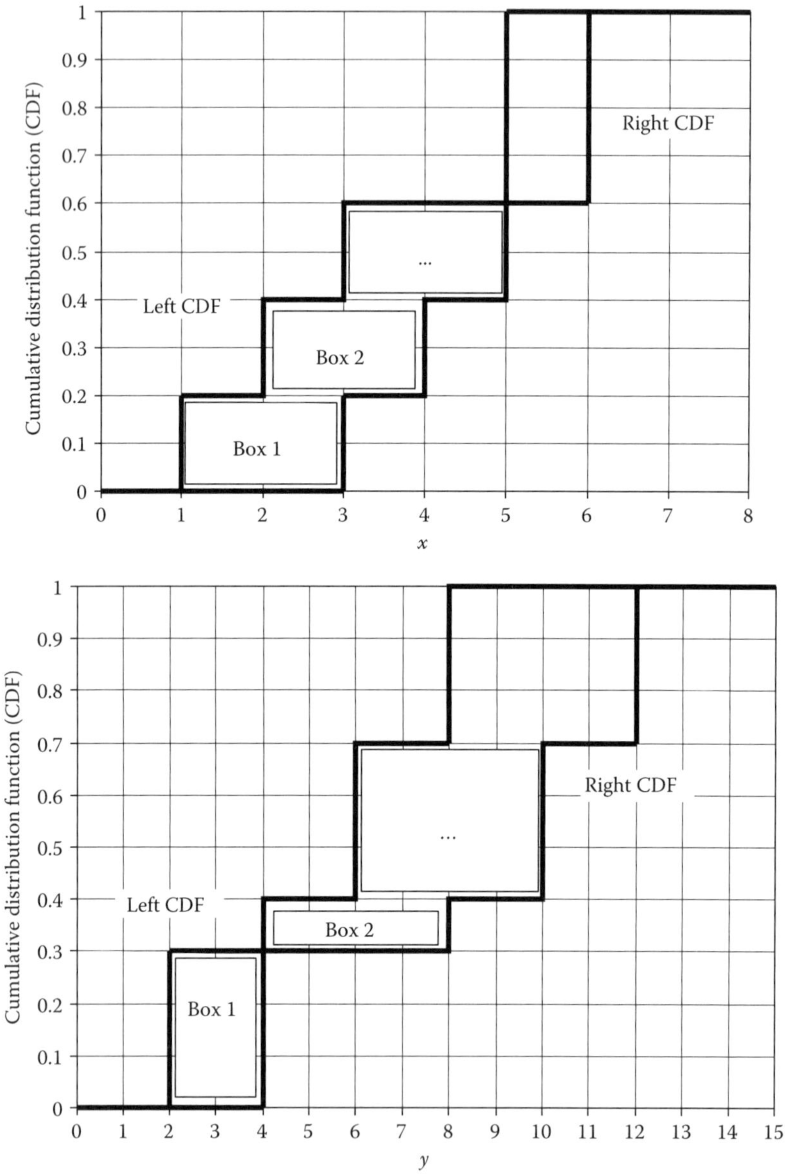

FIGURE 3.21 Bounds on the cumulative distribution functions of *X* and *Y*.

TABLE 3.11
Addition ($Z = X + Y$) Using CDF Bounds with Underlying Independence Expressed as Intervals with Probabilities

Intervals for Y and Their Probabilities	Intervals for X and Their Probabilities			
	$P(1 \le X < 3) =$ **0.2**	$P(2 \le X < 4) =$ **0.2**	$P(3 \le X < 5) =$ **0.2**	$P(5 \le X < 6) =$ **0.4**
$P(2 \le Y < 4) = 0.3$	$P(3 \le Z < 7) =$ 0.06	$P(4 \le Z < 8) =$ 0.06	$P(5 \le Z < 9) =$ 0.06	$P(7 \le Z < 10) =$ 0.12
$P(4 \le Y < 8) = 0.1$	$P(5 \le Z < 11) =$ 0.02	$P(6 \le Z < 12) =$ 0.02	$P(7 \le Z < 13) =$ 0.02	$P(9 \le Z < 14) =$ 0.04
$P(6 \le Y < 10) =$ 0.3	$P(7 \le Z < 13) =$ 0.06	$P(8 \le Z < 14) =$ 0.06	$P(9 \le Z < 15) =$ 0.06	$P(11 \le Z < 16) =$ 0.12
$P(8 \le Y < 12) =$ 0.3	$P(9 \le Z < 15) =$ 0.06	$P(10 \le Z < 16) =$ 0.06	$P(11 \le Z < 17) =$ 0.06	$P(13 \le Z < 18) =$ 0.12

TABLE 3.12
Probability Bounds for the Addition ($Z = X + Y$)

Addition Result of $Z = X + Y$	Left Bound	Right Bound
2	0	0
3	0.06	0
4	0.12	0
5	0.20	0
6	0.22	0
7	0.42	0.06
8	0.48	0.12
9	0.64	0.18
10	0.70	0.30
11	0.88	0.32
12	0.88	0.34
13	1.00	0.42
14	1.00	0.52
15	1.00	0.64
16	1.00	0.82
17	1.00	0.88
18	1.00	1.00
19	1.00	1.00
20	1.00	1.00

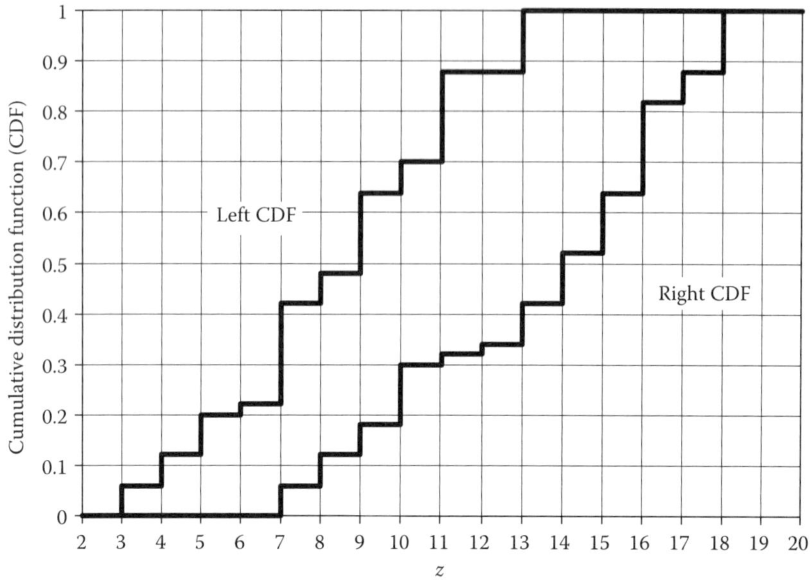

FIGURE 3.22 Bounds on the cumulative distribution functions of $Z = X + Y$.

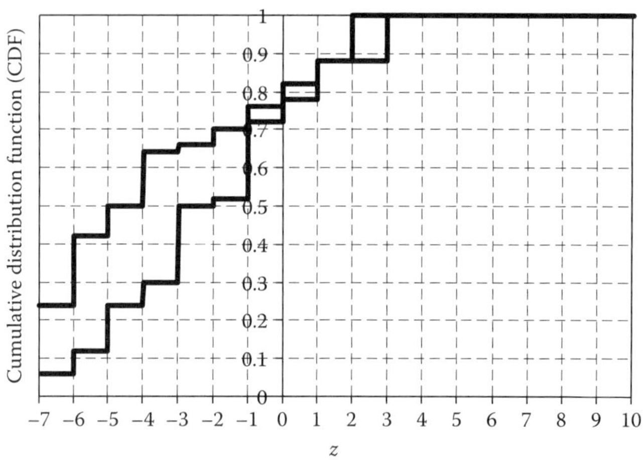

FIGURE 3.23 Bounds on the cumulative distribution functions of $Z = X - Y$.

each Z value, such as z, the left bound was constructed as the cumulative sum of all interval probabilities for Z in Table 3.11, where the lower (left) limits of the intervals are less than or equal to z. The right bound for Z can be constructed in a similar manner as the cumulative sum of interval probabilities from Table 3.11, where the upper (right) values of the intervals are less than or equal to z. The resulting probability bounds of $Z = X + Y$ are shown in Figure 3.22. Other arithmetic operations, such as subtraction, multiplication, and division, can be performed in a manner similar to the above process for addition.

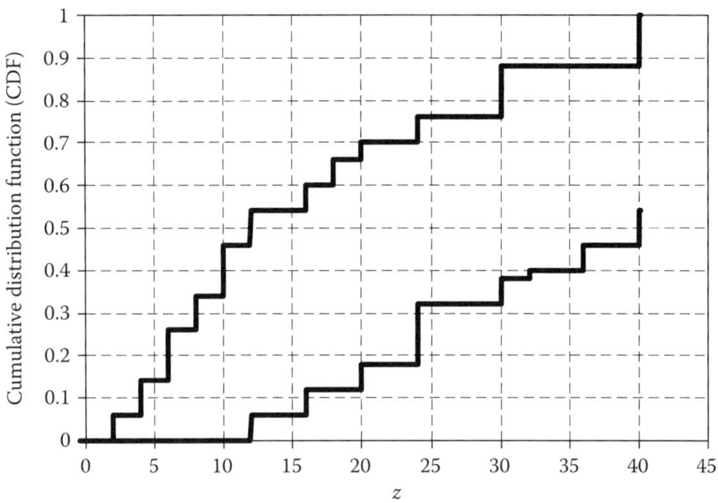

FIGURE 3.24 Bounds on the cumulative distribution functions of $Z = X \times Y$.

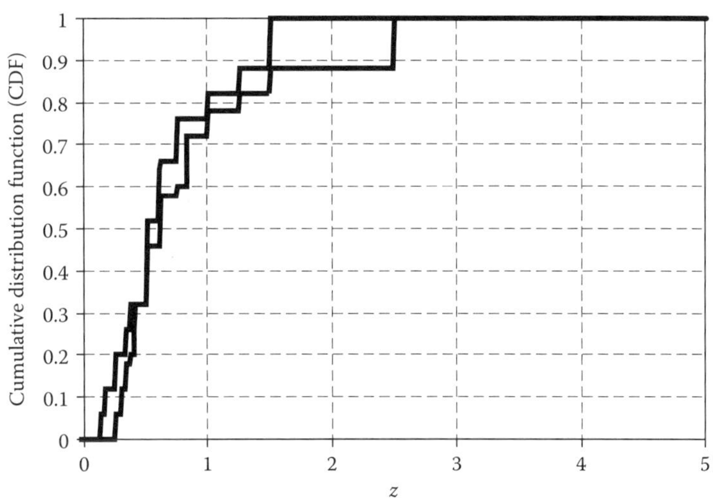

FIGURE 3.25 Bounds on the cumulative distribution functions of $Z = X \div Y$.

The results for the cases of addition, multiplication, and division are provided in Figure 3.23 to Figure 3.25.

3.9 FUZZY MEASURES AND FUZZY INTEGRALS

A fuzzy measure is defined in Chapter 2 as a function μ from a family of subsets C of a universal set X to the interval [0, 1]. Commonly, C is the power set of X, i.e., $P(X)$. This mapping can be expressed mathematically as

$$\mu : C \rightarrow [0,1] \tag{3.160}$$

This function must have the following properties in addition to continuity from above and continuity from below:

$$\mu(\varnothing) = 0 \quad \text{if } \varnothing \in C \tag{3.161a}$$

$$\mu(X) = 1 \quad \text{if } X \in C \tag{3.161b}$$

$$\mu(A) \leq \mu(B) \quad \text{if } A \,\&\, B \in C \text{ and } A \subseteq B \tag{3.161c}$$

Another function f can be introduced on the elements of X as follows:

$$f : X \rightarrow [0,1] \tag{3.162}$$

For this function f, an α-cut and a strict α-cut can be defined, similar to fuzzy sets, as the following respective sets:

$$^{\alpha}f = \{x: f(x) \geq \alpha\} \tag{3.163a}$$

$$^{\alpha+}f = \{x: f(x) > \alpha\} \tag{3.163b}$$

A primary difference between μ and f is that in addition to quantifying two different attributes, the former relates to a subset of X and the latter refers to the elements of X.

A fuzzy integral of f based on the μ values attached to the power set can be defined as

$$\int_X f d\mu = \sup_{\alpha \in [0,1]} [a \wedge \mu(X \cap {}^{\alpha}f)] \tag{3.164}$$

where sup = *supremum* of A, defined as the least upper bound of the set, and $\wedge$ = minimum operator. Since $X \cap {}^{\alpha}f = {}^{\alpha}f$, the equation can be simplified to the following by also dropping the X below the integral:

$$\int f d\mu = \sup_{\alpha \in [0,1]} [\alpha \wedge \mu({}^{\alpha}f)] \tag{3.165}$$

The integral provided in Equations 3.164 and 3.165 is called a Sugeno integral or a fuzzy integral (Wang and Klir, 1992). This integral can be computed over a subset A of X as follows:

$$\int_A f d\mu = \sup_{\alpha \in [0,1]} [a \wedge \mu(A \cap {}^{\alpha}f)] \tag{3.166}$$

EXAMPLE 3.9 QUALITY OF CHINESE CUISINE

Wang and Klir (1992) provide examples of using fuzzy integrals to measure the quality of Chinese cuisine based on an assumed universal set X that includes three elements: taste (T), smell (S), and appearance (A). The appearance is assumed to consist of color, shape, and general arrangement of a dish. Therefore, the universal set is

$$X = \{T, S, A\} \tag{3.167}$$

The following fuzzy measure (μ) of importance in defining the quality of Chinese cuisine was subjectively defined based on the power set X:

$$\mu(\{T\}) = 0.7 \tag{3.168a}$$

$$\mu(\{S\}) = 0.1 \tag{3.168b}$$

$$\mu(\{A\}) = 0 \tag{3.168c}$$

$$\mu(\{T, S\}) = 0.9 \tag{3.168d}$$

$$\mu(\{T, A\}) = 0.8 \tag{3.168e}$$

$$\mu(\{S, A\}) = 0.3 \tag{3.168f}$$

$$\mu(\{T, S, A\}) = \mu(X) = 1 \tag{3.168g}$$

$$\mu(\varnothing) = 0 \tag{3.168h}$$

These importance measures were subjectively assessed and assigned values based on intuition, but interestingly are not additive, e.g.,

$$\mu(\{T, S\}) \neq \mu(\{T\}) + \mu(\{S\}) \tag{3.169a}$$

$$\mu(\{S, A\}) \neq \mu(\{S\}) + \mu(\{A\}) \tag{3.169b}$$

An expert was asked to assess four dishes. The expert provided the following assessment for the first dish:

$$f\{T\} = 0.9 \tag{3.170a}$$

$$f\{S\} = 0.6 \tag{3.170b}$$

$$f\{A\} = 0.8 \tag{3.170c}$$

The quality of the first dish (Q_1) can be assessed using the following fuzzy integral based on Equation 3.164:

$$Q_1 = \int_X f d\mu = \sup_{\alpha \in [0,1]} [a \wedge \mu({}^{\alpha}f)] \tag{3.171a}$$

or by substituting values from Equation 3.170a to c, Q_1 becomes

$$\begin{aligned} Q_1 = \int_X f d\mu &= [0.6 \wedge \mu({}^{0.6}f)] \vee [0.8 \wedge \mu({}^{0.8}f)] \vee [0.9 \wedge \mu({}^{0.9}f)] \\ &= [0.6 \wedge \mu(X)] \vee [0.8 \wedge \mu(\{T, A\})] \vee [0.9 \wedge (\{T\})] \\ &= [0.6 \wedge 1] \vee [0.8 \wedge 0.8] \vee [0.9 \wedge 0.7] \\ &= 0.8 \end{aligned} \tag{3.171b}$$

where ${}^{\alpha}f$ can be evaluated using Equation 3.163a. The second dish was assessed to have $f(T) = 1$, $f(S) = f(A) = 0$, and has the following quality value:

$$\begin{aligned} Q_2 &= [1 \wedge \mu({}^{1}f)] \vee 0 \\ &= \mu(\{T\}) \\ &= 0.7 \end{aligned} \tag{3.172}$$

The third dish was assessed to have $f(T) = f(S) = 1$, $f(A) = 0$, and has the following quality value:

$$\begin{aligned} Q_3 &= [1 \wedge \mu(\{T, S\})] \vee 0 \\ &= \mu(\{T, S\}) \\ &= 0.9 \end{aligned} \tag{3.173}$$

The fourth dish was assessed to have $f(T) = f(S) = f(A) = 1$ and has the following quality value:

$$
\begin{aligned}
Q_4 &= [1 \wedge \mu(\{X\})] \\
&= \mu(\{X\}) \\
&= 1
\end{aligned}
\tag{3.174}
$$

EXERCISE PROBLEMS

3.1. Develop a creative nested or relational hierarchy for various measures covered in this chapter, including at least the classical measure, monotone measure, belief measure, plausibility measure, fuzzy measure, fuzzy integrals, probability measure, and possibility measure. Your hierarchical representation should clearly show under what conditions would one type reduce to another type.

3.2. Two judges classified a case to three possible motivations as provided in the following table in the form of basic assignments m_1 and m_2:

Subset (Motivation)	Judge 1 m_1	Judge 2 m_2
Greed (G)	0.05	0.15
Love (L)	0.10	0.05
Self defense (S)	0.15	0.05
$G \cup L$	0.25	0.15
$G \cup S$	0.15	0.20
$L \cup S$	0.05	0.30
$G \cup L \cup S$	Not provided	Not provided

Compute the belief measures for judges 1 and 2. Compute the basic assignment for the combined judgment and the corresponding belief measure.

3.3. Two experts classified a bird species to three possible causes for its population decline, as provided in the following table in the form of basic assignments m_1 and m_2:

Subset (Cause)	Expert 1 m_1	Expert 2 m_2
Changes in land use (C)	0.10	0.15
Hunting (H)	0.15	0.15
Disease (D)	0.15	0.15
$C \cup H$	0.25	0.15
$C \cup D$	0.15	0.10
$H \cup D$	0.15	0.20
$C \cup H \cup D$	Not provided	Not provided

Compute the belief measures for Experts 1 and 2. Compute the basic assignment for the combined judgment and the corresponding belief measure.

3.4. Two experts classified the bird species of problem 3.3 to three possible causes for its population decline, as provided in the following table in the form of basic assignments m_1 and m_2:

Subset (Cause)	Expert 1 m_1	Expert 2 m_2
Changes in land use (C)	0.10	0.10
Hunting (H)	0.20	0.30
Disease (D)	0.10	0.10
$C \cup H$	0.20	0.10
$C \cup D$	0.10	0.10
$H \cup D$	0.20	0.10
$C \cup H \cup D$	Not provided	Not provided

Compute the belief measures for Experts 1 and 2. Compute the basic assignment for the combined judgment and the corresponding belief measure.

3.5. Three experts classified a bird species to three possible causes for its population decline, as provided in the following table in the form of basic assignments m_1, m_2, and m_3:

Subset (Cause)	Expert 1 m_1	Expert 2 m_2	Expert 3 m_3
Changes in land use (C)	0.05	0.05	0.05
Hunting (H)	0.25	0.30	0.50
Disease (D)	0.05	0.10	0.05
$C \cup H$	0.25	0.05	0.01
$C \cup D$	0.05	0.05	0.05
$H \cup D$	0.25	0.20	0.01
$C \cup H \cup D$	Not provided	Not provided	Not provided

Compute the belief measures for Experts 1, 2, and 3. Provide procedures for computing the basic assignment for the combined judgment and the corresponding belief measure, and demonstrate the procedures using the above table.

3.6. Three sensory sources are used to identify targets. Using a formulation similar to Table 3.4 of Example 3.3, determine the combined probability using the following assignment:

Subset (Evidence)	Measures Assignment (m_1)	 Assignment (m_2)
Fighter (F)	0.4	0.10
Reconnaissance (R)	0.2	0.0
Bomber (B)	0.0	0.80
Universal set (X)	0.4	0.10

3.7. Assume that the x's are users or intruders of information network with specific nested characteristics, such as x_1 = a legitimate user from an authorized IP address, x_2 = an intruder from an authorized IP address or an intruder from an IP address that is masked by an authorized IP address, x_3 = a legitimate user with an authorized access protocol (username and password), and x_4 = an intruder from an authorized IP access protocol. The nested structure of Figure 3.5 can be used to construct evidence-gathering methods and their probabilities of affirmative detection, false detection, affirmative nondetection, and false nondetection. These probabilities can be constructed as basic assignments to meet the requirements of the theory of evidence. They are then used to compute belief and plausibility measures for having any of the events $A_1 \ldots A_5$ of Figure 3.5. Develop the mathematical formulation needed for this application with an illustrative example.

3.8. Using Table 3.5, compute the joint probability of the two events A and B that have the following probabilities:

Probability A = likely

Probability B = seldom

Treat the above probabilities as fuzzy sets and use the α-cut method to compute the probability of A and B, assuming that they are independent events. Express your result using a fuzzy set and linguistically based on Table 3.5. What are the limitations of such a hybrid use of linguistic probabilities and fuzzy sets?

3.9. The accident probability at a new intersection is of interest to a traffic engineer. The engineer subjectively estimated the weekly accident probability as follows:

Weekly Accident Probability	Subjective Probability of Accident Probability
0.1	0.3
0.2	0.4
0.4	0.2
0.6	0.05
0.8	0.04
0.9	0.01

Solve the following:

a. What is the average accident probability based on the prior information?
b. Given an accident in the first week of traffic, update the distribution of the accident probability.
c. What is the new average accident probability based on the posterior information?

d. Given an accident in the first and second weeks and no accidents in the third week of traffic, update the distribution of the accident probability.
e. What is the average accident probability after the second week?
f. Given no additional accidents for weeks 4, 5, 6, 7, 8, 9, and 10, update the distribution and average accident probability. Plot your results.

3.10. Develop an inference spreadsheet for assessing the condition of a piece of equipment based on three factors that are related to equipment condition as follows:

Factor No.	Logical Relationship between Factor and Condition
1	Necessary
2	Sufficient
3	Maybe necessary

3.11. Plot an interval distribution for the following random variable:

Logormal probability distribution:
Mean value = [3000, 4000] psi
Standard deviation = [300, 300]

3.12. Plot an interval distribution for the following random variable:

Logormal probability distribution:
Mean value = [3000, 4000] psi
Standard deviation = [300, 400] psi, i.e., coefficient of variation = 0.10

3.13. Plot an interval distribution for the following random variable:

Exponential probability distribution:
Mean value = [3000, 4000]

3.14. Reproduce the example results shown in Figure 3.23 for the subtraction operation, i.e., $Z = X - Y$.

3.15. Reproduce the example results shown in Figure 3.24 for the multiplication operation, i.e., $Z = X \times Y$.

3.16. Reproduce the example results shown in Figure 3.25 for the division operation, i.e., $Z = X \div Y$.

3.17. Redo the example shown in Table 3.11 and Table 3.12 and Figure 3.22 for the addition operation assuming (1) perfect positive dependence and (2) unknown dependence between X and Y. Hint: Construct the probabilities in Table 3.11 so that they are on a diagonal for the perfect dependence case, and determine bounds on the probabilities for the unknown dependence case.

3.18. Redo the example shown in Figure 3.24 for the product operation assuming unknown dependence between X and Y.

3.19. The value of a line of designer pants to potential users depends on five elements: fabric feel to skin, color, fit, appearance, and price. Subjectively assign fuzzy measures of importance to these five elements. Using the following assessment for these elements, assess the value of a line of pants: 0.8, 0.7, 0.5, 0.7, and 0.8. Reevaluate the value for another line design that has the following assessment for these elements: 0.5, 0.9, 0.9, 0.9, and 0.3.

4 Uncertainty Measures

4.1 INTRODUCTION

Ignorance and uncertainty types as described in Chapter 1 were used to develop methods for data and information encoding and expression, and uncertainty-based synthesis of information as discussed in Chapters 2 and 3, respectively. Engineers and scientists often need means to quantify uncertainty. Uncertainty measures can be viewed as similar to scales or measures used to quantify physical quantities, such as temperature, pressure, or dimensions; however, they are unique in that they measure conceived or abstract notions rather than a physical quantity. The objective of this chapter is to present methods for measuring uncertainty contents in the form of basic models.

4.2 UNCERTAINTY MEASURES: DEFINITION AND TYPES

A *measure of uncertainty* of some conceived type represented within a given mathematical theory (e.g., probability theory, possibility theory, Dempster–Shafer theory, etc.) is a function (u) that assigns to each representation of evidence in the theory (μ according to Equation 3.6) (e.g., a probability distribution, a possibility distribution, a body of evidence in Dempster–Shafer theory, etc.) a nonnegative real number. Intuitively, numbers obtained by this function should be inversely proportional to the strength and consistency in evidence, as expressed in the theory employed: the stronger and more consistent the evidence, the smaller the amount of uncertainty. An uncertainty measure can be defined formally as a function that maps the set (U) of all uncertainty functions μ, as defined by Equation 3.6, to the nonnegative real line (R_+) as follows:

$$u: U(\mu) \rightarrow R_+ \tag{4.1}$$

where μ is given by Equation 3.6.

It should be mentioned that measures of uncertainty have been almost exclusively investigated in terms of *disjunctive variables*. A disjunctive variable has at any given time a single value, but we are often uncertain about it due to limited evidence. Probability theory, possibility theory, Dempster–Shafer theory, and the various theories of imprecise probabilities allow us to describe different types of evidence regarding disjunctive variables. Some examples of disjunctive variables are the age of a person, the weight of an object, the speed of a car, the humidity at a given place in the Earth, and the arrival time of a flight.

Distinct from disjunctive variables are *conjunctive variables* (Yager, 1987b). They are characterized by simultaneously assuming multiple values from a given universal set. Some examples of conjunctive variables are friends of a person, the time period a person spent waiting for a flight, components that form a compound, and books written by an author. Since uncertainty theories for conjunctive variables are virtually undeveloped, all results presented in this chapter are based on the assumption that we deal with disjunctive variables.

Uncertainty measures are distinguished from one another by the mathematical representation employed and by the type of uncertainty involved. Although each uncertainty measure should make sense on intuitive grounds, it is even more important that it satisfies certain axiomatic requirements. In the rest of this section, the most fundamental requirements are described. Since the mathematical form of each of these requirements depends on the uncertainty theory employed, they are described in generic terms, independent of the various uncertainty calculi. The following requirements are essential in the sense that they apply to all uncertainty theories (Klir and Smith, 1999):

- *Subadditivity*: The amount of uncertainty in a joint representation of evidence (defined on a Cartesian product) cannot be greater than the sum of the amounts of uncertainty in the associated marginal representations of evidence.
- *Additivity*: The two amounts of uncertainty considered under subadditivity become equal if and only if the marginal representations of evidence are noninteractive according to the rules of the uncertainty calculus involved.
- *Range*: The range of uncertainty is [0, M], where M depends on the cardinality of the universal set involved and on the chosen unit of measurement.
- *Continuity*: Any measure of uncertainty must be a continuous function.
- *Expansibility*: Expanding the universal set by alternatives that are not supported by evidence must not affect the amount of uncertainty.
- *Branching/consistency*: When uncertainty can be computed in several ways, all intuitively acceptable, the results must be the same (consistent).

The remaining two requirements are applicable only to some theories of uncertainty:

- *Monotonocity*: When evidence can be ordered in the uncertainty theory employed (as in possibility theory), the relevant uncertainty measure must preserve this ordering.
- *Coordinate invariance*: When evidence is described within the n-dimensional Euclidean space ($n \geq 1$), the relevant uncertainty measure must not change under isometric transformation of coordinates.

Uncertainty measures are available for three uncertainty classes as follows: (1) imprecision or nonspecificity associated with sizes or cardinalities, (2) fuzziness or vagueness associated with imprecision in boundaries, and (3) conflict or strife and discord among various sets. Although the area of uncertainty measures is an active research area and is evolving, many concepts are mature and could be used to some problems in engineering the sciences. Uncertainty measures were provided in this chapter for selected uncertainty types and theories covered in this book. The presentation in this section is limited to three relatively mature uncertainty types: (1) nonspecificity that results from imprecision connected with set sizes, i.e., cardinalities, and can be represented by the Hartley-like measure; (2) uncertainty expressed in terms of conflict among evidential claims, i.e., Entropy-like measures; and (3) fuzziness as a result of nonsharp boundaries of fuzzy sets. Commonly, engineering and science problems simultaneously contain these uncertainty types and other types.

4.3 NONSPECIFICITY MEASURES

A fundamental uncertainty type stems from lack of specificity as a result of providing several alternatives, with one alternative being the true one. This uncertainty type vanishes, and complete certainty is achieved, when only one alternative is possible. Therefore, the nonspecificity uncertainty type results from having imprecision due to alternative sets that have cardinalities greater than 1.

4.3.1 HARTLEY MEASURE

A fundamental measure of uncertainty based on cardinality was conceived as a measuring of nonspecificity by Hartley (1928). This fundamental measure can therefore be defined for a *finite* set of all possible alternatives, i.e., universal space X of alternatives under consideration, with only one of the alternatives being correct, although the correct alternative is unknown to us. In the context of possibility theory, the finite set X of conceived alternatives includes only one alternative in a given situation that is true. Assume that, according to given evidence, we know that alternatives in a particular set E (such that $E \subseteq X$) are possible, while those outside E are not possible. This means that, according to the evidence, the true alternative is in set E. This simple evidence can be formalized by defining a *possibility measure*, $r_E(x)$, based on evidence focusing on E:

$$r_E(x) = \begin{cases} 1 & \text{when } x \in E \\ 0 & \text{when } x \in \overline{E} \end{cases} \tag{4.2}$$

for all $x \in X$, and

$$r_E(A) = \max_{x \in A}(r_E(x)) \tag{4.3}$$

for all $A \in P_X$, i.e., A is a subset of X. In this case, only the alternatives that belong to A are considered possible candidates for this true alternative. Moreover, it is convenient to introduce a dual *necessity measure*, Nec_E, via the formula

$$Nec_E(A) = 1 - r_E(\bar{A}) \tag{4.4}$$

for all $A \in P_X$.

Clearly, the larger the set E, the less specific the evidence, and the more uncertain we are about the true alternative. Uncertainty is thus caused in this case by the *nonspecificity* of evidence. Hartley (1928) showed that the only sensible way to measure uncertainty of this nonspecificity type is to use function H, defined as follows:

$$H(r_E(x)) = \log_2(|E|) = \frac{\ln(|E|)}{\ln(2)} \tag{4.5}$$

where $\log_2$ is the logarithm to the base 2, resulting in a measurement unit in *bits*, and ln is the natural logarithm. A *bit* is defined as a single digit in a binary number system and can be viewed as a unit of information equal to the amount of information obtained by learning or resolving which of two equally has occurred. Similarly, the measure of uncertainty associated with any finite set A of possible alternatives can be defined as follows:

$$H : U(r) \to R_+ \tag{4.6}$$

or

$$H(r_E(x)) = \log_2 \sum_{x \in X} r_E(x) = \frac{\ln \sum_{x \in X} r_E(x)}{\ln(2)} \tag{4.7}$$

Bits form the basis for a *byte* in computer language, where a *byte* is a string of binary digits (bits), usually eight, operated on as a basic unit by a digital computer. The logarithm ($\log_2$) of i is given by

$$\log_2(i) = x \tag{4.8a}$$

It is the power to which the base, in this case 2, must be raised to obtain i as provided by

$$2^x = i \tag{4.8b}$$

The Hartley measure (H) reaches its maximum for the universal set X; i.e., the Hartley measure takes values in the following range:

$$0 \le H(r_E(x)) \le \log_2(|X|) \text{ for any } A \subseteq X \tag{4.9}$$

The measure has the properties of being additive, monotonic, and normal (i.e., taking a value of 1 at $|A| = 2$).

The Hartley measure can be defined in the context of a relation (R) defined over all the possible combinations of two universal sets X and Y of discrete elements and of finite sizes, represented as the Cartesian product of X and Y. Two sets can be defined as follows:

$$A_X = \{x \in X \,|\, (x, y) \in R \text{ for some } y \in Y\} \tag{4.10}$$

$$B_Y = \{y \in Y \,|\, (x, y) \in R \text{ for some } x \in X\} \tag{4.11}$$

These sets are called the projections of R on the sets X and Y, respectively. The relation is defined between these two sets A_X and B_Y. The *joint Hartley measure* of R is

$$H(R) = H(X,Y) = \log_2(|R|) \tag{4.12}$$

where $|R|$ is the cardinality if R is defined as the number of ordered pairs in R. The *marginal Hartley measures* associated with subsets of two respective universal sets are

$$H(A_X) = \log_2(|A_X|) \tag{4.13a}$$

$$H(B_Y) = \log_2(|B_Y|) \tag{4.13b}$$

The *conditional Hartley measures* based on Equations 4.12 and 4.13 can be defined as follows:

$$H(A_X|B_Y) = \log_2\left(\frac{|R|}{|B_Y|}\right) \tag{4.14a}$$

$$H(B_Y|A_X) = \log_2\left(\frac{|R|}{|A_X|}\right) \tag{4.14b}$$

The ratio $|R|/|B_Y|$ in Equation 4.14a represents the average number of elements of X that are possible alternatives under the condition that a possible element of Y is known. Therefore, Equation 4.14a measures the average nonspecificity regarding possible choices from X for all possible choices from Y. Similar observations can be made relating to Equation 4.14b. The following general relations can be constructed based on these definitions:

$$H(A_X|B_Y) = H(A_X, B_Y) - H(B_Y) \tag{4.15a}$$

$$H(B_Y|A_X) = H(A_X, B_Y) - H(A_X) \tag{4.15b}$$

$$H(A_X) - H(B_Y) = H(A_X|B_Y) - H(B_Y|A_X) \tag{4.15c}$$

Some relations might display *noninteractive* behavior between X and Y, producing the following relations:

$$H(A_X|B_Y) = H(A_X) \tag{4.16a}$$

$$H(B_Y|A_X) = H(B_Y) \tag{4.16b}$$

$$H(A_X, B_Y) = H(A_X) + H(B_Y) \tag{4.16c}$$

The relations of *interactive* sets in X and Y have the following characteristics:

$$H(A_X|B_Y) < H(A_X) \tag{4.17a}$$

$$H(B_Y|A_X) < H(B_Y) \tag{4.17b}$$

$$H(A_X, B_Y) < H(A_X) + H(B_Y) \tag{4.17c}$$

A function can be defined to indicate the level of Hartley-based information transmission (T), i.e., uncertainty resolution, based on a relation R as follows (Klir and Wierman, 1999):

$$T_H(A_X, B_Y) = H(A_X) + H(B_Y) - H(A_X, B_Y) \tag{4.18}$$

The information transmission takes a value of zero for noninteractive sets; otherwise, it is greater than zero. It can also be computed as follows:

$$T_H(A_X, B_Y) = H(A_X) - H(A_X|B_Y) \tag{4.19a}$$

$$T_H(A_X, B_Y) = H(B_Y) - H(B_Y|A_X) \tag{4.19b}$$

TABLE 4.1
Hartley Measure

Cardinality ($\lvert A \rvert$)	Hartley Measure (H)
1	0
2	1
3	1.584963
4	2
5	2.321928
6	2.584963
7	2.807355
8	3
9	3.169925
10	3.321928
197	7.622052
198	7.629357
199	7.636625
200	7.643856
1000	9.965784
2000	10.96578
10,000	13.28771
1,000,000	19.93157

Example 4.1 Causes of Structural Failures

Structures could fail as a result of several causes that include design errors, construction errors, weather conditions, and extreme loads, among other factors. A set A can be defined to include three factors. The Hartley measure of A is

$$H(r_E(x)) = \log_2(\lvert A \rvert) = \frac{\ln(\lvert A \rvert)}{\ln(2)} = \frac{\ln(3)}{\ln(2)} = 1.584963 \tag{4.20}$$

The uncertainty measure can be interpreted as a measure of diagnostic uncertainty. The level of uncertainty increases as the number of factors is increased. Table 4.1 and Figure 4.1 show the relationship between the number of factors and Hartley measure. The uncertainty can be reduced based on performing tests or analytical studies to eliminate some of the factors that are not relevant. For example, several candidate tests can be conducted to assess these factors. The tests could help in assessing varying numbers of factors ($\lvert F \rvert$) for a set of factors F. The test that has the greatest potential to reduce the uncertainty needs to be identified and considered in conjunction with its cost and potential benefit for implementation. Uncertainty measures, among other information, can therefore be used in making this decision. For example, for a test that resolves a nonempty subset A of the factors, i.e., $\lvert A \rvert \geq 1$, the uncertainty reduction gained can be assessed as follows:

$$H(r_F(x), r_A(x)) = H(\lvert F \rvert - \lvert A \rvert) = \log_2(\lvert F \rvert - \lvert A \rvert) \tag{4.21}$$

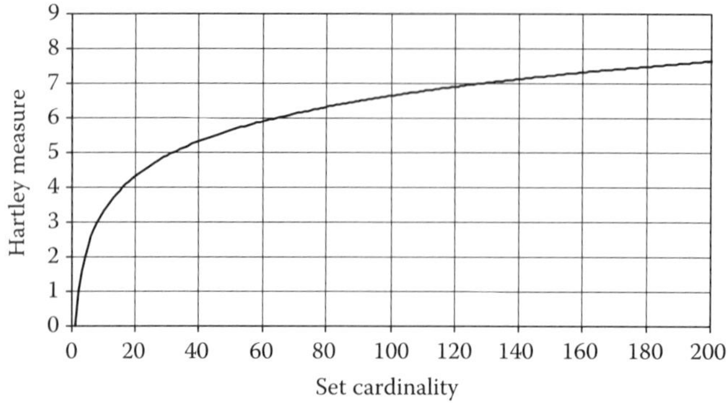

FIGURE 4.1 Hartley measure.

4.3.2 Hartley-Like Measure

The Hartley measure is applicable only to finite sets. Its counterpart for subsets of the n-dimensional Euclidean space $R^n(n \geq 1)$ was suggested by Klir and Yuan (1995a) in terms of a function called a *Hartley-like measure* (HL). The function produces an uncertainly value of 1 for any closed interval of real numbers whose length is 1, a value of 2 in the case of a unit square, a value of 3 in the case of a unit cube, etc. The Hartley-like function measures nonspecificity in situations in which the conceived alternatives are points of an n-dimensional Euclidean space and the evidence is expressed in terms of a bounded and convex subset E of possible points (alternatives) of R^n. The function is defined by

$$HL(E) = \min_{t \in T} \log_2 \left[\prod_{i=1}^{n} (1 + \mu(E_{i_t})) + \mu(E) - \prod_{i=1}^{n} \mu(E_{i_t}) \right] \tag{4.22a}$$

or

$$HL(E) = \min_{t \in T} \ln \left[\prod_{i=1}^{n} (1 + \mu(E_{i_t})) + \mu(E) - \prod_{i=1}^{n} \mu(E_{i_t}) \right] / \ln(2) \tag{4.22b}$$

where μ denotes the Lebesgue measure, T denotes the set of all transformations from one orthogonal coordinate system to another, and E_{i_t} denotes the ith one-dimensional projection of E within the coordinate systems.

For any universal set X and a convex subset of R^n, a normalized Hartley-like measure (NHL) is defined for each convex subset E of X by

$$NHL(E) = \frac{HL(E)}{HL(X)} \tag{4.23a}$$

Clearly, NHL is independent of the chosen unit and satisfies the following condition:

$$0 \le NHL(E) \le 1 \tag{4.23b}$$

4.3.3 Evidence Nonspecificity

The theory of evidence was introduced in Chapter 3. It was demonstrated to classify an element to a family of subsets. The nonspecificity in evidence can be constructed by extending the Hartley measure to each subset, and computing a weighted sum of all the resulting measures of the subsets using the basic assignment as weight factors. The nonspecificity measure (H_e) can therefore be defined for a basic assignment m for a family of subsets, $A_1, A_2, \ldots, A_n \in P_X$, according to the theory of evidence as follows:

$$H_e(m) = \sum_{i=1}^{n} m(A_i)\log_2(|A_i|) \tag{4.24}$$

where m is the evidence body defined in this case as

$$m = \{m(A_1), m(A_2), \ldots, m(A_n)\} \tag{4.25a}$$

or for short

$$m = \{m_1, m_2, \ldots, m_n\} \tag{4.25b}$$

where

$$m_i = m(A_i) \quad \text{for } i = 1, 2, \ldots, n \tag{4.26}$$

Equation 4.24 provides an assessment of the *nonspecificity in evidence*. The nonspecificity in evidence results from associating the basic assignment values to subsets that each can contain more than one element. This uncertainty can be eliminated by making the assignments m to singletons, i.e., individual elements of X.

Example 4.2 Causes of a Bridge Failure during Construction

Example 3.1 dealt with modeling the causes of a bridge failure during construction using the theory of evidence. Three common causes — design error (D), construction error (C), and human error (H) — were considered and used in Table 3.1 for defining an assignment based on the opinion of two experts. Considering the assignments in

TABLE 4.2
Nonspecificity in Evidence Computations

Subset (Failure Cause)	Notation A_i	Expert 1 $m_1(A_i)$	Expert 1 $\lvert A_i \rvert$	Expert 1 $m(A_i)\log_2(\lvert A_i \rvert)$	Expert 2 $m_2(A_i)$	Expert 2 $\lvert A_i \rvert$	Expert 2 $m(A_i)\log_2(\lvert A_i \rvert)$
Design error (D)	A_1	0.1	1	0	0.05	1	0
Construction error (C)	A_2	0.05	1	0	0.1	1	0
Human error (H)	A_3	0.1	1	0	0.15	1	0
$D \cup C$	A_4	0.2	2	0.2	0.25	2	0.25
$D \cup H$	A_5	0.1	2	0.1	0.1	2	0.1
$C \cup H$	A_6	0.05	2	0.05	0.1	2	0.1
$D \cup C \cup H$	A_7	0.4	3	0.633985	0.25	3	0.396240625

Table 3.1 by the two experts, Equation 4.24 can be evaluated as shown in Table 4.2. Therefore, the nonspecificity in evidence for the first expert can be computed as

$$H_e(m_1) = \sum_{i=1}^{n} m(A_i)\log_2(|A_i|)$$

$$= \frac{\ln(1)}{\ln(2)}(0.1+0.05+0.1) + \frac{\ln(2)}{\ln(2)}(0.2+0.1+0.05) + \frac{\ln(3)}{\ln(2)}(0.4) \quad (4.27)$$

$$= 0.983985$$

Similarly, for the second expert the nonspecificity in evidence is

$$H_e(m_2) = 0.846240625 \quad (4.28)$$

Based on these results, expert 1 in this case is providing more nonspecificity in evidence than expert 2. The two evidence bodies can be combined as discussed in Chapter 2 using a combination rule, and the $H_e(m_{12})$ for the combined evidence can be computed.

4.3.4 Nonspecificity of Graded Possibility

A nonspecificity uncertainty that is associated with more than one subset in the family of sets can be defined within the framework of possibility theory of Section 3.6. If all focal elements are nested, then it is convenient to replace function H with a special function, H_p. Assuming that $X = \{x_1, x_2, \ldots, x_n\}$ and ordering $r(x_i) \geq r(x_{i+1})$ of possibility values for all $i = 1, 2, \ldots, n$, the possibilistic measure of nonspecificity H_p is defined by the formula

$$H_p(r) = \sum_{i=2}^{n} [(r(x_i) - r(x_{i+1})) \log_2(i)] \tag{4.29}$$

where $r(x_{n+1}) = 0$ by convention. $H_p(r)$ is also referred to as the *U-uncertainty.*

4.3.5 Nonspecificity of Fuzzy Sets or U-Uncertainty

Higashi and Klir (1983) provided the following nonspecificity measure, also called *U-uncertainty*, for a normal, finite fuzzy set (A):

$$U(A) = \int_0^1 \log_2(|{}^{\alpha}A|)d\alpha \tag{4.30a}$$

where $|{}^{\alpha}A|$ = the cardinality of the α-cut of A for finite sets. Formally, $U(A)$ defined by Equation 4.30a is equivalent to $H_p(r)$ defined by Equation 4.29 under the fuzzy set interpretation of possibility theory.

For fuzzy intervals or numbers on the real line, the measure is

$$HL(A) = \int_0^1 \log_2(1 + \mu({}^{\alpha}A))d\alpha \tag{4.30b}$$

where $\mu({}^{\alpha}A)$is the Lebesgue measure of ${}^{\alpha}A$.

Example 4.3 Fuzzy-Based Nonspecificity for a Symmetric Fuzzy Number

The triangular fuzzy number $A[a_L, a_m, a_R]$, as represented mathematically in Equation 2.82,

$$A(x) = \begin{cases} \dfrac{x - a_L}{a_m - a_L} & \text{for } a_L \le x \le a_m \\ \dfrac{x - a_R}{a_m - a_R} & \text{for } a_m \le x \le a_R \\ 0 & \text{otherwise} \end{cases} \tag{4.31a}$$

can be simplified for a symmetric triangular fuzzy number for the following case:

$$a_m = \frac{a_L + a_R}{2} \tag{4.31b}$$

The α-cut for a symmetric triangular fuzzy number can be expressed as

$$ {}^{\alpha}A = \left[a_L + \alpha\left(\frac{a_L + a_R}{2} - a_L\right), a_R - \alpha\left(a_R - \frac{a_L + a_R}{2}\right)\right] \quad \text{for } \alpha \in [0,1] \quad (4.32a) $$

or

$$ {}^{\alpha}A = \left[a_L + \alpha\left(\frac{a_R - a_L}{2}\right), a_R - \alpha\left(a_R - \frac{a_R - a_L}{2}\right)\right] \quad \text{for } \alpha \in [0,1] \quad (4.32b) $$

The *HL*-uncertainty can be evaluated according to Equation 4.30b as follows:

$$ HL(A) = \frac{1}{\ln(2)} \int_0^1 \ln(1 + \mu({}^{\alpha}A))d\alpha \quad (4.33) $$

where $\mu({}^{\alpha}A)$ is the Lebesgue measure of ${}^{\alpha}A$. The Lebesgue measure of ${}^{\alpha}A$ is the length of the interval as follows:

$$ \mu({}^{\alpha}A) = (a_R - a_L) - 2\left(\frac{a_R - a_L}{2}\right)\alpha = (1-\alpha)(a_R - a_L) \quad (4.34) $$

Substituting Equation 4.32 into Equation 4.33 produces

$$ \begin{aligned} HL(A) &= \frac{1}{\ln(2)} \int_0^1 \ln(1 + (1-\alpha)(a_R - a_L))d\alpha \\ &= \frac{1}{(a_R - a_L)\ln(2)}([1 + (a_R - a_L)]\ln[1 + (a_R - a_L)] - (a_R - a_L)) \end{aligned} \quad (4.35) $$

For nonsymmetric triangular fuzzy numbers, the same Equation 4.35 applies since

$$ \begin{aligned} {}^{\alpha}A &= [a_L + \alpha(a_m - a_L), a_R - \alpha(a_R - a_m)] \\ \mu({}^{\alpha}A) &= (a_R - a_L) - \alpha(a_m - a_L) - \alpha(a_R - a_m) \\ &= (a_R - a_L) - \alpha a_m + \alpha a_L - \alpha a_R + \alpha a_m \\ &= (a_R - a_L) + \alpha a_L - \alpha a_R \\ &= (a_R - a_L) + \alpha(a_L - a_R) \\ &= (1-\alpha)(a_R - a_L) \end{aligned} \quad (4.36) $$

Equation 4.36 is the same as Equation 4.34.

For a trapezoidal fuzzy number $A = [a_L, a_{mL}, a_{mR}, a_R]$, the α-cut of A is

$$^{\alpha}A = [a_L + \alpha(a_{mL} - a_L), a_R - \alpha(a_R - a_{mR})] \tag{4.37}$$

The corresponding cardinality is

$$\mu(^{\alpha}A) = (a_R - a_L) - \alpha(a_R - a_L - (a_{mR} - a_{mL})) \tag{4.38}$$

The HL-uncertainty can be evaluated according to Equation 4.30b as follows:

$$HL(A) = \frac{1}{\ln(2)} \int_0^1 \ln(1 + \mu(^{\alpha}A))d\alpha \tag{4.39}$$

or

$$HL(A) = \frac{1}{\ln(2)} \int_0^1 (1 + (a_R - a_L) - \alpha(a_R - a_L - (a_{mR} - a_{mL}))) \tag{4.40}$$

The following expression can be obtained:

$$\begin{aligned} HL(A) = \frac{1}{\ln(2)} \frac{1}{(a_R - a_L) - (a_{mR} - a_{mL})} \times \\ ([1 + (a_R - a_L)]\ln[1 + (a_R - a_L)] - [1 + (a_{mR} - a_{mL})] \times \\ \ln[1 + (a_{mR} - a_{mL})] - [(a_R - a_L) - (a_{mR} - a_{mL})]) \end{aligned} \tag{4.41}$$

It can be shown that Equation 4.41 reduced to Equation 4.36 by substituting $a_{mL} = a_{mR} = a_m$.

4.4 ENTROPY-LIKE MEASURES

Probability mass functions are used to provide likelihood measures associated with all possible outcomes. Assuming that there are n possible outcomes, the probability assignment (P) can be expressed as p_i, $i = 1, 2, \ldots, n$. The uncertainty in this case has two aspects: (1) nonspecificity due to the existence of more than one possible outcome and (2) conflict as described by the probability distribution provided by the probability mass function. The Hartley measure as provided in the previous section is well suited for the former aspect, but it does not cover the latter. The

Shannon entropy was developed to measure the conflict uncertainty associated with a probability assignment for finite sets (Shannon, 1948). The Shannon entropy was extended to measure uncertainty based on basic assignments in evidence theory. The basic Shannon entropy and its extensions to evidence theory are discussed in this section.

4.4.1 Shannon Entropy for Probability Distributions

Shannon (1948) provided an uncertainty measure for *conflict* that arises from a probability mass function. This measure is commonly known as the *entropy measure*, or the *Shannon entropy measure*. The entropy measure, $S(p)$, for a probability distribution of a random variable with discrete values defined over $x \in X$ is given by

$$S(p) = -\sum_{i=1}^{n} p_i \log_2(p_i) \tag{4.42}$$

This entropy measure takes on values larger than 0. Its value is zero if $p_i = 1$ for exactly an $i \in \{1, 2, \ldots, n\}$, and is maximum for equally likely outcomes of $p_i = 1/n$ for all i. Equation 4.42 can be expressed as

$$S(p(x)\,|\,x \in X) = -\sum_{x \in X} p(x) \log_2\left(1 - \sum_{y \neq x} p(y)\right) \tag{4.43}$$

The term $\sum_{y \neq x} p(y)$ in this equation expresses the total evidential claim pertaining to all alternatives that are different from alternative x. This evidential claim fully conflicts with the evidential claim $p(x)$. Therefore, Equation 4.43 measures conflict between

$$p(x) \text{ and } \sum_{y \neq x} p(y) \tag{4.44}$$

where $\sum_{y \neq x} p(y)$ in Equation 4.44 is expressed via its monotone transformation $-\log_2\left[1 - \sum_{y \neq x} p(y)\right]$.

Equation 4.43 measures the mean (expected) value of the conflict among the evidential claims expressed by a probability distribution function for a finite set of

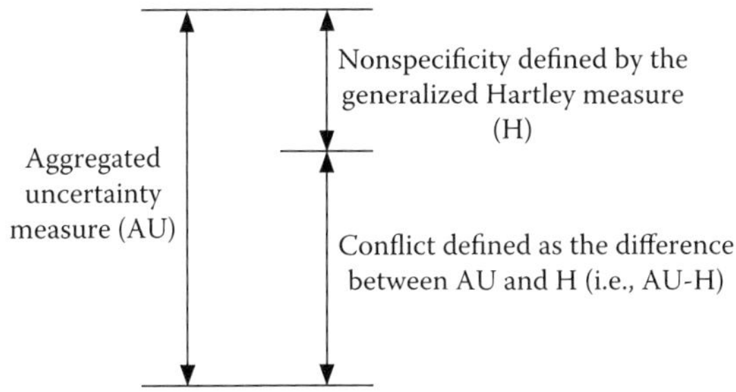

FIGURE 4.2 Relationships among aggregated uncertainty, nonspecificity, and conflict.

mutually exclusive alternatives. This type of uncertainty whose amount is measured by the Shannon entropy is thus conflict.

Although the Shannon entropy for the equally likely case produces the same numerical value as the Hartley measure according to Equation 4.7, the two measures are fundamentally different in that the Hartley measure provides an estimate of the nonspecificity type of uncertainty and the Shannon entropy provides an estimate of the conflict type of uncertainty. The relationship between the Hartley measure and the measure of conflict is illustrated in Figure 4.2. The total of the two types is called the aggregated uncertainty, which is discussed in Section 4.4.4.

It is obvious that the Shannon entropy is applicable only to finite sets of alternatives. At first sight, it seems suggestive to extend it to probability density functions, f, on R (or, more generally, on R^n, $n \geq 1$), by replacing in Equation 4.42 p with f and the summation with integration. This entropy is called the Boltzmann (1894) entropy as given by (Harr, 1987)

$$B(f) = -\int_a^b f_X(x)\log_2(f_X(x))dx \tag{4.45}$$

where a = lower limit, b = upper limit, and f_X = probability density function on X. However, there are several reasons why the resulting functional does not qualify as a measure of uncertainty: (1) it may be negative, (2) it may be infinitely large, (3) it depends on the chosen coordinate system, and most importantly, (4) the limit of the sequence of its increasingly more refined discrete approximations diverges (Klir and Wierman, 1999). These problems can be overcome by the following modified functional:

$$B(f(x), f'(x)|x \in R) = \int_R f_X(x)\log_2\left(\frac{f_X(x)}{f'_X(x)}\right)dx \tag{4.46}$$

which involves two probability density functions, f and f'. Uncertainty is measured by B in relative rather than absolute terms.

TABLE 4.3
Probability Mass Function in Example 4.4

Element (x_i)	Probability (case a)	Equally Likely Probabilities (case b)
x_1	0.2	1/3
x_2	0.4	1/3
x_3	0.4	1/3

When f in Equation 4.46 is a joint probability density function on R^2 and f' is the product of the two marginal distributions of f, we obtain the information transmission:

$$B(f_{XY}(x,y), f_X(x), f_Y(y) \big| x \in R, y \in R) = \int_R \int_R f_{XY}(x,y) \log_2 \left(\frac{f_{XY}(x,y)}{f_X(x) f_Y(y)} \right) dxdy \tag{4.47}$$

Example 4.4 Computational Aspects of Shannon Entropy

For the probability mass function provided in Table 4.3 (case a), the Shannon entropy can be computed as follows:

$$\begin{aligned} S(P) &= -\frac{1}{\ln(2)} \sum_{i=1}^{n} p_i \ln(p_i) \\ &= -\frac{1}{\ln(2)} (0.2 \ln(0.2) + 2(0.4) \ln(0.4)) \\ &= 1.5219293 \end{aligned} \tag{4.48}$$

The Shannon entropy becomes largest for equally likely elements (i.e., case b of Table 4.3). For this case, the maximum value is

$$\begin{aligned} S(P) &= -\frac{1}{\ln(2)} \sum_{i=1}^{n} p_i \ln(p_i) \\ &= -\frac{1}{\ln(2)} \left(3 \left(\frac{1}{3} \right) \ln \left(\frac{1}{3} \right) \right) \\ &= 1.584963 \end{aligned} \tag{4.49}$$

The results in Equation 4.49 match the value in Table 4.1 for the Hartley measure using $n = 3$, although they measure totally different types of uncertainty.

4.4.2 Discrepancy Measure

This measure is used in expert opinion elicitation (Ayyub, 2001). An expert can be used to estimate a probability distribution function (P) expressed as p_i, $i = 1, 2, \ldots, n$. This function is an estimate of a true, yet unknown, probability distribution function (S) expressed as s_i, $i = 1, 2, \ldots, n$. The *discrepancy* measure is between the true and provided probability values as given by

$$S_D(S,P) = -\sum_{i=1}^{n} s_i \log_2\left(\frac{s_i}{p_i}\right) \tag{4.50}$$

This *discrepancy measure* (S_D) is based on the Shannon entropy measure. It can be used to obtain assessments of opinions obtained from a set of experts with equal circumstances and conditions, although equal circumstances and conditions might not be attainable. It provides an assessment of the degree of surprise that someone would experience, if an estimate p_i, $i = 1, 2, \ldots, n$, is obtained, whereas the real values are s_i, $i = 1, 2, \ldots, n$ (Cooke, 1991).

4.4.3 Entropy Measures for Evidence Theory

4.4.3.1 Measure of Dissonance

Dissonance is a state of contradiction between claims, beliefs, or interests (Yager, 1983). The measure of dissonance, D, can be defined based on evidence theory as follows:

$$D(m) = -\sum_{i=1}^{n} m(A_i)\log_2(Pl(A_i)) \tag{4.51}$$

where $m(A_i) > 0$; $\{A_1, A_2, \ldots, A_n\}$ = a family set of subsets, i.e., focal points, that contains some or all elements of the universal set X; $m(A_i)$ = a basic assignment that is interpreted either as the degree of evidence supporting the claim that a specific element belongs to the subset A_i but not to any special subset of A_i, or as the degree of belief that such a claim is warranted; $\sum_{i=1}^{n} m(A_i) = 1$; and $Pl(A_i)$ = plausibility measure, which represents the total evidence or belief that the element of concern belongs to the set A_i or to any other sets that intersect with A_i, as provided by Equation 3.29.

4.4.3.2 Measure of Confusion

The measure of confusion characterizes the multitude of subsets supported by evidence as well as the uniformity of the distribution of strength of evidence among the subsets. The greater the number of subsets involved and the more uniform the distribution, the more confusing the presentation of evidence (Klir and Folger, 1988). The measure of confusion, C, is defined as

$$C(m) = -\sum_{i=1}^{n} m(A_i)\log_2(Bel(A_i)) \tag{4.52}$$

where $Bel(A_i)$ = belief measure, which represents the total evidence or belief that the element of concern belongs to the subset A_i as well as to the various special subsets of A_i, as provided by Equation 3.28.

EXAMPLE 4.5 UNCERTAINTY MEASURES ASSOCIATED WITH EXPERT OPINIONS

The assignment provided in Example 4.2 by expert 1 is used to demonstrate the computations of the measures of dissonance. Table 4.4 summarizes the computations. The measure of dissonance, D, can be computed as follows based on the values in Table 4.4:

$$\begin{aligned} D(m) &= -\sum_{i=1}^{n} m(A_i)\log_2(Pl(A_i)) \\ &= 0.032192809 + 0.025728659 + \ldots + 0 \\ &= 0.165471137 \end{aligned} \tag{4.53}$$

Similar computations can be made for the measure of confusion as follows based on the values in Table 4.5:

$$\begin{aligned} C(m) &= -\sum_{i=1}^{n} m(A_i)\log_2(Bel(A_i)) \\ &= 0.332192809 + 0.216096405 + \ldots + 0 \\ &= 1.473189622 \end{aligned} \tag{4.54}$$

TABLE 4.4
Dissonance Computations for Example 4.5

Subset (Failure Cause)	Notation				
	A_i	$m(A_i)$	Pl	$\log_2(Pl(A_i))$	$m(A_i)\log_2(Pl(A_i))$
Design error (D)	A_1	0.1	0.8	–0.32193	–0.032192809
Construction error (C)	A_2	0.05	0.7	–0.51457	–0.025728659
Human error (H)	A_3	0.1	0.65	–0.62149	–0.062148838
$D \cap C$	A_4	0.2	0.9	–0.152	–0.030400619
$D \cap H$	A_5	0.1	0.95	–0.074	–0.007400058
$C \cap H$	A_6	0.05	0.9	–0.152	–0.007600155
$D \cap C \cap H$	A_7	0.4	1	0	0

TABLE 4.5
Confusion Computations for Example 4.5

Subset (Failure Cause)	Notation				
	A_i	$m(A_i)$	Bel	$\log_2(Bel(A_i))$	$m(A_i)\log_2(Bel(A_i))$
Design error (D)	A_1	0.1	0.1	–3.32193	–0.332192809
Construction error (C)	A_2	0.05	0.05	–4.32193	–0.216096405
Human error (H)	A_3	0.1	0.1	–3.32193	–0.332192809
$D \cap C$	A_4	0.2	0.35	–1.51457	–0.302914635
$D \cap H$	A_5	0.1	0.3	–1.73697	–0.173696559
$C \cap H$	A_6	0.05	0.2	–2.32193	–0.116096405
$D \cap C \cap H$	A_7	0.4	1	0	0

4.4.4 Aggregate and Disaggregate Uncertainty in Evidence Theory

An aggregate uncertainty (AU) in evidence theory measures the combined nonspecificity and conflict provided by a given body of evidence. The function AU is defined as a mapping from the set of all belief measures (B) to the nonnegative real line (R_+) as follows:

$$AU : B \rightarrow R_+ \tag{4.55}$$

The measure is given by (Klir and Wierman, 1999)

$$AU(Bel) = \max_{P_{Bel}} \left[-\sum_{x \in X} p_x \log_2(p_x) \right] \tag{4.56}$$

where P_{Bel} is the set of all probability distributions (p_x) defined over X that satisfy the following two constraints:

$$p_x \in [0,1] \text{ for all } x \in X \text{ and } \sum_{x \in X} p_x = 1 \tag{4.57a}$$

$$Bel(A) \leq \sum_{x \in A} p_x \leq 1 - Bel(\bar{A}) \text{ for all } A \subseteq X \tag{4.57b}$$

The evaluation of Equation 4.56 can be performed using an iterative process to obtain the maximum value over the set of all probability distributions. The iterative process is summarized in the following seven steps, and illustrated using the example at the end of the section (Harmanec et al., 1996):

Input: A frame of discernment *X*, and a belief function *Bel* on *X*.
Output: *AU*(*Bel*), and p_x for $x \in X$, that satisfies Equation 4.57a and b.

Step 1. Find a nonempty set $A \subseteq X$ such that $Bel(A)/|A|$ is maximal. If more than one set have maximal $Bel(A)/|A|$, use the set with the maximum cardinality.

Step 2. For $x \in A$, define $p_x = Bel(A)/|A|$.

Step 3. For each $B \subseteq X - A$, compute $Bel(B) = Bel(B \cup A) - Bel(A)$.

Step 4. Define a new *X* as the set difference between the current *X* and *A*, i.e., $X - A$.

Step 5. If *X* from step 4 is nonempty and $Bel(X) > 0$, then go to step 1.

Step 6. If $Bel(X) = 0$ and $X \neq \emptyset$, then define $p_x = 0$ for all $x \in X$.

Step 7. Calculate $AU(Bel) = -\sum_{x \in X} p_x \log_2(p_x)$.

The concept of *uncertainty disaggregation* can be used to determine the components of *AU* as provided in Figure 4.2. Once *AU* is computed as provided in Equation 4.56 and the Hartley measure (*H*) is computed, the measure of conflict can be evaluated as the difference *AU* – *H*. It should be noted that *AU* is not sensitive to changes to *Bel* as used in Equation 4.56. The lack of sensitivity is attributed to the reallocation of this *AU* uncertainty to its two components of nonspecificity and conflict. Klir (2006) provides additional information on disaggregation of uncertainty.

Example 4.6 Aggregate Uncertainty for a Particular Assignment

For a universal set $X = \{a, b, c, d\}$ with the following assignment *m*

$$
\begin{aligned}
m(\{a\}) &= 0.26\\
m(\{b\}) &= 0.26\\
m(\{c\}) &= 0.26\\
m(\{a, b\}) &= 0.07\\
m(\{a, c\}) &= 0.01\\
m(\{a, d\}) &= 0.01
\end{aligned}
$$

$$m(\{b, c\}) = 0.01$$
$$m(\{b, d\}) = 0.01$$
$$m(\{c, d\}) = 0.01$$
$$m(\{a, b, c, d\}) = 0.10$$

The belief values (*Bel*) of a particular set (*A*), and estimates of the probability of singletons comprising *A* as the belief divided by the cardinality for the set (*A*) can be computed as follows:

A	*Bel*(*A*)	$\frac{Bel(A)}{\lvert A \rvert}$
$\{a\}$	0.26	0.26
$\{b\}$	0.26	0.26
$\{c\}$	0.26	0.26
$\{a, b\}$	0.59	0.295
$\{a, c\}$	0.53	0.265
$\{a, d\}$	0.27	0.135
$\{b, c\}$	0.53	0.265
$\{b, d\}$	0.27	0.135
$\{c, d\}$	0.27	0.135
$\{a, b, c\}$	0.87	0.29
$\{a, b, d\}$	0.61	0.203
$\{a, c, d\}$	0.55	0.183
$\{b, c, d\}$	0.55	0.183
$\{a, b, c, d\}$	1	0.25

The maximum $\frac{Bel(A)}{\lvert A \rvert}$ values occur at $A = \{a, b\}$. Therefore, the following probabilities can be assigned:

$$p_a = 0.295 \text{ and } p_b = 0.295$$

These probabilities can be used to update the belief values as follows:

A	*Bel*(*A*)	$\frac{Bel(A)}{\lvert A \rvert}$
$\{c\}$	$Bel(\{a, b, c\}) - Bel(\{a, b\}) = 0.87 - 0.59 = 0.28$	0.28
$\{d\}$	$Bel(\{a, b, d\}) - Bel(\{a, b\}) = 0.61 - 0.59 = 0.02$	0.02
$\{c, d\}$	$Bel(\{a, b, c, d\}) - Bel(\{a, b\}) = 1 - 0.59 = 0.41$	0.205

The maximum $\frac{Bel(A)}{|A|}$ values occur at $A = \{c\}$. Therefore, the following probability can be assigned:

$$p_c = 0.28$$

Using the value $p_c = 0.28$, the $Bel(\{d\})$ can be computed as follows:

$$Bel(\{d\}) = Bel(\{a,b,c,d\}) - Bel(\{a,b\}) - Bel(\{d\}) = 1 - 0.59 - 0.28 = 0.13$$

Evaluating Equation 4.56 produces the following aggregate uncertainty:

$$\begin{aligned} AU(Bel) &= -\sum_{i\in\{a,b,c,d\}} p_i \log_2(p_i) \\ &= -2(0.295)\log_2(0.295) - (0.28)\log_2(0.28) - (0.13)\log_2(0,13) \quad (4.58) \\ &= 1.93598 \end{aligned}$$

4.5 FUZZINESS MEASURE

Fuzziness as represented by fuzzy set theory results from uncertainty in belonging to a set. Fuzzy sets are sets that have imprecise boundaries. For a given fuzzy set A, each element x of the universal set X has a membership value $A(x)$ to A, which may also be viewed as a measure of compatibility between x and the concept represented by A. The membership value is in the range [0, 1], as described in Section 2.4. Each fuzzy set has a unique property of having nonempty intersection with its complement. Yager (1979, 1980b) employed this property to define the following measure of fuzziness:

$$f(A) = |X| - \sum_{x\in X} |A(x) - \bar{A}(x)| \tag{4.59}$$

where f = fuzziness measure of a fuzzy, finite set A; X = universal set; $\bar{A}$ = complement of A; and $A(x)$ membership value of x to A. Fuzziness is measured here by the lack of distinction between a fuzzy set and its complement. Using the definition of the standard complement provided by Equation 2.54, Equation 4.59 can be written as

$$f(A) = |X| - \sum_{x\in X} |2A(x) - 1| = \sum_{x\in X} (1 - |2A(x) - 1|) \tag{4.60a}$$

The fuzziness measure becomes zero as the set becomes crisp with $A(x)$, taking only values of zeros and ones, and reaches maximum at $A(x) = 0.5$ for all $x \in X$. For uncountable sets, the measure is

$$f(A) = \int_X (1 - |2A(x) - 1|)dx \tag{4.60b}$$

Example 4.7 Fuzziness of a Symmetric Fuzzy Number

The symmetric triangular fuzzy number $A = [a_L, a_m, a_R]$, with $X = [a_L, a_R]$, is provided in Equation 4.31a as

$$A(x) = \begin{cases} \dfrac{x - a_L}{a_m - a_L} & \text{for } a_L \le x \le a_m \\ \dfrac{x - a_R}{a_m - a_R} & \text{for } a_m \le x \le a_R \\ 0 & \text{otherwise} \end{cases}$$

where

$$a_m = \frac{a_L + a_R}{2}$$

The measure of fuzziness according to Equation 4.60b is

$$f(A) = \int_X (1 - |2A(x) - 1|)dx \tag{4.61}$$

Substituting Equation 4.31a into Equation 4.61 produces

$$f(A) = \int_{a_L}^{a_m} \left(1 - \left|2\frac{x - a_L}{a_m - a_L} - 1\right|\right)dx + \int_{a_m}^{a_R} \left(1 - \left|2\frac{x - a_R}{a_m - a_R} - 1\right|\right)dx \tag{4.62}$$

The expression provided by Equation 4.62 is applicable to both symmetric and non-symmetric fuzzy numbers.

For a trapezoidal fuzzy number $A = [a_L, a_{mL}, a_{mR}, a_R]$, the measure is

$$
\begin{aligned}
f(A) &= \int_X (1 - |2A(x) - 1|)dx \\
&= \int_{a_L}^{a_{mL}} \left(1 - \left|2\frac{x - a_L}{a_{mL} - a_L} - 1\right|\right)dx + \int_{a_{mL}}^{a_{mR}} (1 - |2(1) - 1|)dx + \\
&\quad \int_{a_{mR}}^{a_R} \left(1 - \left|2\frac{x - a_R}{a_{mR} - a_R}\right| - 1\right)dx \\
&= \int_{a_L}^{a_{mL}} \left(1 - \left|2\frac{x - a_L}{a_{mL} - a_L} - 1\right|\right)dx + \int_{a_{mR}}^{a_R} \left(1 - \left|2\frac{x - a_R}{a_{mR} - a_R}\right| - 1\right)dx
\end{aligned}
\tag{4.63}
$$

4.6 APPLICATION: COMBINING EXPERT OPINIONS

In some applications, expert opinions in the form of subjective probabilities of an event need to be combined into a single value, and perhaps confidence intervals for their use in probabilistic and risk analyses. Cooke (1991) and Rowe (1992) provided a summary of methods for combining expert opinions. The methods can be classified into consensus methods and mathematical methods (Clemen, 1989; Ferrell, 1985). The mathematical methods can be based on assigning equal weights to the experts or different weights. This section provides a summary of classical methods for combining expert opinions.

4.6.1 Consensus Combination of Opinions

A consensus combination of opinions is arrived at through a facilitated discussion among the experts to some agreeable common values, with perhaps a confidence interval or outer quartile values. The primary shortcomings of this method are (1) socially reinforced irrelevance or conformity within a group, (2) dominance of strong-minded or strident individuals, (3) group motive of quickly reaching an agreement, and (4) group-reinforced bias due to common background of group members. The facilitator of an expert opinion elicitation session should play a major role in reducing group pressure, individual dominance, and biases.

4.6.2 Percentiles for Combining Opinions

A p-percentile value (x_p) for a random variable based on a sample is defined as the value of the parameter such that $p\%$ of the data is less than or equal to x_p. On the basis of this definition, the median value is considered to be the 50th percentile value. Aggregating the opinions of experts can be based on computing the 25th, 50th, and 75th percentile values of the gathered opinions. The computation of these values depends on the number of experts providing opinions. Table 3.6 provides a summary of the needed equations for 4 to 20 experts. For example, seven experts provided the following subjective probabilities of an event, which are sorted in decreasing order:

$$\text{Probabilities} = \{1.0\text{E-}02, 5.0\text{E-}03, 5.0\text{E-}03, 1.0\text{E-}03, 1.0\text{E-}03, 5.0\text{E-}04, 1.0\text{E-}04\} \quad (4.64)$$

The median and arithmetic quartile points according to Table 3.6 are respectively given by

$$\text{25th percentile} = 5.0\text{E-}03 \quad (4.65a)$$

$$\text{50th percentile (median)} = 1.0\text{E-}03 \quad (4.65b)$$

$$\text{75th percentile} = 7.5\text{E-}04 \quad (4.65c)$$

4.6.3 Weighted Combinations of Opinions

French (1985) and Genest and Zidek (1986) provided summaries of various methods for combining probabilities and example uses. For E experts with the ith expert providing a vector of n probability values, $p_{1i}, p_{2i}, \ldots, p_{ni}$, for sample space outcomes $A_1, A_2, \ldots, A_n$, the E expert opinions can be combined using weight factors $w_1, w_2, \ldots, w_E$, that sum up to 1, using one of the following selected methods:

1. *Weighted arithmetic average*: The weighted arithmetic mean for outcome j can be computed as follows:

$$\text{Weighted arithmetic mean for outcome } j = M_1(j) = \sum_{i=1}^{E} w_i p_{ji} \quad (4.66)$$

   The weighted arithmetic means are then normalized using their total to obtain the 1-norm probability for outcome for each outcome as follows:

$$\text{1-norm probability for outcome } j = P_1(j) = \frac{M_1(j)}{\sum_{k=1}^{n} M_1(k)} \quad (4.67)$$

2. *Weighted geometric average*: The weighted geometric mean for outcome j can be computed as follows:

$$\text{Weighted geometric mean for outcome } j = M_0(j) = \prod_{i=1}^{E} (p_{ji})^{w_i} \quad (4.68)$$

The weighted geometric means are then normalized using their total to obtain the 0-norm probability for outcome for each outcome as follows:

$$\text{0-norm probability for outcome } j = P_0(j) = \frac{M_0(j)}{\sum_{k=1}^{n} M_0(k)} \tag{4.69}$$

3. *Weighted harmonic average*: The weighted harmonic mean for outcome j can be computed as follows:

$$\text{Weighted harmonic mean for outcome } j = M_{-1}(j) = \frac{1}{\sum_{i=1}^{E} \frac{w_i}{p_{ji}}} \tag{4.70}$$

The weighted harmonic means are then normalized using their total to obtain the –1-norm probability for outcome for each outcome as follows:

$$-1\text{-norm probability for outcome } j = P_{-1}(j) = \frac{M_{-1}(j)}{\sum_{k=1}^{n} M_{-1}(k)} \tag{4.71}$$

4. *Maximum value*: The maximum value for outcome j can be computed as follows:

$$\text{Maximum value for outcome } j = M_{\infty}(j) = \max_{i=1}^{E}(p_{ji}) \tag{4.72}$$

The maximum values are then normalized using their total to obtain the ∞-norm probability for outcome for each outcome as follows:

$$\infty\text{-norm probability for outcome } j = P_{\infty}(j) = \frac{M_{\infty}(j)}{\sum_{k=1}^{n} M_{\infty}(k)} \tag{4.73}$$

5. *Minimum value*: The minimum value for outcome j can be computed as follows:

$$\text{Minimum value for outcome } j = M_{-\infty}(j) = \min_{i=1}^{E}(p_{ji}) \tag{4.74}$$

The minimum values are then normalized using their total to obtain the $-\infty$-norm probability for outcome for each outcome as follows:

$$-\infty\text{-norm probability for outcome } j = P_{-\infty}(j) = \frac{M_{-\infty}(j)}{\sum_{k=1}^{n} M_{-\infty}(k)} \tag{4.75}$$

6. *Generalized weighted average*: The generalized weighted average for outcome j can be computed as follows:

$$\text{Generalized weighted average for outcome } j = M_r(j) = \left(\sum_{i=1}^{E} w_i p_{ji}^r \right)^{1/r} \tag{4.76}$$

The generalized weighted averages are then normalized using their total to obtain the r-norm probability for outcome for each outcome as follows:

$$r\text{-norm probability for outcome } j = P_r(j) = \frac{M_r(j)}{\sum_{k=1}^{n} M_r(k)} \tag{4.77}$$

where for $r = 1, 0, -1, \infty$, and $-\infty$, cases 1 to 5 result, respectively.

Example 4.8 Aggregation of Expert Opinions

Five experts provided the following occurrence probabilities for an event E:

$$P(E) = [0.001, 0.01, 0.002, 0.008, 0.005] \tag{4.78}$$

The experts were assigned the following weight factors based on their abilities as perceived by an analyst:

$$\text{Weight factors} = [0.1, 0.3, 0.25, 0.15, 0.2] \tag{4.79}$$

The weighted, aggregated opinion of the five experts can be computed using applicable methods of Section 4.6.3 as provided in Table 4.6. The probability values provided by the experts are spread out over one order of magnitude. Therefore, the geometric or the harmonic means, in this case, provide a better representation of the aggregated means. Since only one value is one order of magnitude higher than the lowest value, the arithmetic mean is giving a likable answer. Maximum and minimum aggregates provide the extreme values and are useful only for cases where the risk associated with any decision made based on the aggregated result is high, and therefore the aggregation process is desired to represent the best or the worst situation, as the case may be.

TABLE 4.6
Aggregation of Expert Opinions for Example 4.8

			Weighted Averages				
			Arithmetic	Geometric	Harmonic	Maximum	Minimum
No.	$p(i)$	$w(i)$	$[w(i)][p(i)]$	$[p(i)]^{w(i)}$	$w(i)/p(i)$	$p(i)$	$p(i)$
1	0.001	0.1	0.0001	0.50119	100.00	0.0010	0.0010
2	0.01	0.3	0.0030	0.25119	30.00	0.0100	0.0100
3	0.002	0.25	0.0005	0.21147	125.00	0.0020	0.0020
4	0.008	0.15	0.0012	0.48469	18.75	0.0080	0.0080
5	0.005	0.2	0.0010	0.34657	40.00	0.0050	0.0050
Aggregated opinion			**0.0058**	**0.00447**	**0.00319**	**0.0100**	**0.0010**

EXERCISE PROBLEMS

4.1. For the following five possible alternatives represented by the x values in the table, compute the Hartley uncertainty measure for these possible alternatives:

x	10	20	30	40	50
$P_X(x)$	0.40	0.30	0.20	0.05	0.05

Now assume that the above probability mass function applies to these five values; compute the Shannon entropy uncertainty measure. Discuss their meanings.

4.2. For the following five possible alternatives represented by the x values in the table, compute the Hartley uncertainty measure for these possible alternatives:

x	1	2	3	4	5
$P_X(x)$	0.10	0.20	0.40	0.20	0.10

Now assume that the above probability mass function applies to these five values; compute the Shannon entropy uncertainty measure. Discuss their meanings.

4.3. For the following assignment, compute the evidence nonspecificity measure for each expert and for the combined opinion:

Subset A_i	Expert 1 $m_1(A_i)$	Expert 2 $m_2(A_i)$
A	0.05	0.1
B	0.05	0.1
C	0.05	0.1
$A \cup B$	0.2	0.3
$A \cup C$	0.2	0.1
$B \cup C$	0.05	0.1
$A \cup B \cup C$	0.4	0.2

4.4. For the following assignment, compute the evidence nonspecificity measure for each expert and for the combined opinion:

Subset A_i	Expert 1 $m_1(A_i)$	Expert 2 $m_2(A_i)$
A	0.1	0.1
B	0.1	0.1
C	0.1	0.1
$A \cup B$	0.3	0.1
$A \cup C$	0.2	0.1
$B \cup C$	0.1	0.1
$A \cup B \cup C$	0.1	0.4

4.5. Redo Example 4.6 using the following assignment:

$$
\begin{aligned}
m(\{a\}) &= 0.25 \\
m(\{b\}) &= 0.25 \\
m(\{c\}) &= 0.25 \\
m(\{a, b\}) &= 0.05 \\
m(\{a, c\}) &= 0.1 \\
m(\{a, d\}) &= 0.05 \\
m(\{b, c\}) &= 0.01 \\
m(\{b, d\}) &= 0.01 \\
m(\{c, d\}) &= 0.01 \\
m(\{a, b, c, d\}) &= 0.02
\end{aligned}
$$

4.6. For the triangular set defined as A = [10, 15, 20] and the trapezoidal set defined as B = [10, 12, 18, 20], compute their fuzziness uncertainty measures.

4.7. For the triangular set defined as A = [10, 15, 20] and the trapezoidal set defined as B = [10, 12, 18, 20], compute the fuzziness uncertainty measures for outcomes of the following:

a. $A + B$

b. $A - B$

c. $A \times B$
d. A/B
e. $A + B$, with the constraint that $a = b$ where $a \in A$ and $b \in B$
f. $A - B$, with the constraint that $a = b$ where $a \in A$ and $b \in B$
g. $A \times B$, with the constraint that $a = b$ where $a \in A$ and $b \in B$
h. A/B, with the constraint that $a = b$ where $a \in A$ and $b \in B$

Compare and discuss your results.

4.8. Redo Problem 4.7 with $A = [1, 2, 3]$ and $B = [2, 3, 4, 5]$.

4.9. Redo Problem 4.3 and compute the measure of dissonance.

4.10. Redo Problem 4.3 and compute the measure of confusion.

4.11. Redo Problem 4.4 and compute the measure of dissonance.

4.12. Redo Problem 4.4 and compute the measure of confusion.

4.13. Compute the 25th percentile, median, and 75th percentile aggregation of the following expert opinions: [0.1, 0.2, 0.1, 0.3, 0.1, 0.2, 0.15].

4.14. Compute the 25th percentile, median, and 75th percentile aggregation of the following expert opinions: [100, 120, 110, 100, 90, 150, 110, 120, 105].

4.15. Five experts provided the following occurrence probabilities for an event E: [0.001, 0.001, 0.002, 0.003, 0.003].

The experts were assigned the following weight factors based on their abilities as perceived by an analyst: [0.2, 0.2, 0.2, 0.2, 0.2].

Compute the weighted, aggregated opinion of the five experts using all applicable methods of Section 4.7.3. Compare the results from the various methods and discuss. Provide recommendations on the use of various methods.

5 Uncertainty-Based Principles and Knowledge Construction

5.1 INTRODUCTION

The uncertainty measures discussed in Chapter 4 can be used for decision making relating to knowledge construction. Knowledge can be constructed based on information and data using synthesis and cognitive abilities. Uncertainty measures offer the means to analyze information and data content and associated uncertainties. For example, expert opinions are propositions that do not necessarily meet the justified true belief (JTB) requirements of knowledge, and hence can contain both useful information and uncertainties. In combining these opinions, we have a vested interest in using a process that utilizes all the information contents provided by the experts and that can account for the various uncertainties in producing a combined opinion. The uncertainties can include nonspecificity, conflict, confusion, vagueness, biases, varying reliability levels of sources, and other types.

Chapter 4 presents methods for measuring uncertainty. These measures deal with the various uncertainty types. In combining expert opinions, we can develop expert opinion aggregation methods that are either information based or uncertainty based. In fact, information and uncertainty can be argued to represent a duality, since information can be considered useful by a cognitive agent if this information results in reducing its uncertainty under prescribed conditions. Therefore, the amount of relevant information gained by the agent is related and can be measured by the amount of uncertainty reduced. This concept of uncertainty-based information was introduced by Klir (1991a). The various uncertainty measures presented in Chapter 4 can deal with various uncertainty types and offer strengths and weaknesses with commensurate complexity and computational demands. The selection of an appropriate uncertainty measure or combinations thereof is problem dependent, and a trade-off decision between the computational effort needed and the return on this effort in the form of a refined decision or a combination of expert opinions needs to be made.

The objective of this chapter is to present methods for constructing knowledge based on uncertainty and information synthesis. Three *uncertainty-based principles* (also called criteria) can be used to combine expert opinion: (1) the principle of minimum uncertainty, (2) the principle of maximum uncertainty, and (3) the principle of uncertainty invariance. These three principles are described in subsequent sections. The principles of minimum and maximum uncertainty were developed and had great utilities in classical information theory, commonly referred to as *the principles of minimum and maximum entropy*. In addition, the assumption of closed world is examined and discussed, and open-world modeling methods for knowledge construction are provided.

5.2 CONSTRUCTION OF KNOWLEDGE

Decision situations commonly require constructing knowledge from information. Knowledge construction starts with data collection and information gathering that can include various sources and formats as identified in Figure 1.14. Ignorance types need to be identified, and their levels should be assessed in order to quantify and qualify their contribution to modeling the decision situation. Klir (2006) provide analytical methods that can be used for modeling various ignorance types and assessing their magnitudes or levels. Also, uncertainty measures are provided to assess magnitudes or levels of uncertainty. These uncertainty measures can be defined to be nonnegative real numbers and should be inversely proportional to the strength and consistency in evidence as expressed in the theory employed; i.e., the stronger and more consistent the evidence, the smaller the amount of uncertainty. Such uncertainty measures can be constructed to assess collected information, such as opinions rendered by one expert on some issue of interest, or opinions rendered by several experts on the same issue, or collected data and information.

Ignorance models, uncertainty measures, and data collected can be entered into a systematic process to construct knowledge. This knowledge construction process must have a dialectic nature, as schematically depicted in Figure 1.16. Information can be defined as sensed objects, things, places, processes, and information and knowledge communicated by language and multimedia. Information can be viewed as a preprocessed input to our intellect system of cognition, and knowledge acquisition and creation. Information can lead to knowledge through investigation, study, and reflection. However, knowledge and information about the system might not constitute the eventual evolutionary knowledge state about the system as a result of not meeting the justification condition in JTB or the ongoing evolutionary process or both. Knowledge is defined in the context of the humankind, evolution, language and communication methods, and social and economic dialectic processes, and cannot be removed from them. As a result, knowledge would always reflect the imperfect and evolutionary nature of humans that can be attributed to their reliance on their senses for information acquisition; their dialectic processes; and their mind for extrapolation, creativity, reflection, and imagination with associated biases as a result of preconceived notions due to time asymmetry, specialization, and other factors. An important dimension in defining the state of knowledge and truth about a system is nonknowledge or ignorance.

Opinions rendered by experts that are based on information and existing knowledge can be defined as preliminary propositions with claims that are not fully justified or justified with adequate reliability but are not necessarily infallible. Expert opinions are seeds of propositional knowledge that do not meet one or more of the conditions required for the JTB with the reliability theory of knowledge. They are valuable, as they might lead to knowledge expansion, but decisions made based on them sometimes might be risky propositions since their preliminary nature might lead to proving them false by others or in the future.

The relationships among knowledge, information, opinions, and evolutionary epistemology are schematically shown in Figure 1.16. The dialectic processes include communication methods such as languages, visual and audio formats, and other forms. Also, they include economic schools of thought and political and social dialectic processes within peers, groups, colonies, societies, and the world.

Complex decision situations can challenge human ability to construct knowledge from information. Humans as complex, intelligent systems have the ability to anticipate the future and learn and adapt in ways that are not yet fully understood. Engineers and scientists who study or design systems have to deal with complexity more often than ever, hence the interest in the field of complexity. The study of complexity led to developing theories, such as chaos and catastrophe theories. Even if complexity theories do not produce solutions to problems, they can still help us to understand complex systems and perhaps direct experimental studies. Theory and experiment go hand in glove, therefore providing opportunities to make major contributions.

Complexity can be classified into two broad categories: (1) complexity with structure and (2) complexity without structure, as discussed in Chapter 1. The complexity can be dealt with to some extent, but our analytical and cognitive abilities are limited, as also discussed in Chapter 1.

5.3 MINIMUM UNCERTAINTY PRINCIPLE

The *principle of minimum uncertainty* is basically an arbitration basis. It facilitates the selection of meaningful alternatives from solution sets obtained by solving problems in which some of the initial information is inevitably reduced in the solutions to various degrees. According to this principle, we should accept only those solutions in a given solution set for which the loss of the information is as small as possible. This means, in turn, that we should accept only solutions with minimum uncertainty.

Examples of problems for which the principle of minimum uncertainty is applicable are simplification problems and conflict resolution problems of various types, for example, in simplifying a finite-state nondeterministic system by coarsening state sets of its variables. This simplification requires that each state set be partitioned in a meaningful way (e.g., preserving a given order of states) into a given number of subsets. This simplification can usually be accomplished in many different ways. The minimum uncertainty principle allows us to compare the various competing partitions by their amount of relevant uncertainty (predictive, diagnostic, etc.) and consider only those with minimum uncertainty. As another example, let us consider a set of systems, S_1, S_2, …, S_n, that share some variables. These systems may be locally inconsistent in the sense that projections from individual systems into variables they share (e.g., marginal probabilities, marginal bodies of evidence, etc.) are not the same. To resolve the local inconsistencies, we need to replace each system S_i with another system, C_i, such that systems C_1, C_2, …, C_n be locally consistent. This simplification, of course, can be done in many different ways, but the proper way to do that is to minimize the loss of information caused by these replacements. Denoting the relevant uncertainty measure by U, the principle of minimum uncertainty is applicable to deal with this problem by formulating the following optimization problem:

$$\text{Minimize} \sum_{i=1}^{n} [U(C_i) - U(S_i) \tag{5.1}$$

subject to the following three types of constraints:

1. Axioms of the respective theory in which systems S_i and C_i ($i = 1, 2, \ldots, n$) are formulated
2. Equations by which all conditions of local consistency among systems $C_1, C_2, \ldots, C_n$ are defined
3. $U(C_i) \geq U(S_i)$ for all $i = 1, 2, \ldots, n$ to avoid introducing bias

Other examples include conflict resolution problems in cases of gathering evidence in failure classification.

An analogy can be made in engineering and the sciences where there is the need to select among alternative solutions based on information given on a problem for which each solution has a different level of information retention, bias, and error uncertainties. In this analogy, engineers and scientists often encounter a need to fit a curve representing an analytical model to empirical results. This curve fitting commonly involves the computation of model parameters with each set of parameters, leading to various levels of information retention and uncertainty. An optimal solution in this case can be defined as the solution that maximizes information retention, i.e., minimizes uncertainty. This is analogous to the principle of least squares in regression analysis (Ayyub and McCuen, 2003), although it should be noted that the least squares principle is not an information uncertainty principle. Christensen (1985) used maximum and minimum entropy in dealing with curve fitting.

5.4 MAXIMUM UNCERTAINTY PRINCIPLE

The *principle of maximum uncertainty* is essential for any problem that involves *ampliative reasoning* involving the drawing of conclusions that are not entailed in the given premises. In such cases, we should intuitively use all information supported by available evidence, but without unintentionally adding information unsupported by the given evidence. This principle employs the relationship between information and uncertainty by requiring any conclusion resulting from any ampliative inference to maximize the relevant uncertainty within constraints representing given premises. As a result, we fully limit our inference ability by our ignorance when making inferences beyond the premise information domain, and we fully utilize information provided by the premises. This principle therefore provides us with assurances of maximizing our nonreliance on information not contained in the premises.

Let f and $U(f)$ denote, respectively, a relevant uncertainty function (probability or possibility distribution function, basic probability assignment function, etc.) and the associated measure of uncertainty. Then, the principle of maximum uncertainty is operationally formulated in terms of a generic optimization problem of determining f, for which $U(f)$ reaches its maximum under the following constraints:

1. Axioms upon which the uncertainty function f is based
2. Constraints $E_1, E_2, \ldots$, which represent partial information about f (e.g., marginal distributions, lower or upper bounds, values of f for some sets, etc.)

The principle of maximum uncertainty appeals to engineers and scientists since it results in inferences and solutions that do go beyond premises given. For example, whenever we make predictions based on a given scientific model, we employ ampliative reasoning. Similarly, estimating microstates from the knowledge of relevant macrostates and partial knowledge of the microstates (as in image processing and many other problems) requires ampliative reasoning. The problem of the identification of an overall system from some of its subsystems is another example that involves ampliative reasoning, and hence the principle of maximum uncertainty. For example, predictive, scientific models can be viewed as inference models using premises. In system identification, statements on a system or subsystems need to be made based on partial knowledge of the system, hence the need to make sure that our inferences do not go beyond information and premises available to us. In selecting a likelihood distribution, the principle of maximum uncertainty can provide us with the means of complete uncertainty retention and not using information that we do not possess.

Example 5.1 Selection of Distribution Types Based on Selected Constraints Using Uncertainty Measures

The principle of maximum uncertainty can be used to select the distribution type that maximizes uncertainty for given constraints. The entropy uncertainty measure can be used for this purpose. Although the Boltzman measure of Equation 4.45 does not meet all the requirements of a formal uncertainty measure, as discussed in Section 4.4.1, it is used in the example to illustrate the use of the maximum uncertainty principle based on readily available works by Harr (1987). The deficiencies of this measure include lack of invariance with respect to transformation of the coordinate system, taking negative values, and not a limiting case of a finite set. Table 5.1 summarizes distribution types that maximize uncertainty for a selected list of constraints. For example, the constraints $a \le X \le b$ and

$$\int_a^b f_X(x)dx = 1 \tag{5.2}$$

TABLE 5.1
Maximum Entropy Probability Distributions

Constraints	Maximum Entropy Distribution
$\int_a^b f_X(x)dx = 1$ Minimum value = a Maximum value = b	Uniform
$\int_0^\infty f_X(x)dx = 1$ Expected value = μ	Exponential
$\int_{-\infty}^\infty f_X(x)dx = 1$ Expected value = μ Standard deviation = σ^2	Normal
$\int_a^b f_X(x)dx = 1$ Expected value = μ Standard deviation = σ^2 Finite range of a minimum value = a and a maximum value = b	Beta
$\sum_{i=1}^\infty p_i = 1$ where p_i = probability of i independent and identical events occurring in an interval T with an expected rate of occurrence of events of λ	Poisson

can be used to maximize the entropy according to Equation 4.35 as follows:

$$\text{Maximize } B(f) = -\int_a^b f_X(x)\ln(f_X(x))dx \quad (5.3)$$

Using the method of Lagrange multipliers, the following equation can be obtained:

$$-\frac{\partial}{\partial f}(f\ln(f)) + \lambda\frac{\partial}{\partial f}(f) = 0 \quad (5.4)$$

This equation has the following solutions:

$$-1 - \ln(f) + \lambda = 0 \quad (5.5a)$$

$$f_X(x) = e^{\lambda - 1} \quad (5.5b)$$

Since λ is a constant, f must be a constant; i.e., $f = c$, leading to the following expression for f:

$$\int_a^b c\,dx = cx\Big|_a^b = cb - ca = c(b - a) = 1 \quad (5.6a)$$

Therefore,

$$f_X(x) = \frac{1}{b - a} \quad (5.6b)$$

The corresponding entropy is

$$B(f) = \ln(b - a) \quad (5.6c)$$

The cases in Table 5.1 were developed by Reza (1961), Goldman (1968), and Tribus (1969).

Example 5.2 Aggregation of Expert Opinions

This section contains an example use of uncertainty measure for aggregating expert opinions (Lai, 1992; Ayyub and Lai, 1992; Lai and Ayyub, 1994). The example demonstrates the use of uncertainty measure to combine opinions in defining failures.

The measures of dissonance and confusion, which are constructed in the framework of the theory of evidence, are applied herein for aggregating the expert opinions.

Let $\alpha_1, \alpha_2, \ldots, \alpha_N$ be fuzzy failure definitions for some specified failure mode expressed as structural response and degree of belief for failure definitions that are obtained from experts 1, 2, …, N, as shown in Figure 5.1. The vertical axis is called the failure level; however, it can be viewed as a degree of belief for failure occurrence. These definitions can be viewed as functions representing the same failure state expressed by the N experts. The combined failure definition or function can be obtained by aggregating the N expert opinions as shown in Figure 5.1. The aggregated function is noted as α_0 in Figure 5.1. The lower bound, r_L, and the upper bound, r_U, of structural response for the entire ranges of all functions, and some specified structural response r^* within the lower bound and the upper bound, are shown in the figure. In this approach, the values of the N fuzzy failure functions at the specified structural response r^* are interpreted as a basic assignment for experts 1, 2, …, N, i.e., $m(\{1\}) = \alpha_1(r^*)$, $m(\{2\}) = \alpha_2(r^*)$, …, $m(\{N\}) = \alpha_N(r^*)$. Since each basic assignment is given for the corresponding set

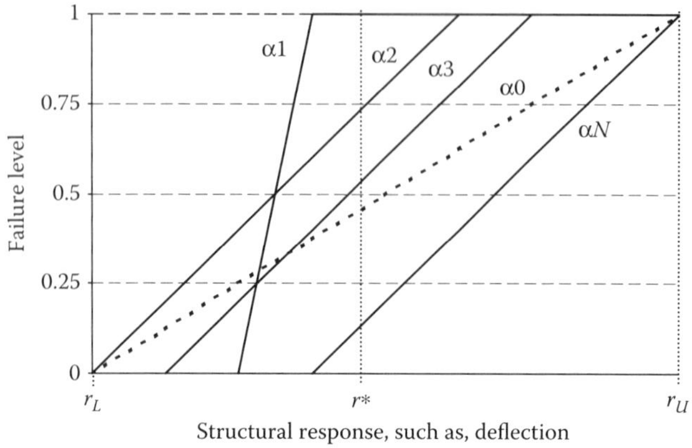

FIGURE 5.1 Fuzzy failures according to several experts.

of individual experts, there is no evidence supporting the unions of any other combinations of expert opinions. This means that the basic assignment corresponds to the sets of singletons only; however, the summation of all the basic assignments is required to be equal to 1.0. Therefore, if the summation of the basic assignment is less than 1.0, i.e., $m(\{1\}) + m(\{2\}) + \ldots + m(\{N\}) < 1.0$, the difference between the summation of the basic assignment and 1 should be distributed to the set of singletons. Since there is no particular preference for any set of individual experts, the difference should be distributed to the sets of singletons by normalization with respect to the summation such that the normalized summation is equal to 1.0.

Once the basic assignments are properly determined, Equations 4.51 and 4.52 are used, respectively, to calculate the measure of dissonance (D) and the measure of confusion (C) for the specified structural responses. It should be noted that the measure of dissonance is equal to the measure of confusion in this case, since the nonzero basic probability assignments exist only for the sets of singletons. Under this circumstance, both the measures are equal to the Shannon entropy (S) (Shannon, 1948). Therefore, the measure of uncertainty can be calculated as the following:

$$D = C = S = -\sum_{i=1}^{N} m(\{i\})\log_2(m(\{i\})) \tag{5.7}$$

where $m(\{i\})$ is the adjusted basic assignment for expert i. It is expected that the maximum measure of uncertainty occurs wherever all the expert opinions and a combined opinion are of the same value at some structural response level, i.e., $\alpha_1(r^*) = \alpha_2(r^*) = \ldots = \alpha_N(r^*) = \alpha_0(r^*)$. Therefore, the closer the experts' opinions and their combination to some common level, the larger the measure of uncertainty. The total measure of uncertainty that is calculated by integrating the measure of uncertainty over the entire range of no common opinion can be treated as some kind of index to measure the uniformity (or agreement) of the experts' opinions. The closer the experts' opinions and the combined opinion to uniformity, the larger the total measure of uncertainty.

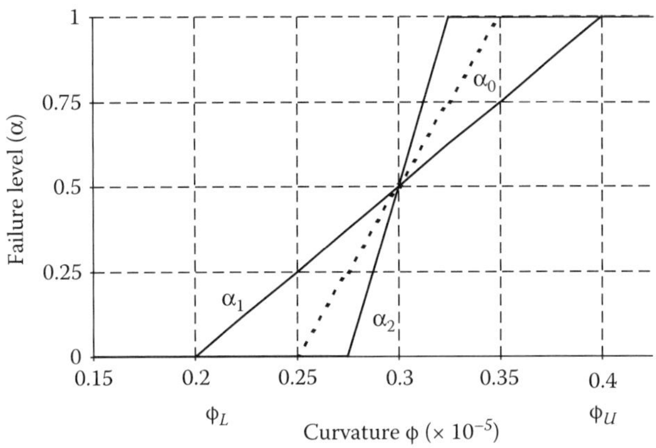

FIGURE 5.2 Fuzzy failures according to two experts.

Therefore, the aggregated linear function α_0 can be obtained by maximizing the total measure of uncertainty.

Now let us examine a particular case by considering the resisting moment vs. curvature relationship of the hull structure of a ship subjected to a hogging moment only. The transition from survival to failure in the crisp case was assumed to be attained at a curvature level of $\phi_f = 0.3 \times 10^{-5}$. In order to illustrate the application of uncertainty measure in aggregating expert opinions, two fuzzy failure definitions are selected as α_1 and α_2 in Figure 5.2. A linear function α_0 of fuzzy failure definition is considered to be the aggregated expert opinion. The lower bound and the upper bound of curvature range for the fuzzy failure function α_0 are also shown as ϕ_L and ϕ_U in the figure. In this example, the two fuzzy failure definitions are expressed by the following equations:

$$\text{Fuzzy definition 1: } \alpha_1 = 5 \times 10^5 \, \phi - 1 \tag{5.8a}$$

$$\text{Fuzzy definition 2: } \alpha_2 = 20 \times 10^5 \, \phi - 5.5 \tag{5.8b}$$

The aggregation failure function is assumed in the following linear form:

$$\alpha_0 = a \, \phi - b \tag{5.9}$$

where a = slope of the linear function of fuzzy failure definition and b = intercept. The slope a and the intercept b can then be derived as:

$$a = \frac{1}{\phi_U - \phi_L} \tag{5.10a}$$

$$b = a \, \phi_L \tag{5.10b}$$

In addition, the aggregated linear function was selected to pass through the point (ϕ, α) = (0.3×10^{-5}, 0.5) since the two fuzzy failure functions proposed by experts pass through the same point. Therefore, the parameters ϕ_L and ϕ_U (or a and b) are related. Only one parameter is needed to uniquely define the function α_0. The lower bound ϕ_L of curvature range is chosen as the independent variable that controls the curve α_0. Once the lower bound ϕ_L is assumed, the upper bound ϕ_U can be calculated using the following equation:

$$\phi_U = \phi_L + 2\,(0.3 \times 10^{-5} - \phi_L) = 0.6 \times 10^{-5} - \phi_L \tag{5.11}$$

The corresponding slope and intercept can then be evaluated using Equation 5.10a and b. The basic probability assignments for all possible sets of expert opinions are shown in the following:

$$m(\{1\}) = \begin{cases} \alpha_1 + \dfrac{1-(\alpha_1+\alpha_2+\alpha_0)}{3} & \text{if } \alpha_1+\alpha_2+\alpha_0 \le 1 \\ \dfrac{\alpha_1}{\alpha_1+\alpha_2+\alpha_0} & \text{if } \alpha_1+\alpha_2+\alpha_0 > 1 \end{cases} \tag{5.12a}$$

$$m(\{2\}) = \begin{cases} \alpha_2 + \dfrac{1-(\alpha_1+\alpha_2+\alpha_0)}{3} & \text{if } \alpha_1+\alpha_2+\alpha_0 \le 1 \\ \dfrac{\alpha_2}{\alpha_1+\alpha_2+\alpha_0} & \text{if } \alpha_1+\alpha_2+\alpha_0 > 1 \end{cases} \tag{5.12b}$$

$$m(\{0\}) = \begin{cases} \alpha_0 + \dfrac{1-(\alpha_1+\alpha_2+\alpha_0)}{3} & \text{if } \alpha_1+\alpha_2+\alpha_0 \le 1 \\ \dfrac{\alpha_0}{\alpha_1+\alpha_2+\alpha_0} & \text{if } \alpha_1+\alpha_2+\alpha_0 > 1 \end{cases} \tag{5.12c}$$

$$m(\{1,2\}) = m(\{2,0\}) = m(\{0,1\}) = m(\{1,2,0\}) = 0 \tag{5.12d}$$

For a particular lower bound ϕ_L, the basic assignments are constructed and the measure of uncertainty can be calculated using Equation 5.7 using integration over the entire range ϕ as follows:

$$D = C = S = -\int_{\phi} \sum_{i=0,1,2} m_{\phi}(\{i\}) \log_2(m_{\phi}(\{i\}))d\phi \tag{5.13}$$

where m_ϕ is the assignment at the ϕ value and the integration is performed over the range ϕ_L to ϕ_U. The results of the total measure of uncertainty for different fuzzy failure definitions are shown in Figure 5.3 and Table 5.2. The aggregated linear function was obtained at the maximum total measure of uncertainty. From the results shown in Figure

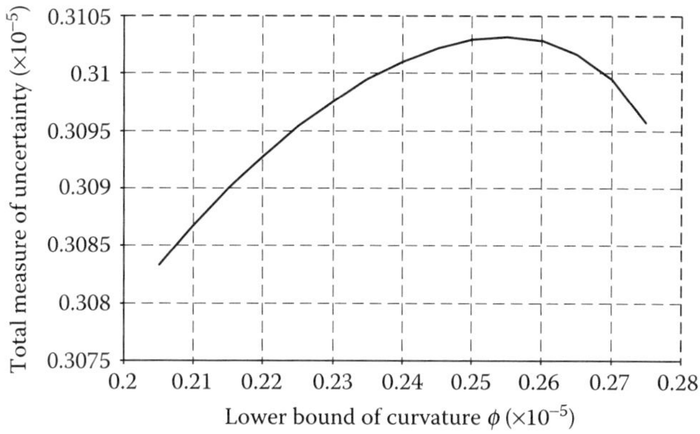

FIGURE 5.3 Total measure of uncertainty for fuzzy failure definitions.

4.3, the maximum total measure of uncertainty occurs where the range of curvature is from 0.255 to 0.345, as indicated in Table 5.2. The resulting aggregated fuzzy failure function is therefore expressed as

$$\alpha_0 = 11.1 \times 10^5\ \phi - 2.83$$

The slope of this aggregated fuzzy failure function ($a = 11.1 \times 10^5$), which is in between the two slopes proposed by the experts ($a = 5 \times 10^5$ for α_1 and 20×10^5 for α_2), is consistent with intuition.

TABLE 5.2
Total Measure of Uncertainty for Fuzzy Failure Definitions

Lower Bound of Curvature for α_0	Slope (a) for α_0	Intercept (b) for α_0	Total Uncertainty Measure (Equation 5.13)
0.200×10^{-5}	5.000×10^{-5}	1.000	0.30795
0.205×10^{-5}	5.263×10^{-5}	1.079	0.30833
0.210×10^{-5}	5.556×10^{-5}	1.167	0.30868
0.215×10^{-5}	5.882×10^{-5}	1.265	0.3090
0.220×10^{-5}	6.250×10^{-5}	1.375	0.30928
0.225×10^{-5}	6.667×10^{-5}	1.500	0.30954
0.230×10^{-5}	7.143×10^{-5}	1.643	0.30976
0.235×10^{-5}	7.692×10^{-5}	1.808	0.30995
0.240×10^{-5}	8.333×10^{-5}	2.000	0.3101
0.245×10^{-5}	9.091×10^{-5}	2.227	0.31022
0.250×10^{-5}	10.000×10^{-5}	2.500	0.3103
0.255×10^{-5}	11.111×10^{-5}	2.833	Max = 0.31032
0.260×10^{-5}	12.500×10^{-5}	3.250	0.31028
0.265×10^{-5}	14.286×10^{-5}	2.786	0.31017
0.270×10^{-5}	16.667×10^{-5}	4.500	0.30995
0.275×10^{-5}	20.000×10^{-5}	5.500	0.30957

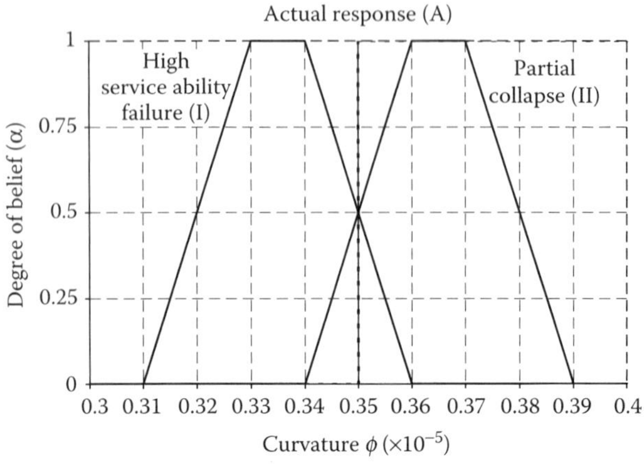

FIGURE 5.4 Two failure events.

Example 5.3 Failure Classification Based on Expert Opinions

The example demonstrates the use of uncertainty measures to classify failure to predefined failure categories (Lai, 1992; Ayyub and Lai, 1992; Lai and Ayyub, 1994).

Consider an actual structural response ϕ_A that is an observed level that can be represented as event A in Figure 5.4. Categories I and II represent serviceability failure and partial collapse, respectively, according to the expert opinions. Category I is called the lower failure category, and Category II is called the higher failure category. Since the magnitude of the structural response ϕ_A is located in the intersection of serviceability failure and partial collapse, confusion exists for the given body of evidence represented by event A and performance categories I and II. Using the measure of confusion, the less distinguishable the two events, the larger the degree of confusion between them. Therefore, if event A is less distinguishable with category I than with category II, event A has a higher level of confusion with category I than with category II. In this case, event A is classified into category I (serviceability failure). One the contrary, if event A has a higher level of confusion with category II, event A is classified into category II (partial collapse).

As an example, consider Figure 5.5; a specified (or observed) curvature level, $\phi_A = 0.35 \times 10^{-5}$, is an assumed actual structural response. Since this level of damage is located in the intersection of two categories, i.e., high serviceability failure and partial collapse, confusion exists in classifying the observed (specified) structural response to a failure category. The measure of confusion is therefore computed herein for the purpose of failure classification. The measure of confusion for each body of evidence can also be evaluated and compared. The results are shown in Table 5.3. It is evident from Table 5.3 that the measure of confusion for high serviceability failure ($C_{A,\mathrm{H}} = 0.9183$) is larger than the measure of confusion for partial collapse ($C_{A,\mathrm{P}} = 0.5586$).

In this case example, six events of structural performance are defined for convenience to track and model structural performance. The following fuzzy events are defined for this purpose as shown in Figure 5.6: complete survival, low serviceability failure,

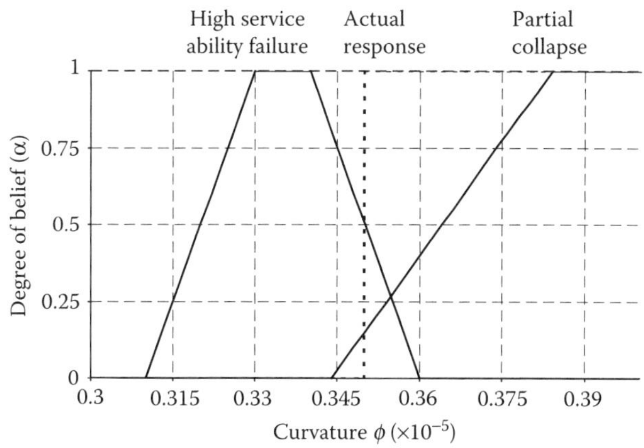

FIGURE 5.5 Computational example using two failure events.

TABLE 5.3
The Measure of Confusion for an Actual Response

	Comparison with the High Serviceability Failure		Comparison with the Partial Collapse	
Parameter	**Actual Response**	**High Serviceability Failure**	**Actual Response**	**Partial Collapse**
Degree of belief at intersection (α)	1.0	0.5	1.0	0.15
Basic assignment (m)	0.6667	0.3333	0.8696	0.1304
Confusion measure (C) According to Equation 5.7	$= \frac{-1}{\ln(2)}$ [0.6667 ln(0.6667) + 0.3333 ln(0.333)] = 0.9183		$= \frac{-1}{\ln(2)}$ [0.8696 ln(0.8696) + 0.1304 ln(0.1304)] = 0.5586	

serviceability failure, high serviceability failure, partial collapse, and complete collapse. These events are not necessarily mutually exclusive, as shown in the figure, since intersections might exist between adjacent events. It is of interest to classify some actual (or observed) structural performance or response to one of these failure categories, i.e., events. If the structural response is located within the range of just one event of structural performance, the structural response is classified to this failure category. If the structural response is located over two or more failure events, confusion in classifying the actual structural response into any of the failure categories results. Therefore, the measure of confusion is used in this case for the purpose of failure classification. These six events were selected for the purpose of illustrating a damage spectrum. The definitions of the six events are not interpreted as "at least"; e.g., although the event *serviceability failure* is commonly interpreted as *at least serviceability failure*, it is not treated as such in this example. Therefore, the failure events are treated as not nested. Lai (1992) examined both nested and nonnested failure events.

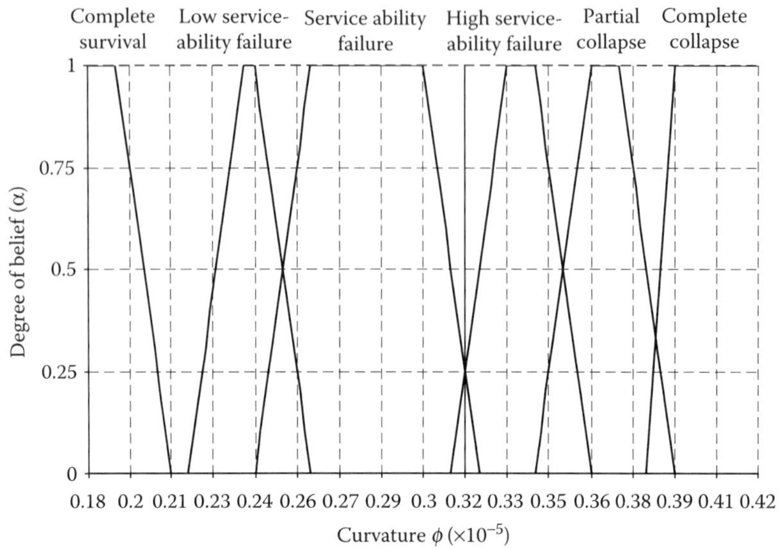

FIGURE 5.6 Six events of structural response.

For the nonnested case, measures of confusion for the six categories are computed as $\phi_A = 0.35 \times 10^{-5}$ and were incrementally increased from the left to the right in Figure 5.6. Since the event with the largest measure of confusion is selected for the structural response classification, the domains of all six classification events can be determined by comparing the degrees of confusion. Event *A* has a confusion measure with each event gradually increasing until a maximum is reached, followed by a decrease as ϕ increases. Figure 5.7 shows the classification of *A* based on the confusion measure as a step function. The numbers in boxes indicate the event classification: 1 = complete survival, 2 = low serviceability failure, 3 = serviceability failure, 4 = high serviceability failure, 5 = partial collapse, and 6 = complete collapse. The confusion measure was computed similar to the case presented in Table 5.3. It is evident from Figure 5.7 that the classification of an event changes from a lower failure category to an adjacent higher failure category at a curvature level located near the intersection of the two adjacent failure categories.

5.5 UNCERTAINTY INVARIANCE PRINCIPLE

The *principle of uncertainty invariance* (also called the *principle of information preservation*) was developed to facilitate meaningful transformations among various uncertainty measures (Brown, 1980; Klir, 1990). It was introduced to facilitate meaningful transformations between the various uncertainty theories. According to this principle, the amount of uncertainty (and the associated uncertainty-based information) should be preserved in each transformation of uncertainty from one mathematical framework to another. Examples of applications of this principle are probability–possibility transformations and probabilistic or possibilistic approximations of general bodies of evidence in Dempster–Shafer theory (DST) or, more generally,

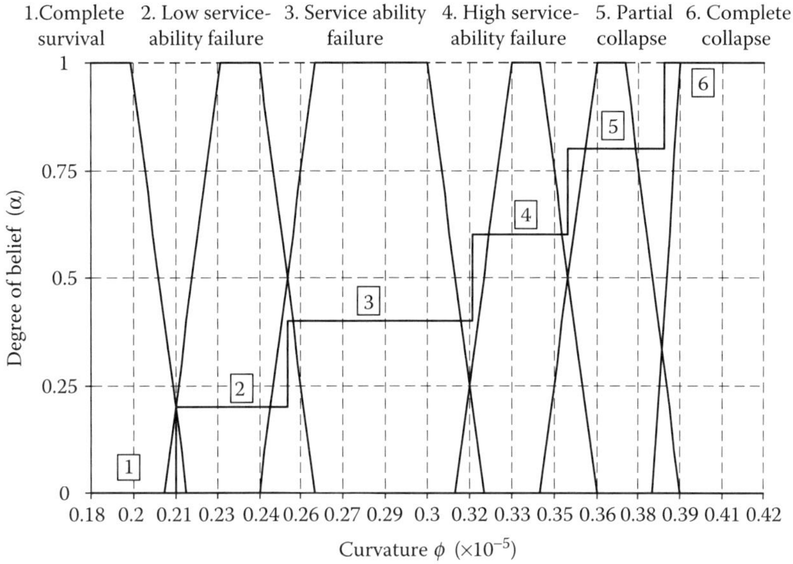

FIGURE 5.7 Classifying an actual response to the six events of structural response.

approximations of imprecise probabilities formalized in a particular theory by their counterparts in a less general theory.

This principle utilizes uncertainty measures that should be carefully constructed in terms of appropriate scale and units to allow for transforming one uncertainty type to another; therefore, once all uncertainties are consistently measured, they can be added, manipulated, and treated using the most convenient theory. The principle of uncertainty invariance was used in probability–possibility transformations in combining objective and subjective information that are represented using probability and possibility theories (Brown, 1980).

Thus far, significant results have been obtained for uncertainty-invariant probability–possibility transformations. It was determined by a thorough mathematical analysis (Geer and Klir, 1992) that these transformations do exist and are unique only under log-interval scales. They are also meaningful, but not unique, under ordinal scales (Klir and Parviz, 1992). In this case, additional, context-dependent requirements may be used to make the transformations unique.

Klir and Wierman (1999) show that a probability mass function $P = (p_1, p_2, \ldots, p_n)$, such that $p_1 \geq p_2 \geq \ldots \geq p_n$, defined on a universal set X of $i = 1, 2, \ldots, n$ elements, can be uniquely transformed to a possibility function $\boldsymbol{r}$ using the uncertainty invariance principle based on the aggregate measure AU. Assume that k unique values exist in $(p_1, p_2, \ldots, p_n)$ that are given k integer values, i.e., $s_1, s_2, \ldots, s_n$, where $k \geq n$. Therefore, the following condition holds:

$$p_1 = p_2 = \ldots = p_{s_1} > p_{s_1+1} = p_{s_1+2} = \ldots = p_{s_2} > \ldots > P_{s_{k-1}+1} = p_{s_{k-1}+2} = \ldots = p_{s_k} \quad (5.14a)$$

The ordinal equivalence requires having k distinct possibility values ($\boldsymbol{r}$) as follows:

$$r_1 = r_2 = \ldots = r_{s_1} > r_{s_1+1} = r_{s_1+2} = \ldots = r_{s_2} > \ldots > r_{s_{k-1}+1} = r_{s_{k-1}+2} = \ldots = r_{s_k} \quad (5.14b)$$

These unique possibility values can be computed as follows:

$$r_{s_j} = \sum_{i=s_{j-1}+1}^{n} p_i \quad (5.14c)$$

where any elements that have the same probabilities are lumped together and assigned the same possibility values, as illustrated in the example at the end of the section.

Example 5.4 Transformation of Probabilities to Possibilities

This example illustrates the use of Equation 5.14c for transforming probabilities to possibilities. Consider the following universal set (Klir and Wierman, 1999):

$$X = \{1,2,3,4,5,6,7,8\} \quad (5.15)$$

with the following probability mass function:

$$P = (0.3, 0.2, 0.2, 0.1, 0.08, 0.04, 0.04, 0.04) \quad (5.16)$$

According to Equation 5.14c, Table 5.4 can be constructed with the possibility distribution $\boldsymbol{r}$, which preserves the uncertainty provided by P, given by

$$r = \{1.0, 0.7, 0.7, 0.3, 0.2, 0.12, 0.12, 0.12\} \quad (5.17)$$

As an illustration, for $j = 3$, r_3 can be computed as

$$r_3 = \sum_{i=2}^{8} p_i = 0.2 + 0.2 + 0.1 + 0.08 + 0.04 + 0.04 + 0.04 = 0.7 \quad (5.18)$$

5.6 METHODS FOR OPEN-WORLD ANALYSIS

The simulated performance of a system depends heavily upon the information available at hand about the problem under consideration. Complete information is difficult to come by and is generally not available even for simple applications. For instance, database systems use the *closed-world assumptions* and introduce *null values* to deal with incomplete information, such as in diagnostics. In this section, methods used to model open-world problems are introduced and demonstrated.

TABLE 5.4
Transforming Probabilities to Possibilities

i	1	2	3	4	5	6	7	8
p_i	0.3	0.2	0.2	0.1	0.08	0.04	0.04	0.04
s_i	s_1		s_2	s_3	s_4			s_5
r_{s_j}	1		0.7	0.3	0.2			0.12
r_i	1	0.7	0.7	0.3	0.2	0.12	0.12	0.12

5.6.1 Statistical Estimators for Sequences and Patterns

5.6.1.1 Laplace Model

An example is used in this section to introduce a class of statistical estimators that are suitable for a bounded open universe for certain applications. In this example, an owner of a processing plant intends to protect the plant from external sabotage by adversaries (e.g., terrorists). Records of previous attacks on this industry show that adversaries have employed attack methods that included four cyber attacks on plants' control systems, two bomb attacks, and two attempts to breach the perimeter through the gate by fraudulent documents. The owner would like to estimate the respective probabilities of observing any of these attacks for the purpose of asset protection. A naively constructed probability distribution based on the closed-world assumption and the empirical evidence at hand produces the following probability mass function:

Attack Type (x)	**Probability ($P(x)$)**	
Cyber attack (C)	4/8 = 0.5	(5.19)
Perimeter breach (P)	2/8 = 0.25	
Bomb attack (B)	2/8 = 0.25	

This poor estimator of Equation 5.19 does not account for the unseen elements, such as potentially using a rifle to attack storage vessels to initiate an explosion.

To account for the unseen elements, Laplace (1825) suggested an estimator that adds 1 for each type of the seen elements (denoted as $k = 3$) and 1 to the collective of unseen elements. Based on this formulation for a sample size of $n = 8$, the following probability mass function is produced:

Attack Type (x)	**Probability ($P(x)$)**	
Cyber attack (C)	$= (4+1)/(k + n + 1)$	
	$= (4 + 1)/12 = 0.4167$	(5.20)
Perimeter breach (P)	$= (2 + 1)/12 = 0.25$	
Bomb attack (B)	$= (2 + 1)/12 = 0.25$	
Unseen elements	$= (0 + 1)/12 = 0.0833$	

5.6.1.2 Add-c Model

The concept of adding 1 can be generalized to adding any constant c, called the add-c model (Gale and Church, 1994; Orlitsky et al., 2003). The concept of a pattern and a sequence is introduced herein to facilitate the presentation of this model. A sequence (S) is defined as follows:

$$S = X_1, X_2, X_3, \ldots, X_n \tag{5.21}$$

where X_i = a possible outcome of some type. The number of types is unknown, but bounded, i.e., a bounded open world. According to Equation 5.19, the attack sequence in this case is assumed to be

$$S = C,\ C,\ P,\ C,\ B,\ B,\ P,\ C \tag{5.22}$$

where C = cyber attack, P = perimeter breach, and B = bomb attack. A pattern (P_S) can be constructed as follows based on Equation 5.22:

$$P_S = 11213321 \tag{5.23}$$

where 1 denotes a cyber attack, 2 denotes a bomb attack, and 3 denotes a perimeter breach. The total number of observed types (k) is 3. The following patterns are valid: 11213211, 12313, and 121121. However, 213, 31212, and 2311 are invalid patterns since a number cannot be introduced for the first time before having already introduced all the numbers preceding it. Denoting n as the sample size, in this case n is 8; n_i = number of observations of each type, and in this case the numbers of observations for the three types are 4, 2, and 2 for cyber attack, perimeter breach, and bomb attack, respectively. The add-c model produces the following probabilities for the ith type and for unseen elements, respectively:

$$P(i\ type) = \frac{n_i(1+c)}{n+nc+c} \quad \text{for } i = 1,2,\ldots,k \tag{5.24a}$$

$$P(Unseen\ Elements) = \frac{c}{n+nc+c} \tag{5.24b}$$

where

$$\sum_{i=1}^{k} n_i = n \tag{5.24c}$$

Where the number of possible elements is large compared to the sample size; add-constant, including add-1 estimators, are lacking. Orlitsky et al. (2003) used an example of studying the DNA (i.e., deoxyribonucleic acid) sequences to demonstrate this deficiency as was initially reported by Gale and Church (1994). Observing the DNA sequences of a large number (n) of animals, and predictably finding that each sequence is unique, each observed sequence has $n_i = 1$. The add-c estimator would assign probability $(1)(1 + c)/(n + nc + c)$ according to Equation 5.24a to each observed sequence, and a probability of $c/(n + nc + c)$ to all unseen sequences. From this assignment, the probability $(n + nc)/(n + nc + c)$ that the estimator assigns to all observed sequences is close to 1, whereas the probability it assigns to all unseen sequences is close to zero, contrary to the truth.

5.6.1.3 Witten–Bell Model

The Witten–Bell model as an enhancement of the Good–Turing model (Chen and Goodman, 1996) was introduced to address the shortcomings of the add-c model. It equates the probabilities of zero-frequency items with one-frequency items, and uses frequency of things observed once to estimate frequency of things we have not observed yet. This distribution allocates uniform probability mass to as yet unseen events by using the number of events that have only been observed once. The probability mass reserved for unseen events is given by

$$P(\mathit{Unseen\ Elements}) = \frac{k}{n+k} \tag{5.25}$$

where k is the number of observed event types, and n is the total number of observed events. This probability equates to the maximum likelihood estimate of a new type event occurring. The remaining probability mass is discounted such that all probability estimates sum to 1, yielding

$$P(\mathit{Each\ Unseen\ Element}) = \frac{k}{z(n+k)} \tag{5.26a}$$

$$P(\mathit{Each\ Observed\ Element}) = \frac{n_i}{(n+k)} \quad \text{for } i = 1, 2, \ldots, k \tag{5.26b}$$

z is the number of element types with zero observed frequency, i.e., requiring the knowledge of and bounding the open world. The summation of Equation 5.26a and b is 1, as expected, i.e.,

$$P(\mathit{All\ Elements}) = z\frac{k}{z(n+k)} + \frac{n_1 + n_2 + \ldots + n_k}{(n+k)} = \frac{k}{(n+k)} + \frac{n}{(n+k)} = 1 \tag{5.27}$$

Example 5.5 Probabilities Using Add-c Model

In this example, the add-c model is illustrated using selected patterns of n sample sizes of 1, 2, 3, 4, 4, 100, and 100, as shown in Table 5.5. The computations in Table 5.5 are performed using Equation 5.19a to c, with $c = 1$. It can be observed from cases 4 and 5 of Table 5.5 that the order of the pattern does not have any effect on the resulting probabilities. This property might not be realistic for some applications. Also, we can observe from cases 6 and 7 that the probability assigned to having a value of 1 or all is, respectively, is close to 1, whereas the probability of having an unseen value element is close to zero.

5.6.2 Transferable Belief Model

In general, an intelligent system must be able to make plausible propositions that may turn out to be incorrect when more information becomes available. The transferable belief model provides a basis for a class of methods for making such propositions when faced with incomplete information. The transferable belief model (TBM) is a nonprobabilistic approach that derives from the Dempster–Shafer mathematical theory of evidence (Shafer, 1976). It is a means for representing quantified degrees of belief. Degrees of belief are obtained from agents providing evidence at a given time within a given frame of discernment. The method is capable of treating inconsistency in data by introducing the open-world assumption. In TBM, a set of all propositions consists of three subsets: (1) a subset of propositions known as *possible propositions* (PP), (2) a subset of propositions known as *impossible propositions* (IP), and (3) a subset of *unknown propositions* (UP). The content of the subsets depends not only on the given problem, but also on the evidence, which is available at a given time. As evidence becomes available, propositions are redistributed between the three sets, as shown in Figure 5.8.

The *closed-world assumption* postulates an empty UP set. The *open-world assumption* admits the existence of a nonempty UP set, and the fact that the truth might be in UP. In this assumption, *unknown* refers to *none of the known propositions*.

The UP set can be considered to be empty where the truth is necessarily in the PP set, and the P_X power set of $2^{|X|}$ is PP. The selection of the type of the world depends on the problem at hand. The closed-world assumption can be selected for a quality condition problem where the condition cannot be other than one or more of the ratings, e.g., poor, poor or good, etc. For diagnosing the degradation underlying causes, the open-world assumption can be a suitable selection since an analyst trying to solve the problem cannot always consider all the possibilities; i.e., one or more underlying causes might exist that are not known to the analyst.

The degree of conflict between two or more evidence sources, k, implies the existence of a proposition not defined in the frame of discernment. In the idealized closed-world assumption, that amount of conflict is redistributed among the known propositions. In the open-world assumption, the degree of conflict corresponds to the amount of belief allocated to the proposition that *none of the known* propositions has the truth. One must keep in mind that the actual underlying physics might be something else other than the causes considered; i.e., the solution is in the set $\emptyset$ = UP and not in the set X = PP.

Table 5.5
Probabilities Using the Add-c Model

Case	Sample Size (n)	Pattern	Probabilities
1	1	1	$P(i=1)=\frac{1(1+1)}{1+1(1)+1}=\frac{2}{3}$ $P(Unseen\ Element)=\frac{1}{1+1(1)+1}=\frac{1}{3}$
2	2	11	$P(i=1)=\frac{2(1+1)}{2+2(1)+1}=\frac{4}{5}$ $P(Unseen\ Element)=\frac{1}{2+2(1)+1}=\frac{1}{5}$
3	3	111	$P(i=1)=\frac{3(1+1)}{3+3(1)+1}=\frac{6}{7}$ $P(Unseen\ Element)=\frac{1}{3+3(1)+1}=\frac{1}{7}$
4	4	1112	$P(i=1)=\frac{3(1+1)}{4+4(1)+1}=\frac{6}{9}$ $P(i=2)=\frac{1(1+1)}{4+4(1)+1}=\frac{2}{9}$ $P(Unseen\ Element)=\frac{1}{4+4(1)+1}=\frac{1}{9}$
5	4	1121	$P(i=1)=\frac{3(1+1)}{4+4(1)+1}=\frac{6}{9}$ $P(i=2)=\frac{1(1+1)}{4+4(1)+1}=\frac{2}{9}$ $P(Unseen\ Element)=\frac{1}{4+4(1)+1}=\frac{1}{9}$
6	100	$\overbrace{111\ldots1}^{100}$	$P(i=1)=\frac{100(1+1)}{100+100(1)+1}=\frac{200}{201}$ $P(Unseen\ Element)=\frac{1}{100+100(1)+1}=\frac{1}{201}$ $P(any\ i)=\frac{1(1+1)}{100+100(1)+1}=\frac{2}{201}$

Continued.

Table 5.5 *(Continued)*
Probabilities Using the Add-c Model

Case	Sample Size (n)	Pattern	Probabilities
7	100	$\overbrace{1234\ldots100}^{100}$	$\sum_{i=1}^{100} P(i) = \frac{200}{201}$ $P(\textit{Unseen Element}) = \frac{1}{100+100(1)+1} = \frac{1}{201}$

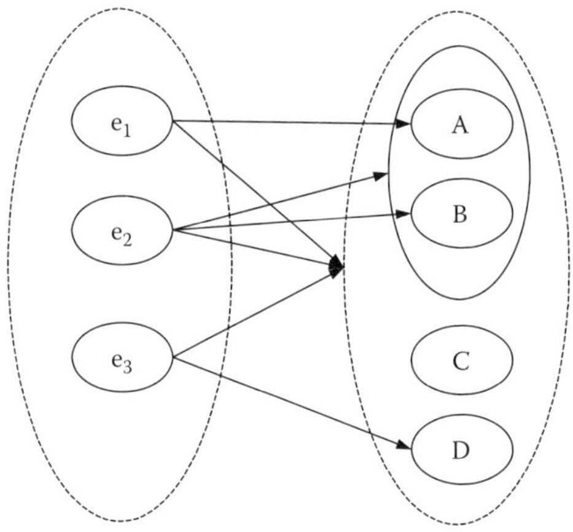

FIGURE 5.8 Mapping of evidence space.

5.6.3 Open-World Assumption Mathematical Framework

The known possible propositions set PP is based on X and is a finite set of elementary propositions. The set Ø is defined as the null or impossible event. In the Dempster–Shafer framework, the mass of the null set, $m(\emptyset)$, is defined as zero when belief functions are normalized, and correspondingly $Bel(X) = 1$. In contrast, under an open-world assumption, the mass of the null set may be nonzero if the frame of discernment X does not contain the truth (Smets, 1998). A set of focal elements can be defined based on $m(A) > 0$ as follows:

$$F(m) = \{A \subseteq X \,|\, m(A) > 0\} \tag{5.28}$$

The elements of $F(m)$ are called the *focal elements* of m. Shafer (1976) initially imposed a normality condition for belief structures, i.e., $\emptyset \notin F(m)$. Smets (1998) proposed to relax this condition and to interpret $m(\emptyset)$ as the part of belief committed to the assumption that none of the hypotheses in X might be true, to allow for an open-world assumption. If, however, the truth is known with absolute certainty to lie in X, i.e., closed-world assumption, then the normality condition can be justified.

The belief and plausibility functions satisfy the following conditions: (1) $Bel(\emptyset) = 0$, (2) $Bel(X) = 1 - m(\emptyset) \leq 1$, and (3) $Bel(B) \leq Pl(B)$. By definition, $Bel(\emptyset) = 0$, even though $m(\emptyset)$ might be positive. If the frame of discernment X is defined such that it included the unknown propositions set X, then this would lead to the same belief function as with the open-world assumption, if one takes care to never allocate any masses to propositions of X that did not include $\emptyset$.

5.6.4 Evidential Reasoning Mechanism, Belief Revision, and Diagnostics

In evidence–hypothesis reasoning, an *evidence space E* is a set of mutually exclusive and collectively exhaustive *evidential elements* that can arise from a source of evidence, e.g., the set of all possible results of a laboratory test. A *hypothesis space H* is a set containing all the mutually exclusive and collectively exhaustive hypotheses possible in the situation under consideration. Evidence–hypothesis reasoning is a mapping from an *evidence space E* to a *hypothesis space H*, which describes the relationships between evidence and hypothesis subsets. An example mapping is shown in Figure 5.7.

Evidence usually exists in two forms as either a linguistic observation such as "high rusted member" or a measured parameter such as "chloride ion concentration rate equals 0.6 kg Cl^-/m^3 (1 lb Cl^-/yd)." Accordingly, the handling of evidence–hypothesis reasoning differs depending on the particularities of an application.

The evidence–hypothesis reasoning mechanism is the task of inferring the belief in some hypotheses by collecting relevant evidence for or against these hypotheses. The inexact relationships among hypotheses and evidence are classified depending on the nature of evidence, i.e., measurement or observation. Linguistic hypothesis–evidence reasoning manipulates if–then rules to manifest the uncertainty associated with hypothesis–evidence relationships. Numerical hypothesis–evidence reasoning deals with computations based on measurements, where the inexact relationships between evidence and hypotheses are presented by two-dimensional plots.

Information is subject to change due to inherent uncertainty in information, or because of the various in ignorance types, or due to an environment that is volatile and dynamic. Current nonmonotonic reasoning systems cannot adequately treat changes in information. Once a change in the knowledge base, however minor, is performed, one must begin from scratch to deal with a problem at hand as a result of evidence fusion being computationally nonmonotonic, with perhaps consequentially changing system architecture. Belief revision methods can be used to deal with changing information (Dubois and Prade, 1997).

Diagnostics is the task of inferring plausible explanations based on a body of evidence, or to decide which explanation accounts for given evidence. Observations of distress and results of laboratory tests can be considered evidence for possible degradation underlying causes. The problem is then to infer the belief in the possible underlying causes producing the observed evidence. A classical method of diagnostic analysis is based on Bayesian analysis. In this case, the relations between evidence and underlying causes are described by conditional probabilities. Since mechanical, physical, and chemical processes of degradation can act in a synergistic manner, assigning a degradation cause might not be a clear-cut case. Since distresses might bear on a set of causes rather than on an individual cause and since evidences are not infallible, one can conclude that the Bayesian theorem is not an appropriate tool for diagnostic problems. In addition, the Bayesian theorem postulates an exhaustive frame of discernment that constitutes a complete set of well-defined causes. The reality is that the actual underlying causes might be something else other than the defined or identified causes. An open-world assumption might be more appropriate for such cases, as described in an earlier section. In addition, the Bayesian theorem can be generalized within the framework of evidence theory where conditional probabilities are replaced by belief functions (Pearl, 1990, 1992; Smets, 1992, 1993). Another generalization is obtained by extending the generalized Bayesian theorem to handle all types of belief functions, i.e., precise, interval, and fuzzy.

EXERCISE PROBLEMS

5.1. Describe the method of Lagrange multipliers used in the derivation of Equation 5.4, and rederive Equation 5.6 in detail.

5.2. Show that the maximum entropy probability distribution that corresponds to the following constraints is the exponential distribution, as provided in Table 5.1:

$$\int_{0}^{\infty} f_X(x)dx = 1$$

Mean value = μ.

5.3. Show that the maximum entropy probability distribution that corresponds to the following constraints is the normal distribution, as provided in Table 5.1:

$$\int_{-\infty}^{\infty} f_X(x)dx = 1$$

Mean value = μ.
Standard deviation = σ.

5.4. Derive the values in Table 5.2.

5.5. Derive the values in Table 5.3.

5.6. For the universal set $X = \{1, 2, 3, 4, 5, 6, 7, 8\}$ with the probability mass function $P = \{0.2, 0.2, 0.2, 0.2, 0.08, 0.04, 0.04, 0.04\}$, compute the corresponding uncertainty invariant possibility distribution r.

5.7. Develop a computational problem to illustrate the open-world assumption.

5.8. Redo Example 5.5 using c values of 2, 3, 4, 5, 10, 20, and 100. What are your observations and conclusions?

5.9. Redo Example 5.5 using the Witten–Bell model with a bounded open world of 100, i.e., $z + k = 100$.

6 Uncertainty Propagation for Systems

6.1 INTRODUCTION

Historically, engineers and scientists have built analytical models to represent natural and human-made systems using a combination of physical laws and empirical tools based on observing system attributes of interest (called system output variables). The intent of these models is to relate output variables (also called criterion or dependent variables) for a system with a set of other input variables (called predictor or independent variables). The input variables are either physically controllable or uncontrollable. For example, a structural engineer might observe the deflection of a bridge as an output of an input, such as a load at the middle of its span. By varying the intensity of the load, the deflection changes. Empirical test methods would vary the load incrementally, and the corresponding deflections are measured, thereby producing a relationship such as

$$y = g(x) \tag{6.1}$$

where x = input variable, y = output variable, and g = a function that relates input to output. In general, a system might have several input variables that can be represented as a vector $\boldsymbol{X}$, and several output variables that can be represented by a vector $\boldsymbol{Y}$. A schematic representation of this model is shown in Figure 6.1. Sometimes, the system is viewed as a whole entity without any knowledge on how the input variables are processed within the system to produce the output variables. This black box view of the system has the advantage of shielding an analyst from the physics governing the system, and providing the analyst with the opportunity to focus on relating the output to the input within some range of interest for the underlying variables. The primary assumptions according to this model in this case are (1) causal relationships exist between input and output variables as defined by the function g, and (2) the effect of time, i.e., time lag or time prolongation within the system, is accounted for by methods of measurement of input and output variables.

For complex engineering systems or natural systems, the numbers of input and output variables might be large with varying levels of importance. In such cases, a system engineer would be faced with the challenge of identifying the most significant variables, and how they should be measured. Establishing a short list of variables might be a most difficult task, especially for novel systems. Some knowledge of the physics of the system might help in this task of system identification. Then, the

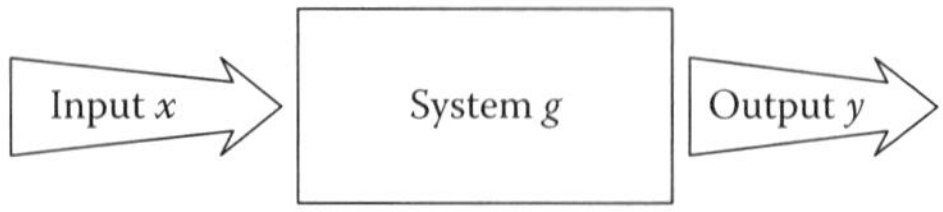

FIGURE 6.1 Black box system model.

analyst needs to decide on the nature of time relation between input and output by addressing questions such as the following:

- Is the output instantaneous as a result of the input?
- If the output lags behind the input, what is the lag time? Are the lag times for the input and output related, i.e., exhibiting nonlinear behavior?
- Does the function g depend on time, number of input applications, or magnitude of input?
- Does the input produce an output and linger within the system, affecting future outputs?

These questions are important for the purpose of defining the model, its applicability range, and validity.

Many models in engineering and the sciences build on knowing the underlying physical laws and utilize methods such as material mechanics and constraints such as boundary conditions to relate input to output variables. The numerical computations might be based on finite element methods that are used to model the entire system. These models can also be viewed according to Equation 6.1. The model complexity can be increased by considering nonlinearity in behavior and other special considerations, such as bifurcation, instability, logic rules, and across-discipline or across-physics interactions. The objective of this chapter is to present methods for propagating uncertainty in input–output systems represented by Equation 6.1.

6.2 FUNDAMENTAL METHODS FOR PROPAGATING UNCERTAINTY

6.2.1 Analytic Probabilistic Methods

Many problems in engineering and the sciences deal with a dependent variable that is a function of one or more independent random variables. In this section, analytical tools for determining the probabilistic characteristics of the dependent random variable based on given probabilistic characteristics of independent random variables, and a functional relationship between them, are provided. The discussion in this section is divided into the following cases: (1) probability distributions for dependent random variables, (2) mathematical expectations, and (3) approximate methods. Cases 1 and 2 result in complete distributions and moments without any approximations, respectively, whereas the third case results in approximate moments.

6.2.1.1 Probability Distributions for Dependent Random Variables

A random variable X is defined as a mapping from a sample space of an engineering system or experiment to the real line of numbers. This mapping can be one-to-one mapping or many-to-one mapping. If Y is defined to be a dependent variable in terms of a function $Y = g(X)$, then Y is also a random variable. Assuming that both X and Y are discrete random variables, for a given probability mass function of X, i.e., $P_X(x)$, the objective herein is to determine the probability mass function of Y, i.e., $P_Y(y)$. This objective can be achieved by determining the equivalent events of Y in terms of the events of X based on the given relationship between X and Y, i.e., $Y = g(X)$. For each value y_i, all of the values of x that result in y_i should be determined, say $x_{i_1}, x_{i_2}, \ldots, x_{i_j}$. Therefore, the probability mass function of Y is given by

$$P_Y(y_i) = \sum_{k=1}^{i} P_X(x_{i_k}) \tag{6.2}$$

If X is continuous but Y is discrete, the probability mass function for Y is given by

$$P_Y(y_i) = \int_{R_e} f_X(x)dx \tag{6.3}$$

where R_e = the region of X that defines an equivalent event to the value $Y = y_i$.

If X is continuous with a given density function $f_X(x)$ and the function $g(X)$ is continuous, then $Y = g(X)$ is a continuous random variable with an unknown density function $f_Y(y)$. The density function of Y can be determined by performing the following four steps:

1. For any event defined by $Y \leq y$, an equivalent event in the space of X needs to be defined.
2. $F_Y(y) = P(Y < y)$ can then be calculated.
3. $f_Y(y)$ can be determined by differentiating $F_Y(y)$ with respect to y.
4. The range of validity of $f_Y(y)$ in the Y space should be determined.

Formally stated, if X is a continuous random variable, $Y = g(X)$ is differentiable for all x, and $g(X)$ is either strictly (i.e., monotonically) increasing or strictly (i.e., monotonically) decreasing for all x; then $Y = g(X)$ is a continuous random variable with the following density function:

$$f_Y(y) = f_X(g^{-1}(y))\left|\frac{\partial g^{-1}(y)}{\partial y}\right| \tag{6.4}$$

where $g^{-1}(y) = x$. Equation 6.4 can be derived by developing an expression for the cumulative distribution function of Y as

$$F_Y(y) = \int_{x \le g^{-1}(y)} f_X(x)dx = \int_{-\infty}^{g^{-1}(y)} f_X(x)dx \tag{6.5}$$

This result can be expressed as

$$F_Y(y) = \int_{-\infty}^{y} f_X(g^{-1}) \frac{\partial g^{-1}}{\partial y} dy \tag{6.6}$$

By taking the derivative $\dfrac{\partial F_Y(y)}{\partial y}$, Equation 6.4 results. The derivative $\dfrac{\partial g^{-1}(y)}{\partial y}$ is known as the Jacobian of the transformation (or inverse) and can be alternatively determined as follows:

$$\frac{\partial g^{-1}(y)}{\partial y} = \frac{1}{\dfrac{\partial g(x)}{\partial x}}\Bigg|_{\text{evaluated at } x=g^{-1}(y)} \tag{6.7}$$

Then the limits on Y should be determined based on the limits of X and $Y = g(X)$. If the inverse $g^{-1}(y)$ is not unique (say $= x_1, x_2, \ldots, x_m$), then the density function of Y is determined as follows:

$$f_Y(y) = \sum_{i=1}^{m} f_X[g_i^{-1}(y)] \left| \frac{\partial g_i^{-1}(y)}{\partial y} \right| \tag{6.8}$$

where

$$g_i^{-1}(y) = x_i \tag{6.9}$$

The following cases are selected special functions of single and multiple random variables that are commonly used, where the resulting variable (Y) can have known distribution types for some cases:

1. For multiple independent random variables $\boldsymbol{X} = (X_1, X_2, \ldots, X_n)$, where the function $g(\boldsymbol{X})$ is given by

$$Y = g(\mathbf{X}) = a_0 + a_1 X_1 + a_2 X_2 + \ldots + a_n X_n \tag{6.10}$$

$a_0, a_1, a_2, \ldots, a_n$ are real numbers, and the mean value and variance of Y are

$$\mathrm{E}(Y) = a_0 + a_1 \mathrm{E}(X_1) + a_2 \mathrm{E}(X_2) + \ldots + a_n \mathrm{E}(X_n) \tag{6.11}$$

and

$$\mathrm{Var}(Y) = \sum_{i=1}^{n}\sum_{j=1}^{n} a_i a_j Cov(X_i, X_j) \tag{6.12}$$

where $Cov(X_i,X_j)$ = covariance of X_i and X_j. It should be noted that $Cov(X_i,X_i) = \mathrm{Var}(X_i) = \sigma_{X_i}^2$. Equation 6.12 can be expressed in terms of the correlation coefficient as follows:

$$\mathrm{Var}(Y) = \sum_{i=1}^{n}\sum_{j=1}^{n} a_i a_j \rho_{X_i X_j} \sigma_{X_i} \sigma_{X_j} \tag{6.13}$$

where $\rho_{X_i X_j}$ = correlation coefficient of X_i and X_j. If the random variables of the vector $\mathbf{X}$ are statistically uncorrelated, then the variance of Y is

$$\mathrm{Var}(Y) = \sum_{i=1}^{n} a_i^2 \mathrm{Var}(X_i) \tag{6.14}$$

2. In Equations 6.11 to 6.14, if the random variables $X_1, X_2, X_3, \ldots, X_n$ have normal probability distributions, then Y has a normal probability distribution with a mean and variance as given by Equations 6.11 to 6.14.
3. If X has a normal distribution, and $Y = g(X) = exp(X)$, then Y has a lognormal distribution.
4. If $Y = X_1 X_2 X_3 \ldots X_n$, the arithmetic multiplication of $X_1 X_2 X_3 \ldots X_n$ with lognormal distributions, then Y has a lognormal distribution.
5. If $X_1 X_2 X_3 \ldots X_n$ are independent random variables that have Poisson distributions with the parameters $\lambda_1, \lambda_2, \ldots, \lambda_n$, respectively, then $Y = X_1 + X_2 + \ldots + X_n$ has a Poisson distribution with the parameter $\lambda = \lambda_1 + \lambda_2 + \ldots + \lambda_n$.

Example 6.1 Mean and Variance of a Linear Function

Consider the following function:

$$Z = 2X + 5Y + 10 \tag{6.15}$$

where X, Y, and Z are random variables. The means of X and Y are 3 and 5, respectively, and the standard deviations are 1 and 2, respectively. The random variables X and Y are assumed to be uncorrelated. Therefore, the mean of Z, μ_Z, is computed using Equation 6.11 as

$$\mu_Z = 2(3) + 5(5) + 10 = 41 \tag{6.16}$$

The variance of Z, σ_Z^2, is computed using Equation 6.14 as

$$\sigma_Z^2 = 2^2(1^2) + 5^2(2^2) = 104 \tag{6.17}$$

The standard deviation of Z is $\sqrt{104} = 10.20$, and the coefficient of variation is $\frac{10(20)}{41} = 0.25$.

Example 6.2 Probability Density Function of a Nonlinear Function

For the following nonlinear function,

$$Y = a\,X^2 + b \tag{6.18}$$

where a and b are constants and X has an exponential distribution with a parameter λ, the probability density function of Y is developed in this example.

In order to use Equation 6.8 to determine $f_Y(y)$, we need to determine X as a function of Y and the derivative of X with respect to Y. The inverse of the function in Equation 6.18 is

$$X = \pm\sqrt{\frac{Y-b}{a}} \tag{6.19}$$

The derivative of X with respect to Y is

$$\frac{dX}{dY} = \pm\frac{1}{2\sqrt{a(Y-b)}} \tag{6.20}$$

Therefore, the density function of Y can be determined based on the density function of the exponential distribution and by substituting in Equation 6.8:

$$f_Y(y) = \frac{1}{2\sqrt{a(y-b)}}\left[f_X\left(+\sqrt{\frac{y-b}{a}}\right) + f_X\left(-\sqrt{\frac{y-b}{a}}\right)\right] \tag{6.21a}$$

or

$$f_Y(y) = \frac{\lambda}{2\sqrt{a(y-b)}}\left[\exp\left(-\lambda\sqrt{\frac{y-b}{a}}\right) + \exp\left(+\lambda\sqrt{\frac{y-b}{a}}\right)\right] \quad (6.21b)$$

The resulting function $f_Y(y)$ is not an exponential density function.

6.2.1.2 Mathematical Expectations

In certain engineering applications, we might be interested in knowing only the moments, specifically the mean and variance, of a random variable *Y*, based on known probabilistic characteristics of a random variable *X* and a function $Y = g(X)$. In such cases, mathematical expectation (E) is very effective in achieving this objective.

The mathematical expectation of an arbitrary function $g(X)$ of a continuous random variable *X* is given by

$$E[g(X)] = \int_{-\infty}^{+\infty} g(x) f_X(x) dx \quad (6.22)$$

The corresponding equation for a discrete random variable is

$$E[g(X)] = \sum_{i=1}^{n} g(x_i) P_X(x_i) \quad (6.23)$$

Mathematical expectation can be used to determine the moments of *Y* for a given probabilistic characteristics of *X* and the function $Y = g(X)$. The mean (or expected value) can be determined by the direct use of Equations 6.22 and 6.23. The variance can be determined, respectively, for continuous and discrete *X* random variables as

$$\mathrm{Var}(Y) = \mathrm{Var}[g(X)] = \int_{-\infty}^{+\infty} [g(X) - E[g(X)]]^2 f_X(x) dx \quad (6.24)$$

and

$$\mathrm{Var}(Y) = \mathrm{Var}[g(X)] = \sum_{i=1}^{n} [g(x_i) - E[g(X_i)]]^2 P_X(x_i) \quad (6.25)$$

If we consider a special case where the function, $g(X)$, is the following linear function,

$$Y = g(X) = a\,X + b \tag{6.26}$$

where a and b = real numbers, mathematical expectation can be used to determine, respectively, the mean and variance of Y as follows:

$$\mathrm{E}(Y) = a\,\mathrm{E}(X) + b \tag{6.27}$$

$$\mathrm{Var}(Y) = a^2\,\mathrm{Var}(X) \tag{6.28}$$

Equations 6.27 and 6.28 are valid regardless of the distribution type of X. Also, they can be used for both cases of discrete and continuous X. Based on Equation 6.26, it can be stated that mathematical expectation, E(.), is a linear operator. However, the variance, Var(.), is not a linear operator. Equations 6.26 to 6.28 can be generalized top produce Equations 6.11 to 6.14.

6.2.1.3 Approximate Methods

The closed-form solutions for the distribution types of dependent random variables, as well as mathematical expectation, provide solutions for the simple cases of functions of random variables. Also, they provide solutions for simple distribution types or a mixture of distribution types for the independent random variables. For cases that involve a more general function $g(\boldsymbol{X})$ or a mixture of distribution types, these methods are not suitable for obtaining solutions due to the analytical complexity according to these methods. Also, in some engineering applications, precision might not be needed. In such cases, approximate methods based on Taylor series expansion, with or without numerical solutions of needed derivatives, can be used. The use of Taylor series expansion, in this section, is divided into the following two headings: (1) single random variable X and (2) multiple random variables, i.e., a random vector $\boldsymbol{X}$.

6.2.1.3.1 Single Random Variable X

The Taylor series expansion of a function $Y = g(X)$ about the mean of X, i.e. E(X), is given by

$$\begin{aligned} Y = g[E(X)] + [X - E(X)]\frac{dg(X)}{dX} + \frac{1}{2}[X - E(X)]^2\frac{d^2g(X)}{dX^2} \\ + \ldots + \frac{1}{k!}[X - E(X)]^k\frac{d^kg(X)}{dX^k} + \ldots \end{aligned} \tag{6.29}$$

in which the derivatives are evaluated at the mean of X. Truncating this series at the linear terms, the *first-order mean and variance* of Y can be obtained by applying the mathematical expectation and variance operators. The first-order (approximate) mean is

$$\mathrm{E}(Y) \approx g(\mathrm{E}(X)) \tag{6.30}$$

The first-order (approximate) variance is

$$\mathrm{Var}(Y) \approx \left(\frac{dg(X)}{dX}\right)^2 \mathrm{Var}(X) \tag{6.31}$$

Again the derivatives in Equations 6.30 and 6.31 are evaluated at the mean of X.

6.2.1.3.2 Random vector $\boldsymbol{X}$

The Taylor series expansion of a function $Y = g(\boldsymbol{X})$ about the mean values of $\boldsymbol{X}$, i.e., $E(X_1)$, $E(X_2)$, …, $E(X_n)$, is given by

$$\begin{aligned} Y = {} & g[E(X_1),E(X_2),\ldots,E(X_n)] + \sum_{i=1}^{n}[X_i - E(X_i)]\frac{\partial g(\boldsymbol{X})}{\partial X_i} \\ & + \sum_{i=1}^{n}\sum_{j=1}^{n}\frac{1}{2}[X_i - E(X_i)][X_j - E(X_j)]\frac{\partial^2 g(\boldsymbol{X})}{\partial X_i \partial X_j} + \ldots \end{aligned} \tag{6.32}$$

in which the derivatives are evaluated at the mean values of $\boldsymbol{X}$. Truncating this series at the linear terms, the *first-order mean and variance* of Y can be obtained by applying the mathematical expectation and variance operators. The first-order (approximate) mean is

$$E(Y) \approx g[E(X_1), E(X_2), \ldots, E(X_n)] \tag{6.33}$$

The first-order (approximate) variance is

$$\mathrm{Var}(Y) \approx \sum_{i=1}^{n}\sum_{j=1}^{n}\frac{\partial g(\boldsymbol{X})}{\partial X_i}\frac{\partial g(\boldsymbol{X})}{\partial X_i}Cov(X_i, X_j) \tag{6.34a}$$

in which the derivatives are evaluated at the mean values of $\boldsymbol{X}$, i.e., $E(X_1)$, $E(X_2)$, …, $E(X_n)$. For uncorrelated random X variables, Equation 6.34a can be simplified as follows:

$$\mathrm{Var}(Y) \approx \sum_{i=1}^{n}\left(\frac{\partial g(\boldsymbol{X})}{\partial X_j}\right)^2 \mathrm{Var}(X_i) \tag{6.34b}$$

in which the derivatives are evaluated at the mean values of X_i.

Example 6.3 Stress in a Beam

The stress, F, in a beam subjected to an external bending moment M is

$$F = \frac{My}{I} \tag{6.35}$$

where y is the distance from the neutral axis of the cross section of the beam to the point where the stress is calculated, and I is the centroidal moment of inertia of the cross section. Assume that M and I are random variables with means μ_M and μ_I, respectively, and variances σ_M^2 and σ_I^2, respectively. Also assume that y is not a random variable. The mean value of the stress, μ_F, based on first-order approximations (Equation 6.33) is

$$\mu_F \approx \frac{\mu_M y}{\mu_I} \tag{6.36}$$

The first-order variance of the stress, σ_F^2, is (see Equation 6.34b)

$$\sigma_F^2 \approx \left(\frac{y}{\mu_I}\right)^2 \sigma_M^2 + \left(\frac{\mu_M y}{\mu_I^2}\right)^2 \sigma_I^2 \tag{6.37}$$

6.2.2 Simulation Methods

The interest in simulation methods started in the early 1940s for the purpose of developing inexpensive techniques for testing engineering systems by imitating their real behavior. These methods are commonly called Monte Carlo simulation techniques. The principle behind the methods is to develop an analytical model, which is computer based, that predicts the behavior of a system. Then, the model is evaluated using data measured from a system, and then the model predicts the behavior of the system, usually for many simulation runs. Each evaluation (or simulation cycle) is based on a certain randomly selected set of conditions for the input parameters of the system. Certain analytical tools are used to ensure the random selection of the input parameters according to their respective probability distributions for each evaluation. As a result, several predictions of the behavior are obtained. Then, statistical methods are used to evaluate the moments and distribution type for the system's behavior.

The analytical and computational steps that are needed for performing Monte Carlo simulation are:

1. Definition of the system using an input–output numerical model
2. Generation of random numbers
3. Generation of random variables using transformation methods from random numbers to random variates

4. Evaluation of the model multiple times (N simulation cycles)
5. Statistical analysis of the resulting behavior
6. Study of simulation efficiency and convergence

The definition of the system should include its boundaries, input parameters, output (or behavior) measures, architecture, and models that relate the input parameters and architecture to the output parameters. The accuracy of the results of simulation is highly dependent on having an accurate definition for the system. All critical parameters should be included in the model. The definition of the input parameters should include their statistical or probabilistic characteristics, i.e., knowledge of their moments and distribution types. It is common to assume the architecture of the system in Monte Carlo simulation to be nonrandom. However, modeling uncertainty can be incorporated in the analysis in the form of bias factors and additional variability, e.g., coefficients of variation. The results of these generations are a set of specific values for the input parameters. These values should then be substituted in the model to obtain an output measure. By repeating the procedure N times (for N simulation cycles), N response measures are obtained. Statistical methods can now be used to obtain, for example, the mean value, variance, or distribution type for the response. The accuracy of the resulting measures for the behavior is expected to increase by increasing the number of simulation cycles. The convergence of the simulation methods can be investigated by studying their limiting behavior as N is increased. Also, the efficiency of simulation methods can be increased by using variance reduction techniques. Simulation methods and variance reduction techniques are described in more detail in other sources, such as Ayyub and McCuen (2003).

Example 6.4 Warehouse Construction

A warehouse is to be constructed from precast concrete elements that are produced by a nearby precast factory. The following construction tasks are identified for building the warehouse:

A: Excavation of foundations
B: Construction of foundations
C: Construction of precast elements at factory
D: Transportation of precast elements to construction site
E: Assembly of elements at site
F: Construction of roof
G: Exterior and interior finishing

Figure 6.2 shows the logical network for performing these activities. The figure indicates that tasks C and D can be performed in parallel to tasks A and B; i.e., as excavation and construction of the footings are being performed at the site, the precast elements can be constructed at the factory and then transported to the construction site. Table 6.1 shows the means and standard deviations for the completion times of these tasks. Normal probability distributions and statistical noncorrelation are assumed for these times. In this section, simulation is used to compute these moments.

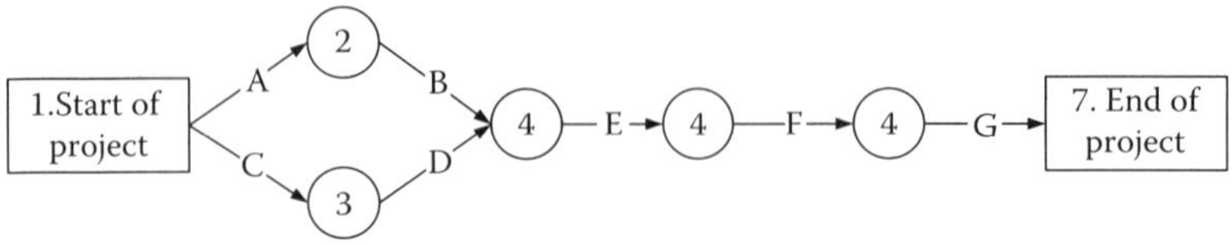

FIGURE 6.2 Construction network.

TABLE 6.1
Moments of Completion Times

Task	Name	Mean (days)	Standard Deviation (days)
A	Foundation excavation	3	1
B	Foundation construction	2	0.5
C	Precast elements construction	4	1
D	Transportation of elements	0.5	0.5
E	Assembly of elements	4	1
F	Roofing	2	1
G	Finishing	3	1

The project completion time, T, is a dependent random variable that is given by

$$T = \max\{A + B, C + D\} + E + F + G \tag{6.38}$$

where A, B, C, D, E, F, and G are random variables that model the times for completing the corresponding tasks. The random numbers used for generating the completion times for tasks A, B, C, D, E, F, and G are shown in Table 6.2. In this example, 20 simulation cycles were used. The random numbers were used to generate the random times as shown in Table 6.3 utilizing the inverse transformation method. Although the simulation processes resulted in some activity times taking negative values, these values were not removed from the analysis, but retained to illustrate the process. Truncated probability distribution could be used to enhance the process and make it more realistic so that such negative values are not encountered. Then, Equation 6.38 was used to compute the completion time for the project for each simulation cycle, as shown in Table 6.3. The mean value and variance of the completion time of the project were computed using values in the last column of Table 6.3. The resulting statistics are

$$\text{Mean value} = 14.59 \text{ days} \tag{6.39}$$

$$\text{Variance} = 5.55 \text{ (days)}^2 \tag{6.40}$$

The approximate mean duration of the project, $\bar{T}$, can be computed as

$$\bar{T} = \max\{3+2, 4+0.5\} + 4 + 2 + 3 = 14 \text{ days} \tag{6.41}$$

TABLE 6.2
Random Numbers Used for Completion Time of Tasks

Task A	Task B	Task C	Task D	Task E	Task F	Task G
0.642707	0.758002	0.547225	0.510713	0.924981	0.44491	0.671304
0.240297	0.092418	0.84715	0.071252	0.98112	0.793358	0.780596
0.169051	0.446979	0.990008	0.079644	0.391058	0.793205	0.276989
0.457609	0.52127	0.606333	0.006137	0.47927	0.121284	0.34367
0.386325	0.395759	0.956544	0.432595	0.723067	0.448813	0.008538
0.313708	0.061922	0.343042	0.230356	0.538481	0.63629	0.211676
0.137571	0.078837	0.471558	0.383158	0.203166	0.500447	0.101354
0.296782	0.610994	0.785467	0.282056	0.282186	0.560465	0.539651
0.908314	0.124274	0.709123	0.508328	0.496352	0.886927	0.720611
0.763968	0.327695	0.506164	0.246872	0.743617	0.275227	0.218178
0.139498	0.935402	0.789508	0.966422	0.440431	0.682035	0.476614
0.220256	0.040641	0.347426	0.282962	0.178687	0.092735	0.96486
0.344963	0.100168	0.963482	0.569873	0.933351	0.64664	0.858627
0.095613	0.791418	0.726318	0.376506	0.872995	0.895403	0.962331
0.22554	0.262949	0.63276	0.550859	0.198235	0.077169	0.08673
0.239485	0.985236	0.212528	0.445724	0.66247	0.32561	0.025242
0.191603	0.108613	0.897544	0.990706	0.933851	0.557361	0.050711
0.94601	0.241317	0.187334	0.015071	0.228146	0.832563	0.816427
0.973859	0.343243	0.19794	0.177672	0.125638	0.099943	0.747989
0.484109	0.214928	0.020997	0.424466	0.893968	0.866459	0.706856

The approximate variance of the project duration, Var(T), was computed as

$$\text{Var}\,(T) = \text{Var(A)} + \text{Var(B)} + \text{Var(E)} + \text{Var(F)} + \text{Var(G)}$$

$$\text{Var}\,(T) = (1)^2 + (0.5)^2 + (1)^2 + (1)^2 + (1)^2 = 4.25\ (\text{days})^2 \tag{6.42}$$

These values differ from the computed values using simulation. The simulation results are also approximate because of the small number of cycles used to compute the results. By increasing the number of simulation cycles, the accuracy of the results is expected to increase. Also, the simulation results in the case of a small number of cycles are dependent on the used random numbers. For example, by using a different set of random numbers, as shown in Table 6.4, the results shown in Table 6.5 are obtained. The statistics in this case are

$$\text{Mean value} = 13.98\ \text{days} \tag{6.43}$$

$$\text{Variance} = 3.67\ (\text{days})^2 \tag{6.44}$$

6.2.3 Vertex Method for Functions of Fuzzy Variables

Some problems in engineering and the sciences can be expressed using functions of fuzzy variables in the same input–output format expressed in Equation 6.1. In this case, the functional relationship is given by

TABLE 6.3
Generated Random Values for Completion Time of Tasks

Task A	Task B	Task C	Task D	Task E	Task F	Task G	Project Completion Time
3.36526	2.3498042	4.1183849	0.5133909	5.4396627	1.8617675	3.443086	16.45958
2.2949296	1.3368915	5.0242746	–0.2334048	6.0778971	2.817944	3.7739874	17.460698
2.0421427	1.933497	6.3270846	–0.2038625	3.7238635	2.8174073	2.4085497	15.073043
2.8937885	2.0266006	4.2693511	–0.7523029	3.9481516	0.8313093	2.5979742	12.297824
2.7115217	1.8680395	5.7122992	0.4152843	4.5916169	1.8716254	0.6144956	13.205321
2.5150499	1.2304214	3.596267	0.1312882	4.0963777	2.3481148	2.1995787	12.389543
1.9086613	1.2934043	3.9288268	0.3516179	3.1698067	2.001117	1.7259547	11.177323
2.4667156	2.1407416	4.7905834	0.2118119	3.4240093	2.1518299	3.0993185	13.677553
4.3306524	1.4230146	4.5504403	0.5104081	3.9908831	3.2104754	3.5842925	16.539318
3.7188585	1.7770724	4.0154067	0.1579639	4.6542217	1.4032765	2.2218555	13.775285
1.9173874	2.7587914	4.8045186	1.4155181	3.8504386	2.472975	2.9415001	15.48495
2.2288943	1.1281585	3.6081638	0.2131524	3.079716	0.6756985	4.8105053	12.387236
2.6014871	1.3596149	5.793009	0.5878512	5.5015181	2.3758197	4.0742012	18.332399
1.6928452	2.4055804	4.601361	0.3428837	5.1407426	3.2559441	4.7787973	18.119729
2.2466222	1.6830249	4.3387303	0.5637742	3.1522166	0.5753632	1.6386032	10.268687
2.2923139	3.0884019	3.2025201	0.4319117	4.4187744	1.5483608	1.0437219	12.391573
2.128135	1.3829576	5.2678487	1.6770614	5.5053989	2.143973	1.3616434	15.955925
4.6076817	1.6491037	3.1123607	–0.5843253	3.2552784	2.9642836	3.9017196	16.378067
4.9412389	1.798406	3.1511551	0.0379101	2.8526594	0.7179441	3.6678706	13.978119
2.9602643	1.6053855	1.9659854	0.4049415	5.2480626	3.109864	3.5438339	16.46741

$$Y = f(X_1, X_2, \ldots, X_n) \tag{6.45}$$

where $X_1, X_2, \ldots, X_n$ = input fuzzy variables, Y = output variable, and f = a continuous function that relates input to output in the n-dimensional rectangular region and has no extreme points in this region, including the boundaries. The α-cut and interval arithmetic can be used as provided in Equations 2.87 to 2.90 (Dong and Shah, 1987; Dong and Wong, 1986a, 1986b, 1986c). An α-cut of a fuzzy number (or set) X_1 can be expressed for convenience as

$$^{\alpha}X_i = [a_i, b_i] \tag{6.46}$$

The ordinates of all the vertices are defined as the combinations of n pairs of endpoints of interval numbers at the α-cut. For example, in the case of $n = 3$, eight vertices (i.e., 2^3) exist as follows:

$$v_1 = (a_1, a_2, a_3) \tag{6.47a}$$

$$v_2 = (a_1, a_2, b_3) \tag{6.47b}$$

TABLE 6.4
A Second Set of Random Numbers Used for Completion Time of Tasks

Task A	Task B	Task C	Task D	Task E	Task F	Task G
0.606883	0.093464	0.125703	0.736216	0.585157	0.033755	0.719628
0.277315	0.682777	0.75993	0.485396	0.288004	0.697372	0.101427
0.725585	0.326737	0.091488	0.718726	0.819744	0.9123	0.910932
0.179915	0.471119	0.07271	0.293896	0.559946	0.441863	0.749723
0.152424	0.240208	0.294833	0.769227	0.786163	0.12152	0.663357
0.168486	0.035771	0.51356	0.880006	0.748794	0.115441	0.953369
0.915682	0.2436	0.610186	0.848375	0.102922	0.009326	0.801494
0.124135	0.682049	0.610019	0.203327	0.081627	0.86644	0.514767
0.342101	0.739733	0.131999	0.569512	0.388688	0.518582	0.204704
0.985961	0.613146	0.914132	0.898415	0.543517	0.091718	0.97092
0.336867	0.616759	0.402409	0.268781	0.913337	0.0987	0.545388
0.583809	0.471045	0.343964	0.278476	0.128413	0.359243	0.341192
0.798033	0.053788	0.467997	0.405734	0.923671	0.587813	0.126547
0.688703	0.028898	0.021365	0.039026	0.483284	0.54659	0.267746
0.959589	0.749079	0.914929	0.72902	0.917082	0.870119	0.652013
0.331024	0.626462	0.697033	0.771629	0.382801	0.702866	0.060994
0.201754	0.233297	0.417021	0.770881	0.034672	0.724181	0.395496
0.633503	0.38085	0.538246	0.326588	0.633842	0.176778	0.346776
0.840578	0.895108	0.071531	0.714916	0.400981	0.243865	0.211002
0.531249	0.46347	0.952944	0.07302	0.345216	0.578557	0.214954

$$v_3 = (a_1, b_2, a_3) \tag{6.47c}$$

$$v_4 = (a_1, b_2, b_3) \tag{6.47d}$$

$$v_5 = (b_1, a_2, a_3) \tag{6.47e}$$

$$v_6 = (b_1, a_2, b_3) \tag{6.47f}$$

$$v_7 = (b_1, b_2, a_3) \tag{6.47g}$$

$$v_8 = (b_1, b_2, b_3) \tag{6.47h}$$

where v_j is the jth vertex. In general, the number of vertices is given by

$$\text{Number of vertices} = 2^n \tag{6.48}$$

Equation 6.45 can be evaluated at the α-cut as follows:

$$^{\alpha}Y\left[\min_{j=1}^{n} f(v_j), \max_{j=1}^{n} f(v_j)\right] \tag{6.49}$$

TABLE 6.5
A Second Set of Generated Random Values for Completion Time of Tasks

Task A	Task B	Task C	Task D	Task E	Task F	Task G	Project Completion Time
3.2707799	1.34004	2.8529732	0.8156949	4.2147164	0.171328	3.5813727	12.578237
2.4095244	2.2375281	4.7058017	0.481743	3.4411537	2.5164544	1.7263661	12.871519
3.5991597	1.775744	2.6681462	0.7893469	4.9142851	3.3552784	4.3467318	17.991199
2.0844111	1.9638631	2.54382	0.2291748	4.1505155	1.8540637	3.6733168	13.72617
1.9739182	1.6473216	3.4610715	0.8680251	4.7929741	0.8324813	3.4212035	13.375756
2.0399005	1.0987845	4.0339016	1.0875602	4.6703954	0.8017872	4.6788003	15.272445
4.3768346	1.6527585	4.2793768	1.0147408	2.7347579	−0.3528514	3.8468091	12.258309
1.8453495	2.2365066	4.2789404	0.0851884	2.6055421	3.1097732	3.0369229	13.116367
2.593705	2.3210996	2.8829455	0.5873924	3.7176888	2.0464701	2.1752441	12.854208
5.1966454	2.1435484	5.366874	1.1363712	4.1090447	0.6695408	4.8949117	17.013691
2.5794094	2.1482732	3.7532931	0.1919199	5.3618234	0.7108243	3.113757	13.914087
3.2112634	1.96377	3.598772	0.2064942	2.8660028	1.6399625	2.5912269	12.272226
3.8344438	1.1952375	3.9198922	0.3809357	5.4304724	2.2215315	1.8570525	14.538738
3.4917621	1.0511701	1.9732268	−0.3812494	3.9582012	2.116783	2.3806975	12.998614
4.7463211	2.3356456	5.3719798	0.8047524	5.3859459	3.1270236	3.3903168	18.985253
2.5633476	2.1610288	4.5154829	0.8719888	3.7023022	2.5322679	1.4531975	13.075239
2.1647959	1.6361127	3.7908605	0.870752	2.1834205	2.5949514	2.7353984	12.175383
3.3407015	1.8485931	4.0957866	0.2755382	4.341604	1.0723773	2.6064022	13.209678
3.9968036	2.6271607	2.5352335	0.7837149	3.7496031	1.3063624	2.1972483	13.877178
3.0782144	1.9542634	5.674461	−0.2269675	3.6021733	2.1978306	2.2108603	13.458358

Example 6.5 Three Fuzzy Variables

The vertex method is illustrated using the following function with three fuzzy variables (Dong and Shah, 1987):

$$Y = f(X_1, X_2, X_3) = X_1(X_2 - X_3) \tag{6.50}$$

For an α-cut that results in the following respective intervals for the three fuzzy variables:

$$\text{The } X_1 \text{ interval} = [1, 2] \text{ with } a_1 = 1 \text{ and } b_1 = 2 \tag{6.51a}$$

$$\text{The } X_2 \text{ interval} = [2, 3] \text{ with } a_2 = 2 \text{ and } b_2 = 3 \tag{6.51b}$$

$$\text{The } X_3 \text{ interval} = [1, 4] \text{ with } a_3 = 1 \text{ and } b_3 = 4 \tag{6.51c}$$

In this case, the eight vertices are

$$v_1 = (1, 2, 1) \quad (6.52a)$$

$$v_2 = (1, 2, 4) \quad (6.52b)$$

$$v_3 = (1, 3, 1) \quad (6.52c)$$

$$v_4 = (1, 3, 4) \quad (6.52d)$$

$$v_5 = (2, 2, 1) \quad (6.52e)$$

$$v_6 = (2, 2, 4) \quad (6.52f)$$

$$v_7 = (2, 3, 1) \quad (6.52g)$$

$$v_8 = (2, 3, 4) \quad (6.52h)$$

The function evaluations can now be performed. For example, for $v_1 = (1, 2, 1)$, the function evaluation is

$$f(v_1) = 1(2 - 1) = 1 \quad (6.53)$$

Then, Equation 6.49 can be used to obtain the final result at this α-cut as follows:

$$^{\alpha}Y = [\min(1,-2,2,-1,2,-4,4,-2), \max(1,-2,2,-1,2,-4,4,-2)] = [-4,4] \quad (6.54)$$

It can be observed from this example that the vertex method has the property of invariance when the form of the expression is changed to

$$Y = f(X_1, X_2, X_3) = X_1X_2 - X_1X_3 \quad (6.55)$$

The same result of [–4, 4] can be obtained using Equation 6.55, although requiring the evaluation of 16 vertices.

6.3 PROPAGATION OF MIXED UNCERTAINTY TYPES

6.3.1 A Fundamental Input–Output System

A fundamental input–output system is introduced in this section to facilitate the introduction and discussion of using methods of propagating uncertainty in systems of the type that can be modeled by Equation 6.1. The system is represented using an algebraic problem set, as was identified by Sandia National Laboratories to be a basic building block for uncertainty propagation in computational mechanics (Oberkampf et al., 2003). The problem set is based on a model structure that is known with certainty and provided as follows:

$$Y = (A + B)^A \tag{6.56}$$

where A and B are the parameters that are independent, and positive and real numbers. This model represents the response Y of a system. Six problem types that reflect various uncertainty representations of A and B are examined and solved in subsequent sections. The solutions presented in this section are based on methods that propagate uncertainties using endpoints of the input intervals to demonstrate the propagation processes. In order to obtain the output interval endpoints, all possible combinations of all values in the input intervals should to propagated using the respective methods, and solutions as output interval endpoints can be obtained through incremental, numerical evaluations throughout the input intervals and using max or min operators. In some of the problems, the endpoints of the output intervals might not correspond to the input interval endpoint evaluations.

6.3.2 Interval Parameters

The parameters in Equation 6.56 in this case are provided in the form of intervals as follows:

$$A = [a_1, a_2] \tag{6.57a}$$

$$B = [b_1, b_2] \tag{6.57b}$$

The interval arithmetic definition of the power of a positive real-valued interval $[b_1, b_2]$ using a positive real-valued power (a) can be defined as

$$[b_1, b_2]^a = [b_1^a, b_2^a] \tag{6.58}$$

Using an interval, positive real-valued power $[a_1, a_2]$, the interval arithmetic definition of the power of a positive real-valued interval $[b_1, b_2]$ is

$$[b_1, b_2]^{[a_1, a_2]} = [b_1^{a_1}, b_2^{a_2}] \tag{6.59}$$

Based on Equations 6.58 and 6.59, the response Y can be computed utilizing interval addition as follows:

$$Y = [[a_1, a_2] + [b_1, b_2]]^{[a_1, a_2]} = [y_1, y_2] \tag{6.60}$$

where

$$y_1 = [a_1 + b_1]^{a_1} \tag{6.61a}$$

$$y_2 = [a_2 + b_2]^{a_2} \tag{6.61b}$$

A pseudo-computational code is as follows:

```
a  =  [al,  ar]
b  =  [bl,  br]
al  =  left(a)
ar  =  right(a)
bl  =  left(b)
br  =  right(b)
cl  = min((ar + br)^ar, (ar + bl)^ar, (al + br)^al,
(al + bl)^al)
cr  = max((ar + br)^ar, (ar + bl)^ar, (al + br)^al,
(al + bl)^al)
c  =  [cl,  cr]
```

Example 6.6 Uncertainty Propagation Using Interval Parameters

This problem is illustrated using the following values for the parameters A and B:

$$A = [0.1,\ 1.0]$$

$$B = [0.0,\ 1.0]$$

The response based on the endpoints can be computed as follows:

$$Y = [[0.1,1.0] + [0.0,1.0]]^{[0.1,1.0]} = [y_1, y_2] = [0.7943282, 2.0]$$

where $y_1 = [0.1 + 0.0]^{0.1} = 0.7943282$, and $y_2 = [1.0 + 1.0]^{1.0} = 2.0$. The endpoint evaluation does not produce the correct value. The minimum value occurs at values not at the ends. The true solutions based on incremental numerical evaluations are $y_1 = 0.692201$, and $y_2 = 2.0$.

6.3.3 A Power as an Interval and a Set of Intervals

The parameters of Equation 6.56 in this case are provided as follows:

$$A = [a_1, a_2] \tag{6.62a}$$

$$B_i = [b_{i1}, b_{i2}] \quad \text{for } i = 1, 2, \ldots, n \tag{6.62b}$$

The information on B is provided based on n independent sources. The universal set of B is defined as the union of the n intervals. Three cases are considered herein based on specific additional information on B.

6.3.3.1 A Consonant or Nested Set of Intervals

The B_i intervals are nested according to the following structure:

$$B_i \subseteq B_{i+1} \quad \text{for } i = 1,2,\ldots,n-1 \tag{6.63}$$

Since the B_i intervals are equally credible, they can be given a basic assignment $m = 1/n$. The belief and plausibility measures, i.e., necessity and possibility, respectively, can be computed as follows:

$$Bel(B_i) = \sum_{all\ B_j \subseteq B_i} m(B_j) \tag{6.64}$$

$$Pl(B_i) = \sum_{all\ B_j \cap B_i \neq \varnothing} m(B_j) \tag{6.65}$$

Equations 6.64 and 6.65 can be evaluated as follows:

i	B_i	$Bel(B_i)$	$Pl(B_i)$
1	B_1	$1/n$	1
2	B_2	$2/n$	1
3	B_3	$3/n$	1
⋮	⋮	⋮	⋮
n	B_n	1	1

(6.66)

Equation 6.66 can now be used to compute the response according to each B_i, and the resulting interval should be associated with the corresponding Bel and Pl.

A pseudo-computational code is as follows:

```
a = [al, ar]
b(1) = [bl1, br1]
b(2) = [bl2, br2]
b(3) = [bl3, br3]
b(4) = [bl4, br4]
al = left(a)
ar = right(a)
```

```
for i = 1 to 4 do begin
bl = left(b(i))
br = right(b(i))
cl = min((ar + br)^ar, (ar + bl)^ar, (al + br)^al,
(al + bl)^al)
cr = max((ar + br)^ar, (ar + bl)^ar, (al + br)^al,
(al + bl)^al)
continue
c = [cl, cr]
```

Example 6.7 Uncertainty Propagation Using a Consonant or Nested Set of Intervals

This case is illustrated herein using the following values for the parameters A and B:

$$A = [0.1, 1.0]$$

$$B_1 = [0.6, 0.8], B_2 = [0.4, 0.85], B_3 = [0.2, 0.9], \text{ and } B_4 = [0.0, 1.0]$$

These intervals are nested as provided in Figure 6.3. The response can be computed using Equations 6.60 and 6.66 as provided in Table 6.6A based on endpoint evaluations as an approximation. The true answer can be obtained using a search algorithm to determine the minimum and maximum of the output intervals by considering all the values in the intervals, called interval evaluation, as opposed to the previous endpoint evaluations. The results in this case are shown in Table 6.6B.

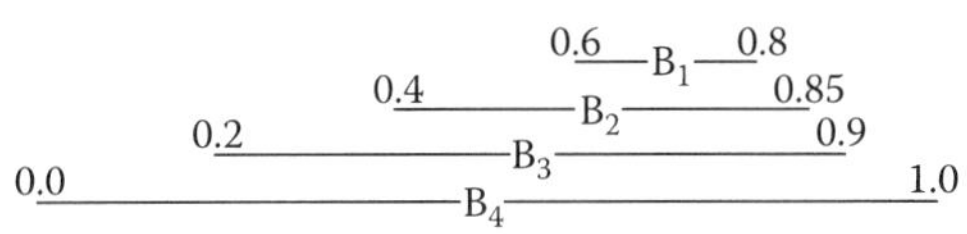

FIGURE 6.3 A consonant nested set of intervals.

TABLE 6.6A
Endpoint Uncertainty Propagation for a Consonant or Nested Set of Intervals

i	B_i	$Bel(B_i)$	$Pl(B_i)$	y_1	y_2
1	[0.6, 0.8]	0.25	1.00	0.9649611	1.80
2	[0.4, 0.85]	0.50	1.00	0.9330329	1.85
3	[0.2, 0.9]	0.75	1.00	0.8865681	1.90
4	[0.0, 1.0]	1.00	1.00	0.7943282	2.00

TABLE 6.6B
Interval Uncertainty Propagation for a Consonant or Nested Set of Intervals

Interval	Output Intervals Based on A = [0.1, 1.0]		a value that produced	b value that produced
B_i	y_1	y_2	y_1	y_2
B_1 = [0.6, 0.8]	0.956196	1.80	0.187900	0.60
B_2 = [0.4, 0.85]	0.897511	1.85	0.268900	0.40
B_3 = [0.2, 0.9]	0.810958	1.90	0.334700	0.20
B_4 = [0.0, 1.0]	0.692201	2.00	0.367900	0.00

6.3.3.2 A Consistent Set of Intervals

The B intervals according to Equation 6.56 are structured such that

$$B_i \cap B_j \neq \phi \quad \text{for } i = 1,2,\ldots,n \text{ and } j = 1,2,\ldots,n \tag{6.67}$$

Similar to the previous case, since the B intervals are equally credible, they can be given a basic assignment $m = 1/n$. The belief and plausibility measures can be computed using Equations 6.64 and 6.65. Then, Equation 6.60 can be used to compute the response according to each B interval, and the resulting interval should be associated with the corresponding *Bel* and *Pl*.

Example 6.8 Uncertainty Propagation Using a Consistent Set of Intervals

This case is illustrated using the following values for the parameters A and B:

$$A = [0.1, 1.0]$$

$$B_1 = [0.6, 0.9], B_2[0.4, 0.8], B_3 = [0.1, 0.7], \text{ and } B_4 = [0.0, 1.0]$$

These intervals have a common range as provided in Figure 6.4. The response can be computed using Equations 6.63, 6.64, and 6.65 as provided in Table 6.7a for all the B intervals based on endpoint evaluations as an approximation. The true answer can be

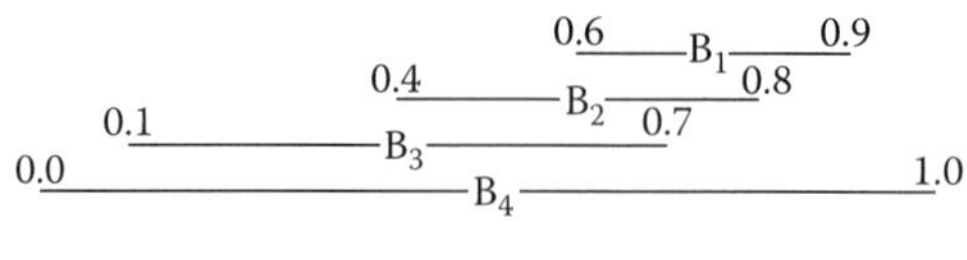

FIGURE 6.4 A consistent set of intervals.

TABLE 6.7A
Endpoint Uncertainty Propagation for a Consistent Set of Intervals

i	B_i	$Bel(B_i)$	$Pl(B_i)$	y_1	y_2
1	[0.6, 0.9]	0.25	1.00	0.9649611	1.90
2	[0.4, 0.8]	0.25	1.00	0.9330329	1.80
3	[0.1, 0.7]	0.25	1.00	0.8513399	1.70
4	[0.0, 1.0]	1.00	1.00	0.7943282	2.00

TABLE 6.7B
Interval Uncertainty Propagation for a Consistent Set of Intervals

Interval	Output Intervals Based on A = [0.1, 1.0]		a value that produced	b value that produced
B_i	y_1	y_2	y_1	y_2
B_1 = [0.6, 0.9]	0.956196	1.90	0.187900	0.60
B_2 = [0.4, 0.8]	0.897511	1.80	0.268900	0.40
B_3 = [0.1, 0.7]	0.756118	1.70	0.357700	0.10
B_4 = [0.0, 1.0]	0.692201	0.692201	2.00	0.367900

TABLE 6.8
Uncertainty Propagation for a Consistent Set of Intervals for the Common Interval B_c

i	B_i	$Bel(B_i)$	$Pl(B_i)$	y_1	y_2
c	[0.6,0.7]	0.25	1.00	0.9649611	1.70

obtained using a search algorithm to determine the minimum and maximum of the output intervals by considering all the values in the intervals, called interval evaluation, as opposed to the previous endpoint evaluations. The results in this case are shown in Table 6.7b. The common range (B_c) among all the B intervals might be of special interest, and its response can be assessed as provided in Table 6.8 based on Table 6.7a.

In cases involving a common range for all intervals, the belief and plausibility of the common range (B_c) were computed based on extension from possibility theory concepts, since B_c is common to all B intervals as follows:

$$Bel(B_c) = \min[Bel(B_i)] \tag{6.68a}$$

$$Pl(B_c) = \max[Pl(B_i)] \tag{6.68b}$$

Equation 6.68a and b is provided as a preliminary solution, and additional investigation is needed in order to qualify it for a particular application.

A pseudo-computational code is as follows:

```
a = [al, ar]
b(1) = [bl1, br1]
b(2) = [bl2, br2]
b(3) = [bl3, br3]
b(4) = [bl4, br4]
al = left(a)
ar = right(a)
for i = 1 to 4 do begin
bl = left(b(i))
br = right(b(i))
cl = min((ar + br)^ar, (ar + bl)^ar, (al + br)^al,
(al + bl)^al)
cr = max((ar + br)^ar, (ar + bl)^ar, (al + br)^al,
(al + bl)^al)
continue
c = [cl, cr]
```

6.3.3.3 An Arbitrary Set of Intervals

In this case, the *B* intervals are provided in any arbitrary structure. Similar to the previous case, since the B_i intervals are equally credible, they can be given a basic assignment $m = 1/n$. The belief and plausibility measures can be computed using Equations 6.64 and 6.65. Then, Equation 6.60 can be used to compute the response according to each B_i, and the resulting interval should be associated with the corresponding *Bel* and *Pl*.

A pseudo-computational code is as follows:

```
a = [al, ar]
b(1) = [bl1, br1]
b(2) = [bl2, br2]
```

```
b(3) = [bl3, br3]
b(4) = [bl4, br4]

al = left(a)
ar = right(a)

for i = 1 to 4 do begin
bl = left(b(i))
br = right(b(i))
cl = min((ar + br)^ar, (ar + bl)^ar, (al + br)^al,
(al + bl)^al)
cr = max((ar + br)^ar, (ar + bl)^ar, (al + br)^al,
(al + bl)^al)
continue
c = [cl, cr]
```

EXAMPLE 6.9 UNCERTAINTY PROPAGATION USING AN ARBITRARY SET OF INTERVALS

This case is illustrated using the following values for the parameters A and B of Equation 6.56:

$$A = [0.1, 1.0]$$

$$B_1 = [0.6, 0.8], B_2 = [0.5, 0.7], B_3 = [0.1, 0.4], \text{ and } B_4 = [0.0, 1.0]$$

These intervals do not have a common range and can be represented as provided in Figure 6.5. The response can be assessed using Equations 6.60, 6.64, and 6.65 as provided in Table 6.9a based on endpoint evaluations as an approximation. The true answer can be obtained using a search algorithm to determine the minimum and maximum of the output intervals by considering all the values in the intervals, called interval evaluation, as opposed to the previous endpoint evaluations. The results in this case are shown in Table 6.9b.

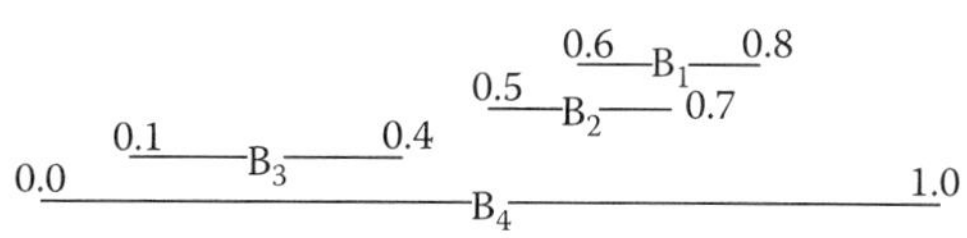

FIGURE 6.5 An arbitrary set of intervals.

TABLE 6.9A
Endpoint Uncertainty Propagation for an Arbitrary Set of Intervals

i	B_i	$Bel(B_i)$	$Pl(B_i)$	y_1	y_2
1	[0.6, 0.8]	0.25	0.75	0.9649611	1.80
2	[0.5, 0.7]	0.25	0.75	0.9502002	1.70
3	[0.1, 0.4]	0.25	0.50	0.8513399	1.40
4	[0.0, 1.0]	1.00	1.00	0.7943282	2.00

TABLE 6.9B
Interval Uncertainty Propagation for an Arbitrary Set of Intervals

Interval	Output Intervals Based on A = [0.1, 1.0]		a value that produced	b value that produced
B_i	y_1	y_2	y_1	y_2
B_1 = [0.6, 0.8]	0.956196	1.80	0.187900	0.60
B_2 = [0.5, 0.7]	0.930174	1.70	0.229800	0.50
B_3 = [0.1, 0.4]	0.756118	1.40	0.357700	0.10
B_4 = [0.0, 1.0]	0.692201	0.692201	2.00	0.367900

6.3.4 Sets of Intervals

In this case, the parameters are provided as follows:

$$A_i = [a_{i1}, a_{i2}] \quad \text{for } i = 1, 2, \ldots, k \tag{6.69a}$$

$$B_i = [b_{i1}, b_{i2}] \quad \text{for } i = 1, 2, \ldots, n \tag{6.69b}$$

The information on A and B is provided based on k and n independent sources, respectively. The universal sets of A and B are defined as the union of the respective k and n intervals. Three cases are considered herein based on specific additional information on A and B.

6.3.4.1 Consonant or Nested Sets of Intervals

The A_i and B_i intervals are nested according to the following structure:

$$A_i \subseteq A_{i+1} \quad \text{for } i = 1, 2, \ldots, k-1 \tag{6.70a}$$

$$B_i \subseteq B_{i+1} \quad \text{for } i = 1, 2, \ldots, n-1 \tag{6.70b}$$

Since the A_i and B_i intervals are equally credible, they can be given the basic assignments $m_A = 1/k$ and $m_B = 1/n$, respectively. The belief and plausibility measures, i.e., necessity and possibility, respectively, can be computed according to Equations 6.64 and 6.65 as follows:

i	A_i	$Bel(A_i)$	$Pl(A_i)$
1	A_1	$1/k$	1
2	A_2	$2/k$	1
3	A_3	$3/k$	1
⋮	⋮	⋮	⋮
n	A_n	1	1

(6.71a)

and

i	B_i	$Bel(B_i)$	$Pl(B_i)$
1	B_1	$1/n$	1
2	B_2	$2/n$	1
3	B_3	$3/n$	1
⋮	⋮	⋮	⋮
n	B_n	1	1

(6.71b)

Equation 6.6 can now be used to compute the response according to each combination of A_i and B_i, and the resulting interval should be associated with the corresponding *Bel* and *Pl* using the intersection relationships from the following rules:

$$Bel(A \cap B) = \min[Bel(A), Bel(B)] \tag{6.72a}$$

$$Bel(A \cup B) \geq \max[Bel(A), Bel(B)] \tag{6.72b}$$

$$Pl(A \cap B) \leq \min[Pl(A), Pl(B)] \tag{6.72c}$$

$$Pl(A \cup B) = \max[Pl(A), Pl(B)] \tag{6.72d}$$

Example 6.10 Uncertainty Propagation Using Consonant or Nested Sets of Intervals

This case is illustrated using the following values for the parameters A and B:

$$A_1 = [0.5, 0.7], A_2 = [0.3, 0.8], \text{ and } A_3 = [0.1, 1.0]$$

$$B_1 = [0.6, 0.8], B_2 = [0.4, 0.85], B_3 = [0.2, 0.9], \text{ and } B_4 = [0.0, 1.0]$$

TABLE 6.10A
Endpoint Uncertainty Propagation for a Consonant or Nested Sets of Intervals

A_i Intervals	(Bel, Pl) =	$B_1 = [0.6, 0.8]$ (0.25, 1.00)	$B_2 = [0.4, 0.85]$ (0.50, 1.00)	$B_3 = [0.2, 0.9]$ (0.75, 1.00)	$B_4 = [0.0, 1.0]$ (1.00, 1.00)
$A_1 = [0.5, 0.7]$	(0.33, 1.00)	$y_1 = 1.048809$ $y_2 = 1.328201$ (0.25, 1.00)	$y_1 = 0.948683$ $y_2 = 1.359040$ (0.33, 1.00)	$y_1 = 0.836666$ $y_2 = 1.389581$ (0.33, 1.00)	$y_1 = 0.707107$ $y_2 = 1.449821$ (0.33, 1.00)
$A_2 = [0.3, 0.8]$	(0.67, 1.00)	$y_1 = 0.968886$ $y_2 = 1.456451$ (0.25, 1.00)	$y_1 = 0.898523$ $y_2 = 1.492750$ (0.50, 1.00)	$y_1 = 0.812252$ $y_2 = 1.528830$ (0.67, 1.00)	$y_1 = 0.696845$ $y_2 = 1.600361$ (0.67, 1.00)
$A_3 = [0.1, 1.0]$	(1.00, 1.00)	$y_1 = 0.964961$ $y_2 = 1.80$ (0.25, 1.00)	$y_1 = 0.933033$ $y_2 = 1.85$ (0.50, 1.00)	$y_1 = 0.886568$ $y_2 = 1.90$ (0.75, 1.00)	$y_1 = 0.794328$ $y_2 = 2.00$ (1.00, 1.00)

TABLE 6.10B
Interval Uncertainty Propagation for a Consonant or Nested Sets of Intervals

	$B_1 = [0.6, 0.8]$		$B_2 = [0.4, 0.85]$		$B_3 = [0.2, 0.9]$		$B_4 = [0.0, 1.0]$	
A_i Intervals	y_1	y_2	y_1	y_2	y_1	y_2	y_1	y_2
$A_1 = [0.5, 0.7]$	1.048809	1.328201	0.948683	1.359040	0.836660	1.389581	0.707107	1.449821
$A_2 = [0.3, 0.8]$	0.968886	1.456451	0.898523	1.492750	0.810958	1.528830	0.692201	1.600361
$A_3 = [0.1, 1.0]$	0.956196	1.80	0.897511	1.85	0.810958	1.90	0.692201	2.00

The B_i intervals are the same as the previous corresponding case. The response can be computed with the Pl, for the resulting interval is an upper bound according to Equation 6.72c, as shown in Table 6.10a based on endpoint evaluations as an approximation. The true answer can be obtained using a search algorithm by evaluating Equation 6.72a and c to determine the minimum and maximum of the output intervals by considering all the values in the intervals, called interval evaluation, as opposed to the previous endpoint evaluations. The results in this case are shown in Table 6.10b.

6.3.4.2 Consistent Sets of Intervals

The A_i and B_i intervals are structured such that

$$A_i \cap A_j \neq \phi \quad \text{for } i = 1,2,\ldots,k \text{ and } j = 1,2,\ldots,k \tag{6.73a}$$

$$B_i \cap B_j \neq \phi \quad \text{for } i = 1,2,\ldots,n \text{ and } j = 1,2,\ldots,n \tag{6.73b}$$

Similar to the previous case, since the A_i and B_i intervals are equally credible, they can be given the basic assignments $m_A = 1/k$ and $m_B = 1/n$, respectively. The belief and plausibility measures can be computed using Equations 6.64 and 6.65. Then, Equation 6.60 can be used to compute the response according to pair of A_i and B_i, and the resulting interval should be associated with the corresponding *Bel* and *Pl* using Equation 6.72a and c.

Example 6.11 Uncertainty Propagation Using Consistent Sets of Intervals

This case is illustrated using the following values for the parameters *A* and *B*:

$$A_1 = [0.5, 1.0], A_2 = [0.2, 0.7], \text{ and } A_3 = [0.1, 0.6]$$

$$B_1 = [0.6, 0.9], B_2 = [0.4, 0.8], B_3 = [0.1, 0.7], \text{ and } B_3 = [0.0, 1.0]$$

The response can be computed using Equations 6.60, 6.64, and 6.65 as shown in Table 6.11a based on endpoint evaluations as an approximation. The true answer can be obtained using a search algorithm to determine the minimum and maximum of the output intervals by considering all the values in the intervals, called interval evaluation, as opposed to the previous endpoint evaluations. The results in this case are shown in Table 6.11b. The common ranges (A_c and B_c) can be treated similarly to the case presented in Section 6.6.2 if needed.

TABLE 6.11A
Endpoint Uncertainty Propagation for a Consistent Sets of Intervals

A_i Intervals	(*Bel*, *Pl*) =	B_1 = [0.6, 0.9] (0.25, 1.00)	B_2 = [0.4, 0.8] (0.50, 1.00)	B_3 = [0.1, 0.7] (0.75, 1.00)	B_4 = [0.0, 1.0] (1.00, 1.00)
A_1 = [0.5, 1.0]	(0.33, 1.00)	y_1 = 1.048809 y_2 = 1.90 (0.25, 1.00)	y_1 = 0.948683 y_2 = 1.80 (0.33, 1.00)	y_1 = 0.774597 y_2 = 1.70 (0.33, 1.00)	y_1 = 0.707107 y_2 = 2.00 (0.33, 1.00)
A_2 = [0.2, 0.7]	(0.67,1.00)	y_1=0.956353 y_2=1.389581 (0.25,1.00)	y_1=0.902880 y_2=1.328201 (0.50,1.00)	y_1=0.786003 y_2=1.265580 (0.67,1.00)	y_1=0.724780 y_2=1.449821 (0.67,1.00)
A_3=[0.1,0.6]	(1.00, 1.00)	y_1 = 0.964961 y_2 = 1.275426 (0.25, 1.00)	y_1 = 0.933033 y_2 = 1.223705 (0.50, 1.00)	y_1 = 0.851340 y_2 = 1.170485 (0.75, 1.00)	y_1 = 0.794328 y_2 = 1.325782 (1.00, 1.00)

TABLE 6.11B
Interval Uncertainty Propagation for a Consonant or Nested Sets of Intervals

	B_1 = [0.6, 0.9]		B_2 = [0.4, 0.8]		B_3 = [0.1, 0.7]		B_4 = [0.0, 1.0]	
A_i Intervals	y_1	y_2	y_1	y_2	y_1	y_2	y_1	y_2
A_1 = [0.5, 1.0]	1.048809	1.90	0.948683	1.80	0.774597	1.70	0.707107	2.00
A_2 = [0.2, 0.7]	0.956353	1.389581	0.897511	1.328201	0.756118	1.265580	0.692201	1.449821
A_3 = [0.1, 0.6]	0.956196	1.275426	0.897511	1.223705	0.756118	1.170485	0.692201	1.325782

6.3.4.3 Arbitrary Sets of Intervals

In this case, the A_i and B_i intervals are provided in any arbitrary structures. Similar to the previous case, since the A_i and B_i intervals are equally credible, they can be given the basic assignments $m_A = 1/k$ and $m_B = 1/n$, respectively. The belief and plausibility measures can be computed using Equations 6.64 and 6.65. Then, Equation 6.60 can be used to compute the response according to the pair of A_i and B_i, and the resulting interval should be associated with the corresponding *Bel* and *Pl* using Equations 6.73a and 6.74c.

6.3.5 A Power as an Interval and as a Set of Intervals

Three cases of Equation 6.56 are considered in this section: (1) a power as an interval and a lognormally distributed parameter, (2) an interval power and an uncertain lognormally distributed parameter, and (3) a set of power intervals and an uncertain lognormally distributed parameter.

6.3.5.1 Power Intervals and Lognormally Distributed Parameter

In this case, the parameters are provided as follows:

$$A = [a_1, a_2] \tag{6.74}$$

$$\ln(B) \sim N(\mu, \sigma) \tag{6.75}$$

where B according to Equation 6.75 is lognormally distributed with the parameters μ and σ. The mean (μ_B) and variance (σ_B^2) of B can be computed from the parameters as follows:

$$\mu_B = \exp\left(\mu + \frac{1}{2}\sigma^2\right) \tag{6.76a}$$

$$\sigma_B^2 = \mu_B^2(\exp(\sigma^2) - 1) \tag{6.76b}$$

Equation 6.76a and b can be inverted as follows:

$$\sigma^2 = \ln\left(1 + \left(\frac{\sigma_B}{\mu_B}\right)^2\right) \tag{6.77a}$$

$$\mu = \ln(\mu_B) - \frac{1}{2}\sigma^2 \tag{6.77b}$$

Monte Carlo simulation can be used to evaluate the response according to the following steps:

- Randomly generate B to obtain b values according to its probability distribution as provided in Equation 6.75.
- Compute the response interval as follows:

$$Y = [[a_1, a_2] + b]^{[a_1, a_2]} = [y_1, y_2] \tag{6.78}$$

where

$$y_1 = [a_1, b]^{a_1} \tag{6.79a}$$

$$y_2 = [a_2 + b]^{a_2} \tag{6.79b}$$

- Repeat the simulation process N times and compute the moments and distribution types of y_1 and y_2.

Example 6.12 Uncertainty Propagation Using a Power as an Interval and a Lognormally Distributed Parameter

For the following parameters,

$$A = [0.1, 1.0]$$

$$\ln(B) \sim N(0.5, 0.5)$$

simulation was used to compute the response. A total of 100 simulation cycles produced the response moments and histograms that show bimodal characteristics, as summarized and shown in Table 6.12 and Figure 6.6.

TABLE 6.12
Uncertainty Propagation for an Interval Power and a Lognormally Distributed Parameter

Moment	B	y_1	y_2
Mean	1.923245	1.062593	2.923245
Standard deviation	1.009403	0.049313	1.009403
Coefficient of variation	0.5248	0.0464	0.3453

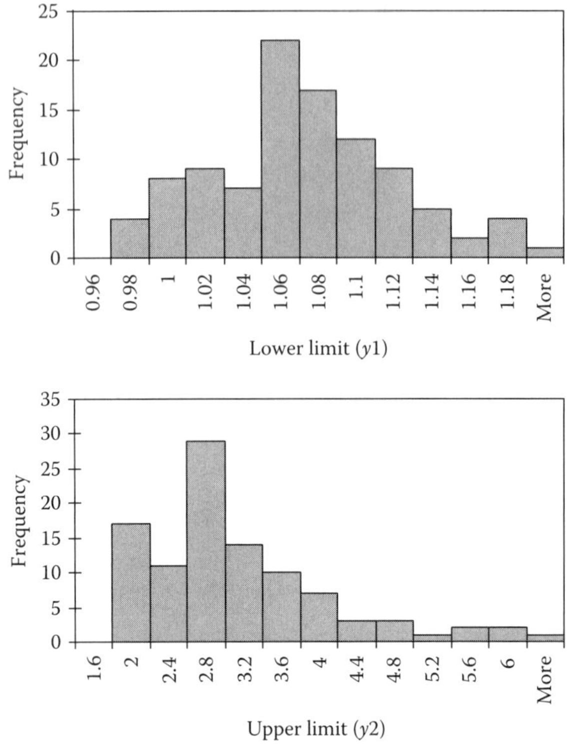

FIGURE 6.6 Uncertainty propagation for an interval power and a lognormally distributed parameter.

6.3.5.2 A Power as an Interval and an Uncertain Lognormally Distributed Parameter

In this case, the parameters are provided as follows:

$$A = [a_1, a_2] \tag{6.80}$$

$$\ln(B) \sim N([\mu_1, \mu_2], [\sigma_1, \sigma_2]) \tag{6.81}$$

The second-order uncertainty provided in characterizing the lognormal parameter can be rolled into the parameters using Monte Carlo simulation to obtain $\ln(B) \sim N(\mu, \sigma)$. Then, the computational procedure presented in the previous section can be used to solve the problem.

6.3.5.3 A Set of Power Intervals and an Uncertain Lognormally Distributed Parameter

The parameters, in this case, are provided as follows:

$$A_i = [a_{i1}, a_{i2}] \quad \text{for } i = 1, 2, \ldots, k \tag{6.82}$$

$$\ln(B_i) \sim N([\mu_{i1}, \mu_{i2}], [\sigma_{i1}, \sigma_{i2}]) \quad \text{for } i = 1, 2, \ldots, n \tag{6.83}$$

The information on A and B is provided based on k and n independent sources, respectively. The universal sets of A and B are defined as the union of the respective k and n intervals. Three cases can be developed as combinations of the computational procedures of previous cases.

EXERCISE PROBLEMS

6.1. The change in the length of a rod due to axial force P is given by

$$\Delta L = \frac{PL}{AE}$$

where L = length of rod, P = applied axial force, A = cross-sectional area of the rod, and E = modulus of elasticity. If P and E are normally distributed with known moments, determine the moments (i.e., mean and variance) of the change in length using first-order approximation for two cases: (1) uncorrelated P and E and (2) correlated P and E. Assume A and L are nonrandom.

6.2. The ultimate moment capacity, M, of an underreinforced concrete rectangular section is given by

$$M = A_s f_y \left(d - \frac{a}{2} \right)$$

where

$$a = \frac{A_s f_y}{0.85\, b f_c'}$$

in which the following are random variables: A_s is the cross-sectional area of the reinforcing steel, f_y is the yield stress (strength) of the steel, d is the distance from the reinforcing steel to the top fibers of the beam, b is the width of the beam, and f_c' is the ultimate stress (strength) of the concrete. If the random variables are assumed to be statistically noncorrelated, determine the first-order mean and variance of the moment capacity.

6.3. For the change in the length of a rod due to axial force P provided in problem 6.1, use 20 simulation cycles to determine the mean and variance of ΔL assuming noncorrelated random variables with the following probabilistic characteristics:

Random Variable	Mean Value	Coefficient of Variation	Distribution Type
P	100 kips	0.35	Lognormal
L	20 in.	0.05	Normal
E	30,000 ksi	0.10	Lognormal
A	1 in.^{2}	0.05	Normal

6.4. For the rod in problem 6.1, study the effect of increasing the number of simulation cycles on the estimated mean and variance of the deformation. Use the following numbers of simulation cycles: 20, 100, 500, 1000, 2000, and 10,000. Provide your results in the form of plots of estimated statistics as a function of the number of simulation cycles.

6.5. For the ultimate moment capacity, M, of an underreinforced concrete rectangular section provided in problem 6.2, assume the random variables to be statistically noncorrelated and determine the mean and variance of the moment capacity using the following information:

Random Variable	Mean Value	Coefficient of Variation	Distribution Type
A_s	0.25 in.^{2}	0.10	Lognormal
f_y	40,000 psi	0.10	Normal
d	20 in.	0.05	Lognormal
b	12 in.	0.05	Normal
f_c'	4000 psi	0.20	Lognormal

Use 100 simulation cycles. Is this a sufficient number of cycles? Why? Discuss.

6.6. Demonstrate that the vertex method property of invariance holds in Example 6.5 based on using the two expressions of the function as follows:

$$Y = f(X_1, X_2, X_3) = X_1(X_2 - X_3)$$

$$Y = f(X_1, X_2, X_3) = X_1X_2 - X_1X_3$$

6.7. Redo Example 6.5 using the following function:

$$Y = f(X_1, X_2, X_3) = X_1^2X_2 - \sqrt{X_3}$$

6.8. Redo Example 6.5 using the following function:

$$Y = f(X_1, X_2, X_3) = X_1 - \sqrt{X_2 X_3}$$

6.9. Redo Example 6.6 using the following values for the parameters A and B:

$$A = [0.1, 2.0]$$

$$B = [0.0, 2.0]$$

6.10. Redo Example 6.7 using the following values for the parameters A and B:

$$A = [0.1, 2.0]$$

$$B_1 = [0.6, 0.8], B_2 = [0.4, 1.0], B_3 = [0.2, 1.5], \text{ and } B_4 = [0.0, 2.0]$$

6.11. Redo Example 6.8 using the following values for the parameters A and B:

$$A = [0.1, 2.0]$$

$$B_1 = [0.6, 0.9], B_2 = [0.4, 1.5], B_3 = [0.1, 1.8], \text{ and } B_4 = [0.0, 2.0]$$

6.12. Redo Example 6.9 using the following values for the parameters A and B:

$$A = [0.1, 2.0]$$

$$B_1 = [0.6, 0.8], B_2 = [0.9, 1.2], B_3 = [0.1, 0.8], \text{ and } B_4 = [0.0, 2.0]$$

6.13. Redo Example 6.10 using the following values for the parameters A and B:

$$A_1 = [0.5, 0.7], A_2 = [0.3, 0.8], \text{ and } A_3 = [0.1, 1.0]$$

$$B_1 = [0.6, 0.8], B_2 = [0.4, 1.0], B_3 = [0.2, 1.5], \text{ and } B_4 = [0.0, 2.0]$$

6.14. Redo Example 6.11 using the following values for the parameters A and B:

$$A_1 = [0.5, 0.7], A_2 = [0.3, 0.8], \text{ and } A_3 = [0.1, 1.0]$$

$$B_1 = [0.6, 0.9], B_2 = [0.4, 1.5], B_3 = [0.1, 1.8], \text{ and } B_4 = [0.0, 2.0]$$

6.15. Provide a solution procedure of Equation 6.56 with uncertain parameters and demonstrate the procedure for the case of arbitrary sets of intervals (Section 6.3.4.3).

6.16. The solution procedures described in previous sections of Equation 6.56 with uncertain parameters are based on computing the ends of the intervals. Reexamine and modify the solution procedure as needed or provide an alternate solution procedure, and demonstrate the procedure for the case of an interval power and a lognormally distributed parameter (Section 6.3.5.1).

6.17. Provide a solution procedure of Equation 6.56 with uncertain parameters and demonstrate the procedure for the case of an interval power and an uncertain lognormally distributed parameter (Section 6.3.5.2).

6.18. Provide a solution procedure of Equation 6.56 with uncertain parameters and demonstrate the procedure for the case of a set of power intervals and an uncertain lognormally distributed parameter (Section 6.3.5.3).

6.19. Select a physics-based input–output system and demonstrate the use of the following cases for the model parameters:
   a. A consonant or nested set of intervals (Section 6.3.3.1)
   b. A consonant set of intervals (Section 6.3.3.2)
   c. An arbitrary set of intervals (Section 6.3.3.3)

6.20. Select a physics-based input–output system and demonstrate the use of the following cases for the model parameters:
   a. Consonant or nested sets of intervals (Section 6.3.4.1)
   b. Consonant sets of intervals (Section 6.3.4.2)
   c. Arbitrary sets of intervals (Section 6.3.4.3)

6.21. Select a physics-based input–output system and demonstrate the use of the following cases for the model parameters:
   a. An interval power and a lognormally distributed parameter (Section 6.3.5.1)
   b. An interval power and an uncertain lognormally distributed parameter (Section 6.3.5.2)
   c. A set of power intervals and an uncertain lognormally distributed parameter (Section 6.3.5.3)

6.22. Use the following input–output regression model with two parameters a and b and demonstrate the use of the following cases for the model parameters:

$$y = aX + b \quad \text{where } 1 \leq X \leq 2$$

   a. A consonant or nested set of intervals (Section 6.3.3.1)
   b. A consonant set of intervals (Section 6.3.3.2)
   c. An arbitrary set of intervals (Section 6.3.3.3)

6.23. Use the following input–output regression model with two parameters a and b and demonstrate the use of the following cases for the model parameters:

$$y = aX + b \quad \text{where } 1 \leq X \leq 2$$

a. Consonant or nested sets of intervals (Section 6.3.4.1)
b. Consonant sets of intervals (Section 6.3.4.2)
c. Arbitrary sets of intervals (Section 6.3.4.3)

6.24. Use the following input–output regression model with two parameters a and b and demonstrate the use of the following cases for the model parameters:

$$y = aX + b \quad \text{where } 1 \leq X \leq 2$$

a. An interval power and a lognormally distributed parameter (Section 6.3.5.1)
b. An interval power and an uncertain lognormally distributed parameter (Section 6.3.5.2)
c. A set of power intervals and an uncertain lognormally distributed parameter (Section 6.3.5.3)

7 Expert Opinions and Elicitation Methods

7.1 INTRODUCTION

Dealing with uncertainty and system complexity requires us to rely on *experts* to address issues. Experts are sought out when encountering uncertainty, ignorance, and complexity. We commonly deal with or listen to experts on a regular basis, such as weather forecasts by weather experts, stock and financial reports by analysts and experts, suggested medication or procedures by medical professionals, policies by politicians, and analyses by world affairs experts. We know from our own experiences that experts are valuable sources of information and knowledge, and can also be wrong in their views rendered to us. Expert opinions, therefore, can be considered to include or constitute nonfactual information. The fallacy of these opinions might disappoint us, but do not surprise us, since issues that require experts tend to be difficult or complex, sometimes with divergent views. The nature of some of these complex issues could only yield views that have subjective truth levels; therefore, they allow for contradictory views that might be all somewhat credible.

Experts, with all their importance and value, can be viewed as double-edged swords. Not only do they bring in a deep knowledge base and thoughts, but also they could provide biases and *pet* theories. The selection of experts, elicitation of their opinions, and aggregation of the opinions should be performed and handled carefully by recognizing uncertainties associated with the opinions, and sometimes with skepticism.

Expert opinion elicitation can be defined as a heuristic process of gathering information and data or answering questions on issues or problems of concern. In this chapter, a focus on occurrence probabilities and adverse consequences of events as primary elements in risk analysis is provided to demonstrate the process presented. Expert opinion elicitation should not be used in lieu of rigorous studies, but should be used to supplement them and to prepare for them, or in cases where data are lacking and cannot be collected.

The expert opinion elicitation process presented in this chapter is a variation of the Delphi technique (Helmer, 1968) with scenario analysis (Kahn and Wiener, 1967) based on uncertainty models (Ayyub, 2001, 2003; Cooke, 1991), social research (Bailey, 1994), U.S. Army Corps of Engineers (USACE) studies (Ayyub et al., 1996), the ignorance, knowledge, information, and uncertainty of Chapter 1, nuclear industry recommendations (NRC, 1997), NUREG/CR-6372 (Budnitz et al., 1997) and NUREG/CR-1563 (Kotra et al., 1996), and the Stanford Research Institute protocol (Spetzler and Stael von Holstein, 1975). Ayyub (2001) provides additional information on expert opinion elicitation.

7.2 TERMINOLOGY

The terminology of Table 7.1 is needed for defining the expert opinion elicitation process. The *expert opinion elicitation* (EE) process is defined as a formal, heuristic process of gathering information and data or answering questions on issues or problems of concern. The EE process requires the involvement of a *leader* of the EE process who is an entity having managerial and technical responsibility for organizing and executing the project, overseeing all participants, and intellectually owning the results.

An *expert* is defined as a skillful person who has a lot of training and has knowledge in some special field. The expert is the provider of an opinion in the process of expert opinion elicitation. An *evaluator* is an expert who has the role of evaluating the relative credibility and plausibility of multiple hypotheses to explain observations. The process involves *evaluators*, who consider available data, become familiar with the views of proponents and other evaluators, question the technical bases of data, and challenge the views of proponents, as well as *observers*, who can contribute to the discussion but cannot provide expert opinion that enters into the aggregated opinion of the experts. The process might require peer reviewers who can provide an unbiased assessment and critical review of an expert opinion elicitation process, its technical issues, and results. Some of the experts might be *proponents*, experts who advocate a particular hypothesis or technical position. In science, a proponent evaluates experimental data and professionally offers a hypothesis that would be challenged by the proponent's peers until proven correct or wrong. *Resource experts* can be used who are technical experts with detailed and deep knowledge of particular data, issue aspects, particular methodologies, or use of evaluators.

The *sponsor* of EE process is an entity that provides financial support and owns the rights to the results of the EE process. Ownership is in the sense of property ownership. A *subject* is a person who might be affected or might affect an issue or question of interest for the process.

A *technical facilitator* (TF) is an entity responsible for structuring and facilitating the discussions and interactions of experts in the EE process, staging effective interactions among experts, ensuring equity in presented views, eliciting formal evaluations from each expert, and creating conditions for direct, noncontroversial integration of expert opinions. A *technical integrator* (TI) is an entity responsible for developing the composite representation of issues based on informed members and sources of related technical communities and experts; explaining and defending composite results to experts and outside experts, peer reviewers, regulators, and policy makers; and obtaining feedback and revising composite results. A *technical integrator and facilitator* (TIF) is an entity responsible for both functions of TI and TF.

7.3 CLASSIFICATION OF ISSUES, STUDY LEVELS, EXPERTS, AND PROCESS OUTCOMES

The NRC (1997) classified issues for expert opinion elicitation purposes into three complexity degrees (A, B, or C), with four levels of study in the expert opinion elicitation process (I, II, III, and IV), as shown in Table 7.2. A given issue is assigned a complexity degree and a level of study that depends on (1) the significance of the

TABLE 7.1
Terminology and Definitions

Term	Definition
Evaluators	Evaluators consider available data, become familiar with the views of proponents and other evaluators, question the technical bases of data, and challenge the views of proponents
Expert	A person with related or unique experience to an issue or question of interest for the process
Expert opinion elicitation (EE) process	A formal, heuristic process of gathering information and data or answering questions on issues or problems of concern
Leader of EE process	An entity having managerial and technical responsibility for organizing and executing the project, overseeing all participants, and intellectually *owning* the results
Observers	Observers can contribute to the discussion but cannot provide expert opinion that enters into the aggregated opinion of the experts
Peer reviewers	Experts that can provide an unbiased assessment and critical review of an expert opinion elicitation process, its technical issues, and results
Proponents	Experts who advocate a particular hypothesis or technical position; in science, a proponent evaluates experimental data and professionally offers a hypothesis that would be challenged by the proponent's peers until proven correct or wrong
Resource experts	Technical experts with detailed and deep knowledge of particular data, issue aspects, particular methodologies, or use of evaluators
Sponsor of EE process	An entity that provides financial support and *owns* the rights to the results of the EE process; ownership is in the sense of property ownership
Subject	A person who might be affected or might affect an issue or question of interest for the process
Technical facilitator (TF)	A person responsible for structuring and facilitating the discussions and interactions of experts in the EE process, staging effective interactions among experts, ensuring equity in presented views, eliciting formal evaluations from each expert, and creating conditions for direct, noncontroversial integration of expert opinions
Technical integrator (TI)	A person responsible for developing the composite representation of issues based on informed members and sources of related technical communities and experts; explaining and defending composite results to experts and outside experts, peer reviewers, regulators, and policy makers; and obtaining feedback and revising composite results
Technical integrator and facilitator (TIF)	A person responsible for both functions of TI and TF

TABLE 7.2
Issue Degrees and Study Levels

a. Issue Complexity Degree		b. Study Level	
Degree	**Description**	**Level**	**Requirements**
A	Noncontroversial Insignificant effect on risk	I	A technical integrator (TI) evaluates and weighs models based on literature review and experience and estimates needed quantities
B	Significant uncertainty Significant diversity Controversial Complex	II	A technical integrator (TI) interacts with proponents and resource experts, assesses interpretations, and estimates needed quantities
C	Highly contentious Significant effect on risk Highly complex	III	A technical integrator (TI) brings together proponents and resource experts for debate and interaction; TI focuses the debate, evaluates interpretations, and estimates needed quantities
		IV	A technical integrator (TI) and technical facilitator (TF) (that can be one person, i.e., TIF) organize a panel of experts to interpret and evaluate, focus discussions, keep the experts' debate orderly, summarize and integrate opinions, and estimate needed quantities

issue to the final goal of the study, (2) the issue's technical complexity and uncertainty level, (3) the amount of nontechnical contention about the issue in the technical community, and (4) important nontechnical consideration, such as budgetary, regulatory, scheduling, public perception, or other concerns. Experts can be classified into five types (NRC, 1997): (1) proponents, (2) evaluators, (3) resource experts, (4) observers, and (5) peer reviewers. These types are defined in Table 7.1.

The study level as shown in Table 7.2 involves a technical integrator (TI) or a technical integrator and facilitator (TIF). A TI can be one person or a team (i.e., an entity) that is responsible for developing the composite representation of issues based on informed members and sources of related technical communities and experts; explaining and defending composite results to experts and outside experts, peer reviewers, regulators, and policy makers; and obtaining feedback and revising composite results. A TIF can be one person or a team (i.e., an entity) that is responsible for the functions of a TI, and structuring and facilitating the discussions and interactions of experts in the EE process, staging effective interactions among experts, ensuring equity in presented views, eliciting formal evaluations from each expert, and creating conditions for direct, noncontroversial integration of expert opinions. The primary difference between the TI and the TIF is in the intellectual responsibility for the study, where it lies with only the TI or with the TIF and the experts,

TABLE 7.3
Guidance on Use of Peer Reviewers

Expert Opinion Elicitation Process	Peer Review Subject	Peer Review Method	Recommendation
Technical integrator	Technical	Participatory	Recommended
and facilitator		Late stage	Can be acceptable
	Process	Participatory	Strongly recommended
		Late stage	Risky: unlikely to be successful
Technical integrator	Technical	Participatory	Strongly recommended
		Late stage	Risky but can be acceptable
	Process	Participatory	Strongly recommended
		Late stage	Risky but can be acceptable

respectively. The TIF also has the added responsibility of maintaining the professional integrity of the process and its implementation.

The TI and TIF processes are required to utilize peer reviewers for quality assurance purposes. Peer review can be classified according to peer review method and according to peer review subject. Two methods of peer review can be performed: (1) participatory peer review, which would be conducted as an ongoing review throughout all study stages, and (2) late-stage peer review, which would be performed as the final stage of the study. The former method allows for affecting the course of the study, whereas the latter one might not be able to affect the study without a substantial rework of the study. The second classification of peer reviews is by peer review subject and has two types: (1) technical peer review, which focuses on the technical scope, coverage, contents, and results, and (2) process peer review, which focuses on the structure, format, and execution of the expert opinion elicitation process. Guidance on the use of peer reviewers is provided in Table 7.3 (NRC, 1997).

The expert opinion elicitation process should preferably be conducted to include a face-to-face meeting of experts that is developed specifically for the issues under consideration. The meeting of the experts should be conducted after communicating to the experts in advance of the meeting background information, objectives, list of issues, and anticipated outcome from the meeting. The expert opinion elicitation based on the technical integrator and facilitator (TIF) concept can result in consensus or disagreement, as shown in Figure 7.1. Consensus can be of four types, as shown in Figure 7.1 (NRC, 1997). Commonly, the expert opinion elicitation process has the objective of achieving consensus type 4; i.e., experts agree that a particular probability distribution represents the overall scientific community. The TIF plays a major role in building consensus by acting as a facilitator. Disagreement among experts, whether it is intentional or unintentional, requires the TIF to act as an integrator by using equal or nonequal weight factors. Sometimes, expert opinions

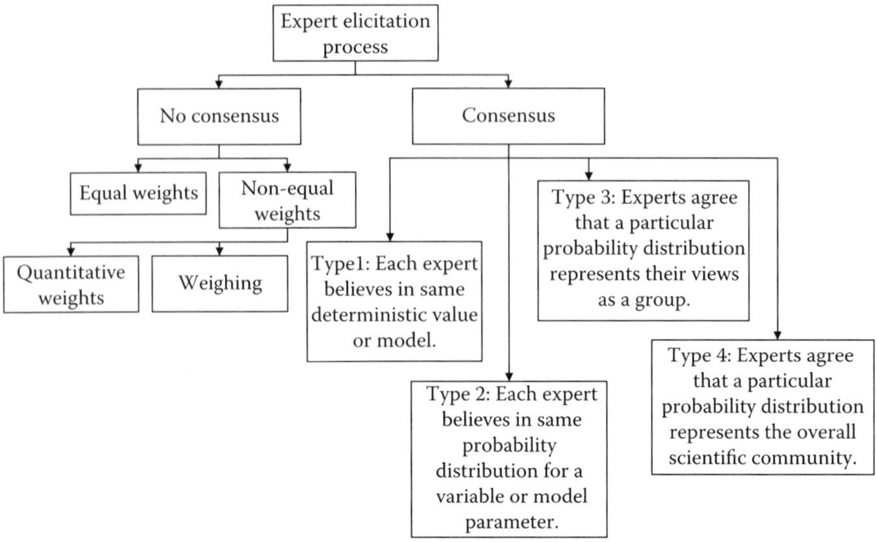

FIGURE 7.1 Outcomes of the expert opinion elicitation process.

need to be weighed for appropriateness and relevance rather than strictly weighted by factors in a mathematical aggregation procedure.

7.4 PROCESS DEFINITION

Expert opinion elicitation was defined as a formal, heuristic process of obtaining information or answers to specific questions about certain quantities, called issues, such as failure rates, failure consequences, and expected service lives. The suggested steps for an expert opinion elicitation process depend on the use of a technical integrator (TI) or a technical integrator and facilitator (TIF), as shown in Table 7.4. Table 7.4 was constructed based on NRC (1997), supplemented with details, and added steps. The details of the steps involved in these two processes are defined in subsequent subsections.

7.5 NEED IDENTIFICATION FOR EXPERT OPINION ELICITATION

The primary reason for using expert opinion elicitation is to deal with uncertainty in selected technical issues related to a system of interest. Issues with significant uncertainty, issues that are controversial or contentious, issues that are complex, and issues that can have a significant effect on risk are most suited for expert opinion elicitation. The value of the expert opinion elicitation comes from its initial intended uses as a heuristic tool, not a scientific tool, for exploring vague and unknown issues that are otherwise inaccessible. It is not a substitute to scientific, rigorous research.

TABLE 7.4
Expert Opinion Elicitation Process

Technical Integration (TI) Process	Technical Integration and Facilitation (TIF) Process
Need identification for expert opinions	Need identification for expert opinions
Select study leader and study level	Select study leader and study level
Select technical integrator	Select technical integrator and facilitator
Identify and select peer reviewers	Identify and select experts and peer reviewers
Define technical issues	Define technical issues
Collect information and analyze	Discuss with experts and refine issues
Perform data diagnostics and summarize results	Train experts for elicitation
Administer peer review	Facilitate expert group discussion
Revise based on reviews	Aggregate results, discuss, and summarize
Document and communicate results	Administer peer review
	Address comments of reviewers
	Document and communicate results

The identification of need and its communication to experts are essential for the success of the expert opinion elicitation process. The need identification and communication should include the definition of the goal of the study and relevance of issues to this goal. Establishing this relevance would make the experts stakeholders, and thereby increase their attention and sincerity levels. The relevance of each issue or question to the study needs to be established. This question-to-study relevance is essential to enhancing the reliability of collected data from the experts. Each question or issue needs to be relevant to each expert, especially when dealing with subjects with diverse views and backgrounds.

7.6 SELECTION OF STUDY LEVEL AND STUDY LEADER

The goal of a study and nature of issues determine the study level as shown in Table 7.2. The study leader can be either a technical integrator (TI), technical facilitator (TF), or a combined technical integrator and facilitator (TIF). The leader of the study is an entity having managerial and technical responsibility for organizing and executing the project, overseeing all participants, and intellectually *owning* the results. The primary difference between the TI and the TIF is in the intellectual responsibility for the study, where it lies with only the TI or with the TIF and the experts, respectively. The TIF also has the added responsibility of maintaining the professional integrity of the process and its implementation. The TI is required to utilize peer reviewers for quality assurance purposes. A study leader should be selected based on the following attributes:

1. An outstanding professional reputation and wide recognition and competence based on academic training and relevant experience

2. Strong communication skills, interpersonal skills, flexibility, impartiality, and ability to generalize and simplify
3. A large contact base of industry leaders, researchers, engineers, scientists, and decision makers
4. Ability to build consensus, and leadership qualities

The study leader does not need to be a subject expert, but should be knowledgeable of the subject matter.

7.7 SELECTION OF PEER REVIEWERS AND EXPERTS

7.7.1 Selection of Peer Reviewers

Peer review can be classified according to peer review method and according to peer review subject. Two methods of peer review can be performed: (1) participatory peer review, which would be conducted as an ongoing review throughout all study stages, and (2) late-stage peer review, which would be performed as the final stage of the study. The second classification of peer reviews is by peer review subject and has two types: (1) technical peer review that focuses on the technical scope, coverage, contents, and results, and (2) process peer review that focuses on the structure, format, and execution of the expert opinion elicitation process. These classifications were discussed in Section 7.3.

Peer reviewers are needed for both the TI and TIF processes. The peer reviewers should be selected by the study leader in close consultation with perhaps the study sponsor. The following individuals should be sought after as peer reviewers:

1. Researchers, scientists, or engineers that have an outstanding professional reputation and widely recognized competence based on academic training and relevant experience
2. Researchers, scientists, and engineers with a general understanding of the issues in other related areas and with relevant expertise and experiences from other areas
3. Researchers, scientists, and engineers who are available and willing to devote the needed time and effort
4. Researchers, scientists, and engineers with strong communication skills, interpersonal skills, flexibility, impartiality, and ability to generalize and simplify

7.7.2 Identification and Selection of Experts

The size of an expert panel should be determined on case-by-case basis. The size should be large enough to achieve a needed diversity of opinion, credibility, and result reliability. In recent expert opinion elicitation studies, a nomination process was used to establish a list of candidate experts by consulting archival literature, technical societies, governmental organization, and other knowledgeable experts (Trauth et al., 1993). Formal nomination and selection processes should establish appropriate criteria for nomination, selection, and removal of experts. For example, the following criteria were used in an ongoing Yucca Mountain seismic hazard analysis (NRC, 1997) to select experts:

1. Strong relevant expertise through academic training, professional accomplishment and experiences, and peer-reviewed publications
2. Familiarity and knowledge of various aspects related to the issues of interest
3. Willingness to act as proponents or impartial evaluators
4. Availability and willingness to commit needed time and effort
5. Specific related knowledge and expertise of the issues of interest
6. Willingness to effectively participate in needed debates, to prepare for discussions, and provide needed evaluations and interpretations
7. Strong communication skills, interpersonal skills, flexibility, impartiality, and ability to generalize and simplify

In this NRC study, criteria were set for expert removal that include failure to perform according to commitments and demands as set in the selection criteria, and unwillingness to interact with members of the study.

The panel of experts for an expert opinion elicitation process should have a balance and broad spectrum of viewpoints, expertise, technical points of view, and organizational representation. The diversity and completeness of the panel of experts is essential for the success of the elicitation process. For example, it can include the following:

1. Proponents who advocate a particular hypothesis or technical position
2. Evaluators who consider available data, become familiar with the views of proponents and other evaluators, question the technical bases of data, and challenge the views of proponents
3. Resource experts who are technical experts with detailed and deep knowledge of particular data, issue aspects, particular methodologies, or use of evaluators

The experts should be familiar with the design, construction, operational, inspection, maintenance, reliability, and engineering aspects of the equipment and components of a facility of interest. It is essential to select people with basic engineering or technological knowledge; however, they do not necessarily need to be engineers. It might be necessary to include one or two experts from management with engineering knowledge of the equipment and components, consequences, safety aspects, administrative and logistic aspects of operation, expert opinion elicitation process, and objectives of this study. One or two experts with a broader knowledge of the equipment and components might be needed. Also, one or two experts with a background in risk analysis and risk-based decision making, and their uses in areas related to the facility of interest, might be needed.

Observers can be invited to participate in the elicitation process. Observers can contribute to the discussion, but cannot provide expert opinion that enters into the aggregated opinion of the experts. The observers provide expertise in the elicitation process, probabilistic and statistical analyses, risk analysis, and other support areas. The composition and contribution of the observers are essential for the success of this process. The observers may include the following:

1. Individuals with a research- or administrative-related background from research laboratories or headquarters of the U.S. Army Corps of Engineers with engineering knowledge of equipment and components of Corps facilities
2. Individuals with expertise in probabilistic analysis, probabilistic computations, consequence computations and assessment, and expert opinion elicitation

A list of names with biographical statements of the study leader, technical integrator, technical facilitator, experts, observers, and peer reviewers should be developed and documented. All attendees can participate in the discussions during the meeting. However, only the experts can provide the needed answers to questions on the selected issues. The integrators and facilitators are responsible for conducting the expert opinion elicitation process. They can be considered to be a part of the observers or experts, depending on the circumstances and the needs of the process.

7.7.3 Items Needed by Experts and Reviewers before the Expert Opinion Elicitation Meeting

The experts and observers need to receive the following items before the expert opinion elicitation meeting:

1. An objective statement of the study.
2. A list of experts, observers, integrators, facilitators, study leader, sponsors, and their biographical statements.
3. A description of the facility, systems, equipment, and components.
4. Basic terminology; definitions should include probability, failure rate, average time between unsatisfactory performances, mean (or average) value, median value, and uncertainty.
5. Failure consequence estimation.
6. A description of the expert opinion elicitation process.
7. A related example on the expert opinion elicitation process and its results, if available.
8. Aggregation methods of expert opinions, such as computations of percentiles.
9. A description of the issues in the form of a list of questions with background descriptions. Each issue should be presented on a separate page with spaces for recording an expert's judgment, any revisions, and comments.
10. Clear statements of expectations from the experts in terms of time, effort, responses, communication, and discussion style and format.

It might be necessary to personally contact individual experts for the purpose of establishing a clear understanding of expectations.

7.8 IDENTIFICATION, SELECTION, AND DEVELOPMENT OF TECHNICAL ISSUES

The technical issues of interest should be carefully selected to achieve certain objectives. In these guidelines, the technical issues are related to the quantitative assessment of failure probabilities and consequences for selected components, subsystems, and systems within a facility. The issues should be selected such that they would have a significant impact on the study results. These issues should be structured in a logical sequence, starting with background statement, followed by questions, and then answer selections or answer format and scales. Personnel with risk analysis background who are familiar with the construction, design, operation, and maintenance of the facility need to define these issues in the form of specific questions. Also, background materials about these issues need to be assembled. The materials will be used to familiarize and train the experts about the issues of interest, as described in subsequent steps.

An introductory statement for the expert opinion elicitation process should be developed that includes the goal of the study and establishes relevance. Instructions should be provided with guidance on expectations, answering the questions, and reporting. The following are guidelines on constructing questions and issues-based social research practices (Bailey, 1994):

1. Each issue can include several questions; however, each question should consist of only one sought-after answer. It is a poor practice to include two questions in one.
2. Questions and issue statements should not be ambiguous. Also, the use of ambiguous words should be avoided. In expert opinion elicitation of failure probabilities, the word *failure* might be vague or ambiguous to some subjects. Special attention should be given to its definition within the context of each issue or question. The level of wording should be kept to a minimum. Also, the choice of the words might affect the connotation of an issue, especially by different subjects.
3. The use of factual questions is preferred over abstract questions. Questions that refer to concrete and specific matters result in desirable concrete and specific answers.
4. Questions should be carefully structured in order to reduce biases of subjects. Questions should be asked in a neutral format, sometimes more appropriately without lead statements.
5. Sensitive topics might require stating questions with lead statements that would establish supposedly accepted social norms in order to encourage subjects to answer the questions truthfully.

Questions can be classified into *open-ended questions* and *closed-ended questions*, as was previously discussed. The format of the question should be selected carefully. The format, scale, and units for the response categories should be selected to best achieve the goal of the study. The minimum number of questions and question order should be selected using the above guidelines.

Once the issues are developed, they should be pretested by administering them to a few subjects for the purpose of identifying and correcting flaws. The results of this pretesting should be used to revise the issues.

7.9 ELICITATION OF OPINIONS

The elicitation process of opinions should be systematic for all the issues according to the steps presented in this section.

7.9.1 Issue Familiarization of Experts

The background materials that were assembled in the previous step should be sent to the experts about 1 to 2 weeks in advance of the meeting with the objective of providing sufficient time for them to become familiar with the issues. The objective of this step is also to ensure that there is a common understanding among the experts of the issues. The background material should include the objectives of the study; a description of the issues and lists of questions for the issues; a description of systems and processes, their equipment and components, the elicitation process, and selection methods of experts; and biographical information on the selected experts. Also, example results and their meaning, methods of analysis of the results, and lessons learned from previous elicitation processes should be made available to them. It is important to break down the questions or issues into components that can be easily addressed. Preliminary discussion meetings or telephone conversations between the facilitator and experts might be necessary in some cases in preparation for the elicitation process.

7.9.2 Training of Experts

This step is performed during the meeting of the experts, observers, and facilitators. During the training the facilitator needs to maintain flexibility to refine wording or even change approach based on feedback from experts. For instance, experts may not be comfortable with "probability," but they may answer on "events per year" or a "recurrence interval." This indirect elicitation should be explored with the experts. The meeting should be started with presentations of background material to establish relevance of the study to the experts, and study goals in order to establish rapport with the experts. Then, information on uncertainty sources and types, occurrence probabilities and consequences, the expert opinion elicitation process, technical issues and questions, and the aggregation of expert opinions should be presented. Also, experts need to be trained on providing answers in an acceptable format that can be used in the analytical evaluation of the failure probabilities or consequences. The experts need to be trained in certain areas, such as the meaning of probability, central tendency, and dispersion measures, especially to experts who are not familiar with the language of probability. Additional training might be needed on consequences, subjective assessment, logic trees, problem structuring tools such as influence diagrams, and methods of combining expert evaluations. Sources of bias, including overconfidence and base rate fallacy, and their contribution to bias and error should be discussed. This step should include a search for any motivational bias of experts, for example, due to previous positions experts have taken in public, experts wanting to influence decisions and funding allocations, preconceived notions that experts will be evaluated by their superiors as a result of their answers, or experts wanting to be perceived as authoritative. These motivational biases, once identified, can sometimes be overcome by redefining the incentive structure for the experts.

7.9.3 Elicitation and Collection of Opinions

The opinion elicitation step starts with a technical presentation of an issue and, by decomposing the issue to its components, discussing potential influences and describing event sequences that might lead to top events of interest. These top events are the basis for questions related to the issue in the next stage of the opinion elicitation step. Factors, limitations, test results, analytical models, and uncertainty types and sources need to be presented. The presentation should allow for questions to eliminate any ambiguity and clarify scope and conditions for the issue. The discussion of the issue should be encouraged. The discussion and questions might result in refining the definition of the issue. Then, a form with a statement of the issue should be given to the experts to record their evaluation or input. The experts' judgment along with their supportive reasoning should be documented about the issue. It is common that experts would be asked to provide several conditional probabilities in order to reduce the complexity of the questions and thereby obtain reliable answers. These conditional probabilities can be based on fault tree and event tree diagrams. Conditioning has the benefit of simplifying the questions by decomposing the problems. Also, it results in a conditional event that has a larger occurrence probability than its underlying events, therefore making the elicitation less prone to biases since experts tend to have a better handle on larger probabilities than very small ones. It is desirable to have the elicited probabilities in the range of 0.1 to 0.9 if possible. Sometimes it might be desirable to elicit conditional probabilities using linguistic terms. If correlation among variables exists, it should be presented to the experts in great detail, and conditional probabilities need to be elicited.

Issues should be dealt with one issue at a time, although sometimes similar or related issues might be considered simultaneously.

7.9.4 Aggregation and Presentation of Results

The collected assessments from the experts for an issue should be assessed for internal consistency, analyzed, and aggregated to obtain composite judgments for the issue. The means, medians, percentile values, and standard deviations need to be computed for the issues. Also, a summary of the reasoning provided during the meeting about the issues needs to be developed. Uncertainty levels in the assessments should also be quantified. Methods for combining expert opinions are provided in previous chapters. The methods can be classified into consensus methods and mathematical methods. The mathematical methods can be based on assigning equal weights to the experts or different weights.

7.9.5 Group Interaction, Discussion, and Revision by Experts

The aggregated results need to be presented to the experts for a second round of discussion and revision. The experts should be given the opportunity to revise their assessments of the individual issues at the end of discussion. Also, the experts should be asked to state the rationale for their statements and revisions. The revised assessments of the experts need to be collected for aggregation and analysis. This step can produce either consensus or no consensus, as shown in Figure 7.1. The selected aggregation procedure might require eliciting weight factors from the experts. In this step the technical facilitator plays a major role in developing a consensus and maintaining the integrity and credibility of the elicitation process. Also, the technical integrator is needed to aggregate the results without biases with reliability measures. The integrator might need to deal with varying expertise levels for the experts, outliers (i.e., extreme views), nonindependent experts, and expert biases.

7.10 DOCUMENTATION AND COMMUNICATION

A comprehensive documentation of the process is essential in order to ensure acceptance and credibility of the results. The document should include complete descriptions of the steps, the initial results, revised results, consensus results, and aggregated results spreads and reliability measures.

Example 7.1 Risk-Based Approval of Personal Flotation Devices

With the introduction of inflatable personal flotation devices (PFDs), the U.S. Coast Guard (USCG) and the PFD industry were faced with limitations inherent within the current PFD approval practice. Inflatable PFDs perform better than inherently buoyant PFDs in some aspects, but they involve new hazards that were not present in the traditional inherently buoyant PFDs. For the approval of inflatable PFDs, it became apparent that in some areas such devices offered performance advantages over inherently buoyant PFDs, but had some disadvantages in other areas. The need to perform equivalency analysis of engineering designs is a common problem for the regulation of engineering systems. Therefore, an improved process for evaluating and comparing PFD performance is needed. The introduction of this concept applied to PFD analysis required the use of expert opinion elicitation to model the relationships between performance variables of PFDs and the probability of the PFDs meeting the needs of a person from the population of potential users, i.e., relationships between the performance levels of a PFD and respective fractions of the population — that their needs will be met at these levels. Example performance measures include (1) freeboard, defined as a distance measured perpendicular to the surface of the water to the lowest point where the PFD user's respiration may be impeded; (2) face plane angle, defined as the angle, relative to the surface of the water, of the plane formed by the most forward part of the forehead and chin of a user floating in the attitude of static balance;

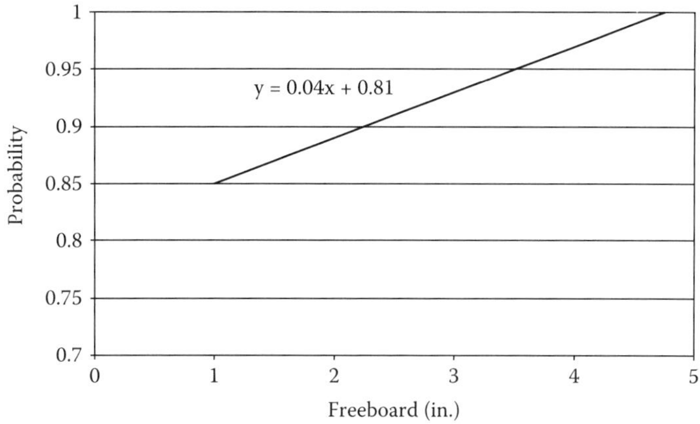

FIGURE 7.2 Probability of meeting the needs of a PFD user and freeboard.

(3) chin support, defined as the PFD device being in direct contact with the jaw line while the subject is in either the vertical upright or relaxed face-up position; (4) torso angle, defined as the angle between a vertical line and a line passing through the shoulder and hip; and (5) turning time, defined as the average time required for a device to turn a facedown wearer to a position in which the wearer's respiration is not impeded and the proportion of test subjects are turned face up. These sample performance measures are used in this example to illustrate the use of expert opinion elicitation to develop relationships between varying performance levels and the respective fractions of the population — that their needs will be met at these levels.

Personal Flotation Device Freeboard (FB)

Freeboard is defined as a distance measured perpendicular to the surface of the water to the lowest point where the user's respiration may be impeded. The objective of freeboard is to minimize the probability of drowning. Greater freeboard means that user movement and water movement are less likely to cause mouth immersion and water inhalation. Figure 7.2 shows a linear relationship between FB and the probability of meeting the needs of a PFD user based on expert opinion elicitation. Defining this linear relationship requires two points that were elicited from experts, as shown in Table 7.5, for the freeboard needed to achieve a probability of 1, the absolute minimum freeboard, and the probability that corresponds to the absolute minimum freeboard.

Personal Flotation Device Face Plane Angle (FPA)

Face plane angle is defined as the angle, relative to the surface of the water, of the plane formed by the most forward part of the forehead and chin of a user floating in the attitude of static balance. The face plane angle's objective is to decrease the probability of drowning. A positive angle is achieved when a user's forehead is higher than his or her chin. Proper face plane angle decreases chances of water inhalation. Figure 7.3 shows a linear relationship between FPA and the probability of meeting the needs of a PFD user based on expert opinion elicitation. Defining this linear relationship

TABLE 7.5
Expert Opinion Elicitation for Freeboard

Values to Define Model	Expert Opinion Collection									Expert Opinion Aggregation				
	Expert 1	Expert 2	Expert 3	Expert 4	Expert 5	Expert 6	Expert 7	Expert 8	Expert 9	Minimum	25th	50th	75th	Maximum
Freeboard at probability of 1	5	5	3.5	4.5	4	4.75	4.75	5	4.75	3.5	4.25	4.75	5	5
Absolute minimum freeboard	0.5	0.5	1	1	0.5	0.75	1	1	1	0.5	0.5	1	1	1
Probability at absolute minimum freeboard	0.85	0.8	0.95	0.8	0.8	0.85	0.8	0.9	0.9	0.8	0.8	0.85	0.9	0.95

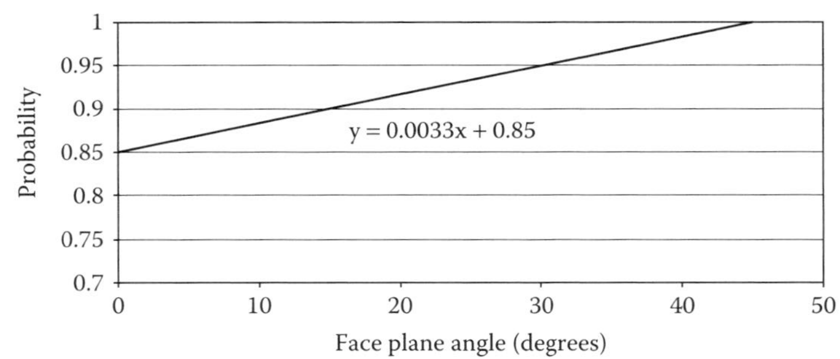

FIGURE 7.3 Probability of meeting the needs of a PFD user and face plane angle.

TABLE 7.6
Expert Opinion Elicitation for Face Plane Angle

	Expert Opinion Collection									Expert Opinion Aggregation				
Values to Define Model	**Expert 1**	**Expert 2**	**Expert 3**	**Expert 4**	**Expert 5**	**Expert 6**	**Expert 7**	**Expert 8**	**Expert 9**	**Minimum**	**25th**	**50th**	**75th**	**Maximum**
Face plane angle at probability of 1	35	90	30	45	25	60	90	45	45	25	32.5	45	75	90
Absolute minimum face plane angle	5	–5	–10	0	–5	3	15	0	15	–10.0	–5	0	10	15
Probability at absolute minimum face plane angle	0.8	0.75	0.9	0.9	0.8	0.9	0.85	0.9	0.5	0.5	0.775	0.85	0.9	0.9

requires two points that were elicited from experts, as shown in Table 7.6, for face plane angle at probability of 1, absolute minimum face plane angle, and the probability at the absolute minimum.

Personal Flotation Device Chin Support (CS)

Chin support is defined as the PFD device is in direct contact with the jaw line while the subject is in either the vertical upright or relaxed face-up position. Chin support is to aid the unconscious or exhausted user from allowing the face to fall in the water and then inhaling water. Chin support is also considered adequate if the device prevents the subject from touching the chin to the chest while the subject is in the relaxed face-up position of static balance. Figure 7.4 shows two cases for the chin support: either provided by the PFD design or not provided by the PFD design. Defining this relationship requires eliciting one value, as shown in Table 7.7, for PFD effectiveness without chin support.

Personal Flotation Device Torso Angle (TA)

Torso angle is the angle between a vertical line and a line passing through the shoulder and hip. A desirable torso angle aids in preventing both mouth immersions due to waves and being tipped facedown by wearer or wave movement. A positive torso angle is achieved when a test participant's hips are forward with respect to his or her shoulders. Figure 7.5 shows a linear relationship between TA and the probability of meeting the needs of a PFD user based on expert opinion elicitation. Defining this linear relationship requires two points that were elicited from experts, as shown in Table 7.8, for torso angle at probability of 1, absolute minimum torso angle, and the probability at the absolute minimum.

Personal Flotation Device Turning Time (TT) from Facedown

Turning time is defined as the average time required for a device to turn a facedown wearer to a position in which the wearer's respiration is not impeded and the proportion of test subjects are turned face up. The faster the turning time on as large a portion of the population as possible, the more likely the PFD is to prevent drowning for an unconscious person. Figure 7.6 shows a linear relationship between TT and the probability of meeting the needs of a PFD user based on expert opinion elicitation. Defining this linear relationship requires two points that were elicited from experts, as shown in Table 7.9, for torso angle at probability of 1, absolute maximum torso angle, and the probability at the absolute maximum.

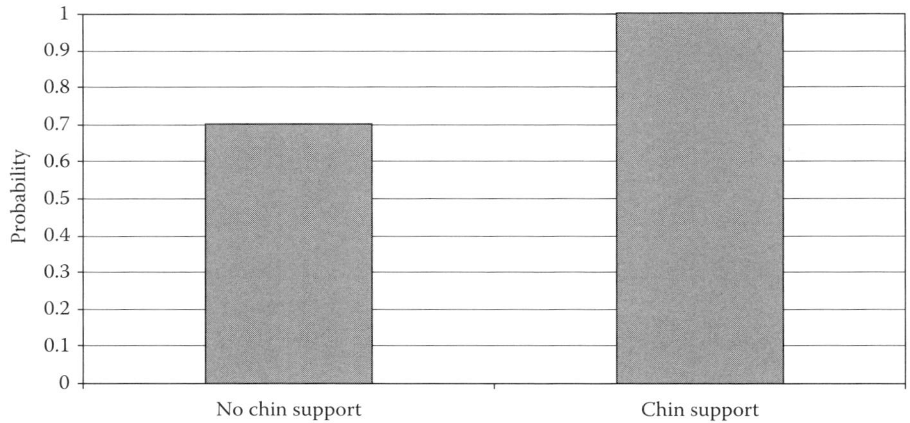

FIGURE 7.4 Probability of meeting the needs of a PFD user without chin support.

TABLE 7.7
Expert Opinion Elicitation for Chin Support

	Expert Opinion Collection									Expert Opinion Aggregation				
Values to Define Model	**Expert 1**	**Expert 2**	**Expert 3**	**Expert 4**	**Expert 5**	**Expert 6**	**Expert 7**	**Expert 8**	**Expert 9**	**Minimum**	**25th**	**50th**	**75th**	**Maximum**
Probability that the PFD is effective with no chin support	0.7	0.6	0.7	0.7	0.5	0.5	0.7	0.7	0.5	0.5	0.55	0.7	0.7	0.7

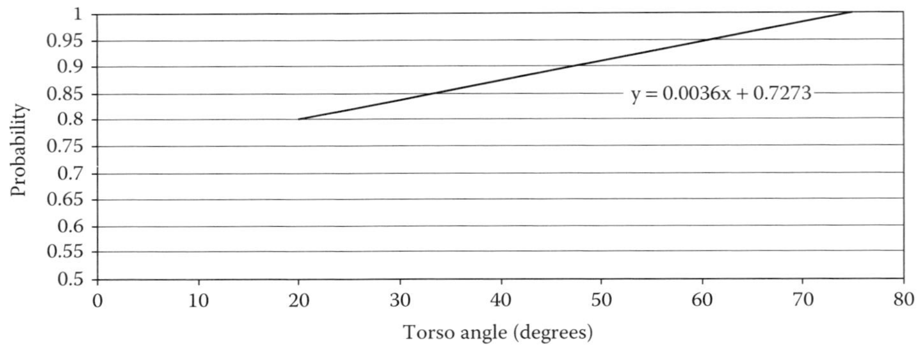

FIGURE 7.5 Probability of meeting the needs of a PFD user and face plane angle.

TABLE 7.8
Expert Opinion Elicitation for Face Plane Angle

	Expert Opinion Collection									Expert Opinion Aggregation				
Values to Define Model	**Expert 1**	**Expert 2**	**Expert 3**	**Expert 4**	**Expert 5**	**Expert 6**	**Expert 7**	**Expert 8**	**Expert 9**	**Minimum**	**25th**	**50th**	**75th**	**Maximum**
Torso angle at probability of 1	85	75	60	45	45	80	60	80	75	45	52.5	75	80	85
Absolute minimum torso angle	30	30	20	20	20	10	15	45	15	10	15	20	30	45
Probability at absolute minimum torso angle	0.75	0.8	0.85	0.9	0.8	0.8	0.85	0.8	0.5	0.5	0.775	0.8	0.85	0.9

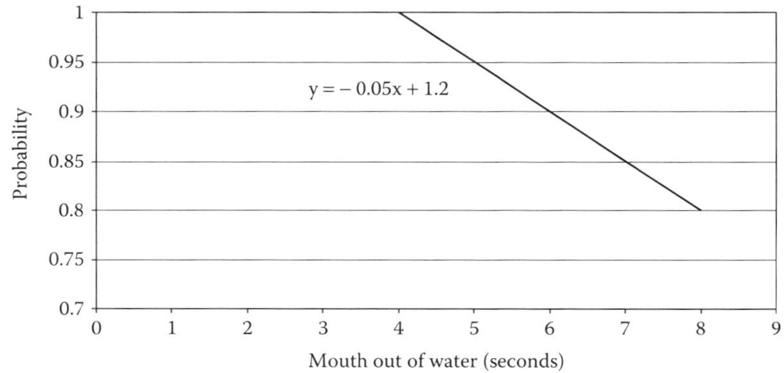

FIGURE 7.6 Probability of meeting the needs of a PFD user and turning time.

TABLE 7.9
Expert Opinion Elicitation for Turning Time

	Expert Opinion Collection									Expert Opinion Aggregation				
Values to Define Model	**Expert 1**	**Expert 2**	**Expert 3**	**Expert 4**	**Expert 5**	**Expert 6**	**Expert 7**	**Expert 8**	**Expert 9**	**Minimum**	**25th**	**50th**	**75th**	**Maximum**
Turning time at probability of 1	2.5	3	3	3	5	5	4	5	5	2.5	3	4	5	5
Absolute maximum turning time	6	8	6.5	8	10	10	7	10	10	6	6.75	8	10	10
Probability at absolute maximum turning time	0.85	0.6	0.5	0.8	0.8	0.75	0.8	0.8	0.9	0.5	0.675	0.8	0.83	0.9

EXERCISE PROBLEMS

7.1. What are the differences between technical facilitator, and technical integrator and facilitator in an expert opinion elicitation process?

7.2. What are the success requirements for selecting experts and developing an expert panel? How many experts would you recommend? For your range on the number of experts, provide guidance in using the lower and upper ends of the range.

7.3. Working in teams, select five classmates as a panel of experts and elicit their opinions on five forecasting issues in engineering. Select these issues such that the classmates can pass the tests of experts on these issues. Perform all the steps of expert opinion elicitation and document your process and results as a part of solving this problem.

7.4. You are asked to form an expert panel and perform expert opinion elicitation about the issues provided below that are concerned with developments by humanity in the current century. In addition to obtaining answers to these questions, you are also being asked to assess the confidence of the participants in their answers on a scale from 1 to 7, corresponding to the highest and smallest confidences, respectively.

a. In your opinion, in what year will the median family income reach twice its present amount?
b. In what year will the percentage of electric automobiles among all automobiles in use reach 50%?
c. In what year will the percentage of intelligent and autonomous (without a driver) automobiles among all automobiles in use reach 50%?
d. By what year will the average life expectancy of a human reach more than 120 years?
e. By what year will it be possible to have commercial carriers to outer space?
f. In what year will a human for the first time travel to Mars, stay at least several days, and return to Earth?

Provide a formal report summarizing the process, listing the experts, and providing answers to these questions.

7.5. Develop a list of communication forecasting issues and elicit opinions similar to that in problem 7.4.

7.6. Develop a list of bioengineering and health forecasting issues and elicit opinions similar to that in problem 7.4.

7.7. Develop a list of power sources and technologies forecasting issues and elicit opinions similar to that in problem 7.4.

7.8. An optimal clearance between the bottom of an overpass bridge and the water surface of a navigation channel needs to be determined to permit safe navigation. A group of seven navigation experts was consulted to offer their opinions about an appropriate design clearance. A formal expert opinion elicitation session resulted in the following opinions:

Clearance Issue	Expert Opinion 1	2	3	4	5	6	7
Clearance (in meters)	50	55	65	70	70	75	80

Aggregate the opinions of the experts by computing the minimum, maximum, 25th percentile, 50th percentile, and 75th percentile values.

7.9. A management consultant is in the process of restructuring the organizational hierarchy of a large corporation. She identified three possible types of organizational structures that are suitable for this large corporation: vertical structure, flat structure, and matrix structure. The selection of a type needs to be based on achieving the highest satisfaction level by employees and their managers. She conducted an expert elicitation session using seven experts and elicited opinions about the best type of structure suitable for the company. The level of satisfaction was measured on a scale of 100 points (lowest level = 0, highest level = 100) with regard to each structure type, as provided in the following table:

Structural Organization Type	Expert Opinions 1	2	3	4	5	6	7
Vertical structure	65	70	70	75	75	80	75
Flat structure	70	85	85	60	75	80	85
Matrix structure	80	70	75	75	90	85	85

Aggregate the opinions of the experts by computing the minimum, maximum, 25th percentile, 50th percentile, and 75th percentile values.

7.10. The probability of performance failure of a newly designed vertical organizational system of a large corporation needs to be assessed by the research and development department of the corporation. The research and development department identified potential sources of this organizational system failure at three management levels: top, middle, and lower management. Nine experts in organizational performances were consulted to offer their opinions and provide probability values. The results are summarized in the following table:

Failure Probability of Vertical Structure	Expert Opinions 1	2	3	4	5	6	7	8	9
Top management level	0.55	0.50	0.45	0.65	0.70	0.65	0.65	0.50	0.65
Middle management level	0.70	0.65	0.65	0.75	0.80	0.70	0.75	0.65	0.70
Lower management level	0.85	0.70	0.85	0.85	0.90	0.80	0.80	0.75	0.80

Aggregate the opinions of the experts by computing the minimum, maximum, 25th percentile, 50th percentile, and 75th percentile values.

8 Visualization of Uncertainty

8.1 INTRODUCTION

Intelligence as the ability to understand and adapt to the environment by using a combination of inherited abilities and learning experiences certainly includes the analysis of uncertainty and making decisions under conditions of uncertainty. Current techniques for visualizing information commonly do not include degrees of certainty (or the degrees and types of ignorance) associated with individual or aggregated information.

For example, for a commander in a battlefield to command, she or he needs to choose. To choose is to decide — almost always on the basis of imperfect information — and momentous decisions require knowledge of threats with a degree of certainty that might not be a requisite for decisions less momentous than waging war. Also, decisions in warfare are commonly made under stress. As warfare grows more complex with networks of ill-defined foes and fronts, battlefield intelligence of both sorts (for decision making and gathering information about the enemy) grows increasingly critical. Sensors and sources, both human and machine, on numerous platforms — manned and unmanned, in the air, on the ground, and under the water — track multitudes of entities and events. Overwhelmed commanders must quickly absorb the flood of data and make life-and-death decisions concerning a fluid and confusing mix of entities: friendly, enemy, civilian, and neutral forces and individuals. The commander must develop a set of goals (or missions) and a mental model of the battle space, and then use this model to make decisions. Forming this mental model, the commander needs to account for the disposition and capability of friendly, enemy, nongovernment organization (NGO), civilian, and neutral forces. Uncertainty is a key element of all of these components of information. However, current information visualization techniques fail to provide the commander with information concerning the degrees of certainty associated with individual or aggregated information elements. Battle space visualization techniques are needed that allow both information and uncertainty to be portrayed effectively and grouped intuitively. Modeling and simulation are promising technologies to host the development of potential visualization technologies, such as intelligent agents, able to assess data uncertainty. Civilian applications can also be constructed to demonstrate this complexity and difficulty.

The present practice of battlefield visualization is illustrated in TRADOC Pam 525-70 (Department of the Army, 1995).

Battle space visualization techniques should allow both information and uncertainty to be portrayed effectively and grouped intuitively. Intelligent agents are promising technologies to host the development of potential visualization technologies to assess data and information uncertainty. Civilian applications can also be constructed to meet societal needs such as Internet *information tagging for uncertainty* and uncertainty visualization of search results.

Ignorance is a key element of data and its processed forms (e.g., information). Current techniques for visualizing information do not help a decision maker sufficiently because they do not include degrees of certainty (or the degrees and types of ignorance) associated with individual or aggregated decision situation entities.

A decision-making process is illustrated in Figure 1.21, showing an ignorance hierarchy, a set of processes that transform data to epiphanies, resulting decisions whose quality may be characterized in a spectrum of appropriate to inappropriate, and the goal of achieving optimum decisions, which are deemed either right (the optimum decision under conditions of certainty — the God's-eye view) or correct (the optimum decision under conditions of uncertainty).

Decision making in the modern age is becoming more difficult. The decision makers' problem is further complicated as each event in a dynamic decision situation is associated with some varying degree of uncertainty. Example dynamic decision situations include waging wars, homeland security applications, disaster management, and air traffic control. In addition, the uncertainty value of potentially important events cannot be assessed in isolation, but must be judged in relation to other events happening at the same time in the same location, as well as events happening at the same time at other locations, to assess conflicting information and confusing situations. Because objects and events in any given decision situation interact, and change rapidly, this degree of complexity can easily overwhelm commanders and other decision makers.

Visualization is any technique for creating images, diagrams, or animations to communicate a message that can be used for the purpose of decision making. A diagram is a simplified and structured visual representation of concepts, ideas, constructions, relations, statistical data, anatomy, etc., used in all aspects of human activities to visualize and clarify a topic. Example diagram types include histograms, probability distributions, tree diagrams, graphs, matrices, networks, flows, and icons. Animation is the technique in which each frame of a film or movie is produced individually, whether generated as a computer graphic, or by photographing a drawn image, or by repeatedly making small changes to a model unit and then photographing the result with a special animation camera. Visualization through visual imagery has been an effective way to communicate both abstract and concrete ideas to enhance decision-making abilities. These methods can be applied to the visualization of uncertainty. The visualization in this case would retain an appropriate level of information and associated uncertainties that are essential to decision makers, and should be tailored for specific applications and types of decision situations under consideration. This chapter provides an introduction to this emerging field of uncertainty visualization.

Example 8.1 Battlefield Uncertainty and Ignorance Types

Symbology in the warfare domain is used for making decisions in battlefields. The primary attributes of objects or units in a battlefield that are communicated by symbology are:

- The identification (ID) attributes of a unit, including:
  - Type
  - Size
  - Intentions
- The position attributes of a unit, including:
  - Location
  - Movement direction
  - Speed

In this section, the ignorance hierarchy provided in Figure 1.19 is examined in terms of the relevance of the ignorance components to the above object attributes communicated by symbology.

The blind ignorance category in Figure 1.19 cannot be represented since information is nonexistent or not adequately processed; however, other attributes, such as conflict and confusion, could be used to infer a state of blind ignorance.

The conscious ignorance with its primary components of inconsistency and incompleteness should be included in proposed visualization methods. Inconsistency results from distortions due to inaccuracy, imprecision, conflict, contradiction, or confusion. Incompleteness results from the absence of knowledge, unknown knowledge (incompleteness in kind), and uncertainty (incompleteness in amount), which can be classified into three types based on its sources of ambiguity (unspecificity and nonspecificity), approximations, and likelihood. Ambiguity arises from having multiple possible outcomes of an event.

Table 8.1 shows a cross-examination of the attributes and selected ignorance types that were judged to be relevant for visualizing ignorance. The table primarily identifies two ignorance types of inconsistency (particularly conflict) and uncertainty in magnitude. In addition to these two primary items, any suggested ignorance symbology should convey the extent (i.e., level) of the ignorance or uncertainty. Movement direction and speed

TABLE 8.1
Battlefield Uncertainty and Ignorance Types

Symbol/Unit Attribute	Uncertainty and Ignorance Types: Inconsistency	Uncertainty and Ignorance Types: Incompleteness
Type	Inconsistency (conflict)	NA
Size	Inconsistency (conflict)	Uncertainty in magnitude
Intentions	Inconsistency (conflict)	NA
Location	NA	Uncertainty in magnitude
Movement direction	NE	NE
Speed	NE	NE

Note: NA = not applicable; NE = not essential and covered by update rates and links.

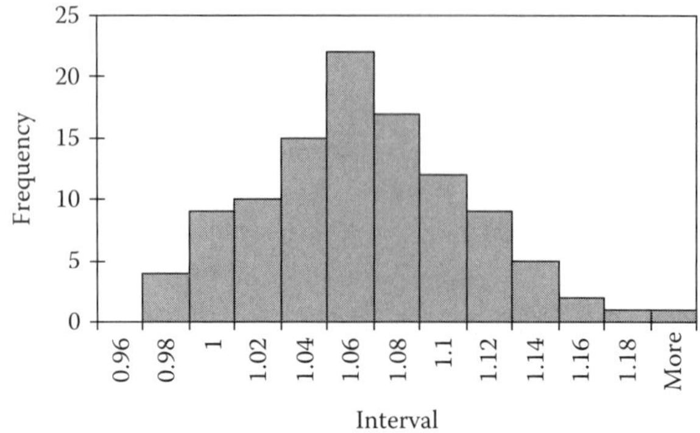

FIGURE 8.1 Frequency histogram.

were judged to be already covered by updates provided by battlefield displays through the update links, and therefore should not be covered by the ignorance symbology.

8.2 VISUALIZATION METHODS

8.2.1 Statistical and Probability-Based Visualization

Traditionally histograms and probability distributions were used to visualize uncertainty represented by random variables. Figure 8.1 shows a histogram that communicates the central tendency, dispersion, skewness, and modal characteristics of a random variable. Figure 8.2 shows a cumulative probability distribution and density functions of a random variable using a mean value of 3000 and a standard deviation of 300. These functions are commonly used to visualize uncertainty.

8.2.2 Point and Global Visualization

Visualization in scientific computing is defined as a method of computing that offers a way for seeing the unseen. It enriches the process of discovery and fosters profound and unexpected insights. Wertheimer (1958), involved with mathematic reasoning in relation with the perception, observed that when elements were gathered into a figure, the figure took on a perceptual salience that exceeded the sum of its parts, leading to what is termed *Gestalt* psychology or theory. It was demonstrated that people extract the global aspects of a scene before smaller (local) details are perceived. The most important aspects of this theory is that in the process of decision making, the solution will never occur unless we add to the given situation some certain contexts and syntheses that eventually change the original meaning and make us cognize the situation and make the correct decision. Such a decision emerges only when our attention has the right direction based on this process.

The global aspects of a scene have processing dominance over local elements. Not only does global perception precede feature-by-feature analysis, but it is also thought to be preattentive in that it occurs without the cognitive effort characteristic of a serial search and analysis of individual elements. The utilizing gestalt theory

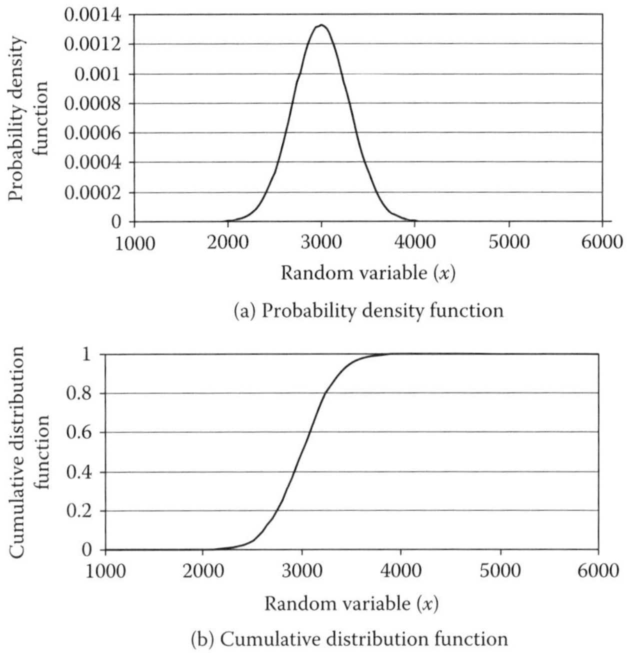

FIGURE 8.2 Probability distributions (mean = 3000, standard deviation = 300). (a) Probability density function. (b) Cumulative distribution function.

of organizing visual presentations supports the emergence of global figures from the chaotic array of discrete elements and allows their perception at very low cognitive cost. In order to apply this approach to the problem of situation awareness and decision making, we *extract a global gestalt from the chaos of moving units in the battlefield*. Wertheimer (1958) has put forward some principles in order to account for the way that humans organize perceptual stimuli. The following three principles are especially important in geometric reasoning and can be helpful in designing the tutoring component of the dynamic environment:

- Principle of proximity, which explains why humans tend to organize elements that are close to each other
- Principle of similarity, which predicts that elements that have a similar structure are perceived together
- Principle of good form, by which, in a complex diagram with multiple configurations, humans tend to perceive only the ones that form closed shapes

The visualization of knowledge (and its components) and ignorance (and its components) must build on these attributes in order to create a system that conveys intuitive meanings leading to correct decisions. We postulate that changes, over time and space in the form of density changes, form such a global gestalt.

Hoffman et al. (1998) used two perceptual domains, spatial and temporal. In the spatial domain, the global preference phenomenon was exploited. Color is used to identify each group (e.g., the traditional red for enemy and blue for friendly). Taking additional advantage of the preattentive global processing ability of humans, the density distribution of troops within a geographical area can be made instantly apparent. Rather than showing the location of each local element, the level of color saturation can represent the density of elements contained within a figure. The user can thus determine troop concentrations with less visual interrogation and cognitive effort because the differential saturation of the figures is immediately apparent. A commander viewing such figures would no doubt come to an immediate, albeit quite different, conclusion as to what is happening. The value in this type of coding becomes more apparent when viewed in a compressed time format. This format is achieved by processing each positional update and saving the rendering as part of an MPEG (Moving Picture Experts Group) movie. The commander can then use the movie to quickly view troop movement and density changes over time.

8.2.3 Use of Colors

Color has a long history of use for visually communicating information. To this end, the American Meteorological Society Interactive Information and Processing Systems Subcommittee for Color Guidelines was formed to poll the meteorological community to determine the most commonly used sets of color assignments that are used in depicting meteorological information. The following recommendations are suggested for the use of colors:

- When selecting color to represent various features or conditions, choose colors that have familiar relationships (Krebs and Wolf, 1979; Rice, 1991; Hoffman, 1991; Travis, 1991). For example, the red, yellow, and green set should be used when depicting dangerous, cautionary, and safe conditions, respectively. Another example relates to the directionality of gray shading used on monochrome satellite imagery: common practice is for light shading to represent high clouds and dark shading to represent low clouds. Bear in mind that across multiple meteorological display products, color may interact with critical values and thereby affect the attention afforded them (Hoffman and Lipton, 1992).
- Use selected colors consistently everywhere they are used (Travis, 1991). For example, if green is used to represent landmasses in one place, it should not be used to represent water bodies in another instance.
- Use time-proven color combinations for color-coding symbols with combinations of different colors (Krebs and Wolf, 1979; Rice, 1991). For example, for a set of eight different colors, the recommended selection is cyan, green, yellow, orange, red, mauve, purple, and blue.
- Avoid color-coding tiny symbols, since color discrimination decreases with the size of the object (Rice, 1991). Also, the use of darker symbols on lighter background usually works better than the reverse (Hoffman and Lipton, 1992).

- Limit the number of different colors that are used in any single visual display or product (Krebs and Wolf, 1979; Hoffman, 1991; Rice, 1991; Grossman, 1992). Four is a desirable limit for complete accuracy in color identification, and more than eight different colors should be avoided.
- Consider the perceived brightness of colors for portraying the relative values of parameters on a quantitative scale (Levkowitz, 1988; Levkowitz and Herman, 1992). For example, the ordered set of five colors brown-red-orange-yellow-white progresses from low to high perceived brightness at its endpoints, while the common spectral set red-orange-yellow-green-blue has its maximum of perceived brightness in the middle (yellow).
- Consider how the color set will map to gray shades if some users will ultimately view the color set in monochrome mode (i.e., monochrome television or hard copy). Pay particular attention to the adequacy of contrast for the monochrome conversion.
- In choosing color palettes, consider how to accommodate the 8% of the population that is color blind. The Human Factors Society recommends (Miller-Jacobs, 1984; Weitzman, 1985; Hoffman, 1991) using blue, green, red, cyan, and yellow-green as an appropriate set of five colors.
- Since visual acuity is best for yellow-green hues, try to use this color for critical information (Hoffman and Lipton, 1992). Conversely, since visual acuity is poor for blue hues, use blue hues for either large areas or large symbols.
- Avoid using red and blue adjacent to each other, because of the three-dimensional illusion that results from what is known as chromostereopsis (Rice, 1991; Tannas, 1992).
- Use bright-colored backgrounds cautiously (Hoffman and Lipton, 1992).
- When color is used to code intensity in an image, such as to create bands of temperature ranges, use perceptually equal steps among the different colors chosen (Levkowitz, 1988; Kaiser and Proffitt, 1989; Travis, 1991).
- Experienced users need to be able to manipulate colors to enhance certain contrasts as they struggle to comprehend the displayed information (Hoffman and Lipton, 1992).

For additional detail on the human factors issues related to applying the color guidelines presented in this report, we recommend Hoffman and Lipton's (1992) essay.

8.2.4 Financial Visualization

Color and intensity are used in financial information visualization, such as *Heatmaps* that organize financial instruments or positions into color-coded cells or spots. Using live data, Heatmaps perform calculations in real-time and display the results as color. They are called Heatmaps because they show what is hot. Users can set visual alerts to highlight important opportunities, critical information, or current results. Heatmaps focus time and attention on the few pieces of information that are most important at this moment.

8.2.5 Icons, Ontology, and Lexicon

Icons, ontology, and lexicon can be used to portray and communicate types of ignorance and uncertainty. In this section, only icons are discussed.

8.2.5.1 Emoticons

Examples of icons used for displaying emotion (*emoticons*) are provided as follows (Nofi, 2000):

:-O Dismay
<G> Grin
:-C Incredulity
:-) Smile
:-J Tongue in cheek
("""") Raised eyebrows
;-) Wink
:-x Kiss
(zzz) Boredom
:-(Frown
O:-) Innocence
[ ] Hug

Emotions can also be displayed using graphical icons; for example, icons could be constructed based on images, as illustrated in Figure 8.3.

8.2.5.2 Ignoricons and Uncerticons

As for ignorance and uncertainty, some conceptual icons were developed for the purposes of this study as a starting point for further development and refinement. These icons are called *ignoricons* and *uncerticons* and are provided in Figure 8.4, which is based on the ignorance hierarchy of Figure 1.19. Visualized data, information, or knowledge having dubious origins and quality could display ignorance and uncertainty levels using these icons in various sizes, colors, and motions to indicate desired attributes. Visualization techniques for ignoricons and others forms need to be identified, selected, and used from glyphs, adding geometry, modifying geometry, and modifying attributes, animation, sonification, and psychovisual approaches. They could also include environmental visualization, surface interpolation, global illumination with radiosity, flow visualization, and figure animation. The visualization methods should satisfy the following principles: (1) apprehension, (2) clarity, (3) consistency, (4) efficiency, (5) necessity, and (6) truthfulness. The following attributes of icons can be used to communicate information on uncertainty and ignorance: (1) form, (2) orientation, (3) color, (4) texture, (5) value, (6) size, (7) position, (8) motion, (9) intensity, (10) shading, and (11) special effects (such as blinking, animation, etc.).

FIGURE 8.3 Images for constructing emotion icons. (By Ziad B. Ayyub, 2005.)

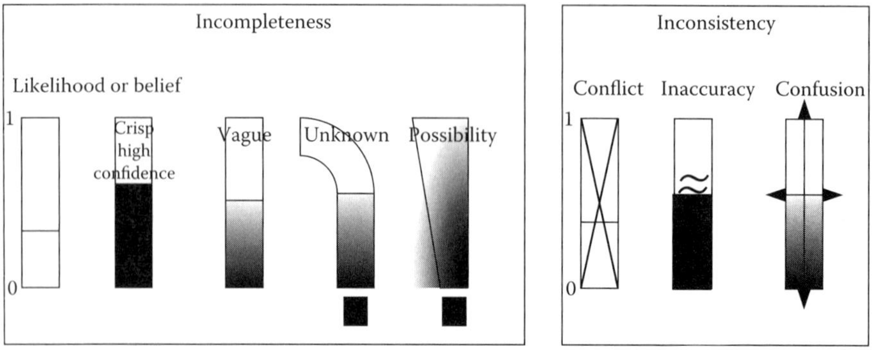

FIGURE 8.4 Suggested ignoricons and uncerticons. (Copyright © BMA Engineering, Inc., 2003. With permission.)

8.3 CRITERIA AND METRICS FOR ASSESSING VISUALIZATION METHODS

8.3.1 Definition of Primary Selection Criteria and Weight Factors

The goal of using a method to portray essential information uncertainty is to effectively and efficiently communicate information by visual means. Suggested visualization methods should be assessed according to selected criteria that represent their effectiveness, efficiency, and consistency in communicating information uncertainty. The following list provides an initial set of criteria:

- Intuitive
- Accurate and precise
- Complete
- Consistent
- Not confusing
- Single/unique meaning
- Fewer errors
- Faster comprehension/decision
- Multiuse or applications or services
- Platform independent
- Established meaning or use

From this initial list of potential assessment criteria, a short list of criteria and subcriteria for evaluating prospective icons for the visualization of uncertainty was assembled as follows:

- Apprehension (mental grasp)
- Clarity
- Consistency

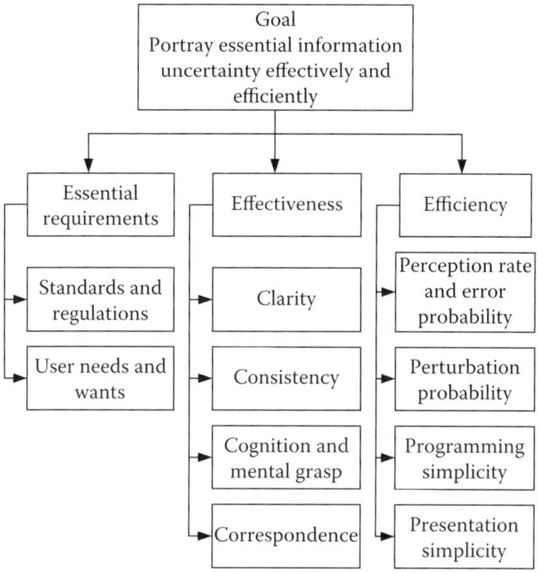

FIGURE 8.5 A criteria hierarchy for assessing methods for information uncertainty portrayal.

- Efficiency
- Necessity
- Truthfulness
- Perturbation probability
- Compatibility with military standards, including MIL-STD-2525B

Using the analytic hierarchy process (AHP), and for the goal of selecting icons for uncertainty in data visualization, the criteria and subcriteria shown in a tree diagram in Figure 8.5 were constructed. The abbreviated labels in the diagram are defined in the figure. Three primary criteria are suggested, represented by the 3E: essentials, effectiveness, and efficiency.

The essentials criterion is defined to encompass exogenous, customer-based requirements for military icons and symbology, and it has two subcriteria: (1) standards and regulations and (2) user needs and wants. The former refers to any military standards and regulations for symbols, icons, and digital displays, such as MIL-STD2525B. The latter encompasses whatever the military user wants or needs with respect to icons used to display uncertainty and ignorance, regardless of military standards and regulations (which may be changed based on new technology, such as outcomes from this project, or user needs, such as needs identified from this project).

The effectiveness criterion (doing the "right thing," in the words of Drucker) focuses on the ability of the icon to impart to the user its intended meaning, while the efficiency criterion (doing "things right," again in the words of Drucker) focuses on the ability of the icon to impart its meaning rapidly and easily. That is, an effective icon is able to map accurately, via the observer's visual sense, its semiotic repre-

TABLE 8.2
Pair-Wise Comparison of Primary Criteria according to the Analytic Hierarchy Process

	Essential Requirements	Effectiveness	Efficiency
Essential requirements	1	1	2
Effectiveness	1	1	2
Efficiency	1/2	1/2	1

Note: 1 = equal; 3 = moderate; 5 = strong; 7 = very strong; and 9 = extremely strong.

sentation into the observer's mind. An efficient icon accomplishes the semiotic mapping, for example, faster or cheaper.

Table 8.2 shows the results of pair-wise comparison of primary criteria according to the analytic hierarchy process in terms to their relative importance to the goal. The table uses an ordinal scale of 1 = equal, 3 = moderate, 5 = strong, 7 = very strong, and 9 = extremely strong. In Figure 8.6, the essentials and effectiveness criteria are judged to be of equal importance, with a weight of 0.40, while the efficiency criterion has a weight of 0.20. It is important to provide the user with symbology that is acceptable under the prevailing rules — although it may be possible (albeit difficult) to change the rules if necessary. But following the rules is insufficient if the resulting icons do not perform as desired or intended — thus the importance of their being effective as well as following military standards. Given that the icons perform as intended, they should do so simply and easily, but their efficiency is not as important as their effectiveness. The AHP inconsistency ratio is satisfactorily below 0.10 (0.0 in this case), as it is for all of the criterion evaluations.

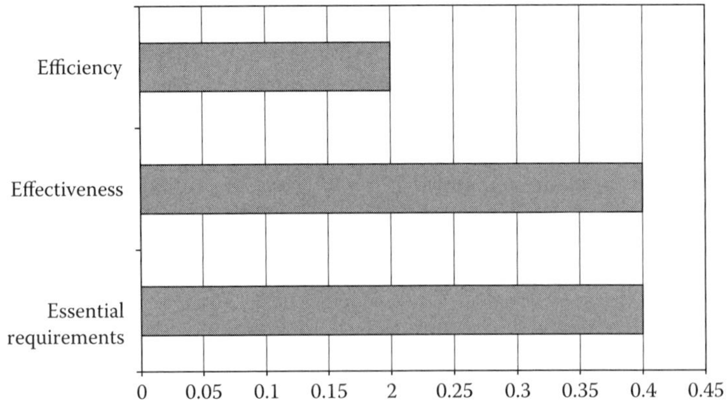

FIGURE 8.6 Weight factors for primary decision criteria.

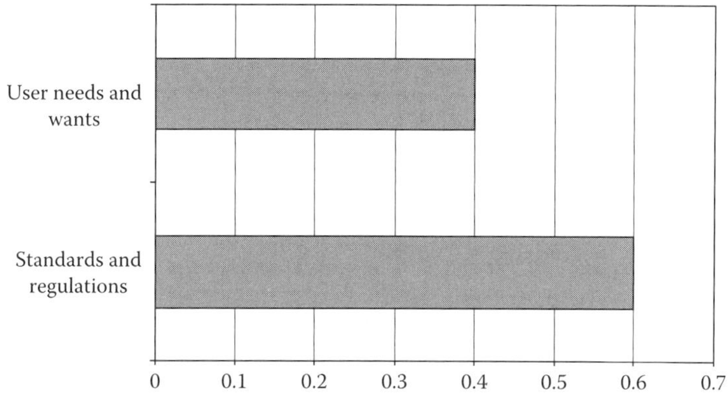

FIGURE 8.7 Weight factors for essential requirements subcriteria.

8.3.2 Definition and Selection Subcriteria and Weight Factors

The essentials subcriteria are evaluated in Figure 8.7. The standards and regulations subcriterion, with a value of 0.60, is judged more important than the user needs and wants subcriterion, with a value of 0.40, because it represents an official constraint, which supersedes the admittedly important objective of satisfying the immediate user.

The effectiveness subcriteria are evaluated in Table 8.3 and Figure 8.8. Echoing the 4C metrics for evaluating diamonds (cut, clarity, color, and carat), we have the 4C subcriteria for evaluating icon effectiveness: clarity, cognition, consistency, and correspondence. Clarity, judged the most important of the four, with a score of 0.39 (rounding), is defined to mean the ability of the icon, on any specified display, to impart the intended meaning without confusion with respect to other symbols or noise. Cognition, deemed second in importance, with a score of 0.28, is the ability of the icon to impart its intended meaning intuitively to the human mind so that extensive training or memorization is not required of typical users. Consistency, with a score of 0.20, is defined as the ability of the icon to maintain its ability to impart

TABLE 8.3
Pair-Wise Comparison of Effectiveness Subcriteria according to the Analytic Hierarchy Process

	Clarity	Cognition	Consistency	Correspondence
Clarity	1	2	2	2
Cognition	1/2	1	2	2
Consistency	1/2	1/2	1	2
Correspondence	1/2	1/2	1	1

Note: 1 = equal; 3 = moderate; 5 = strong; 7 = very strong; and 9 = extremely strong.

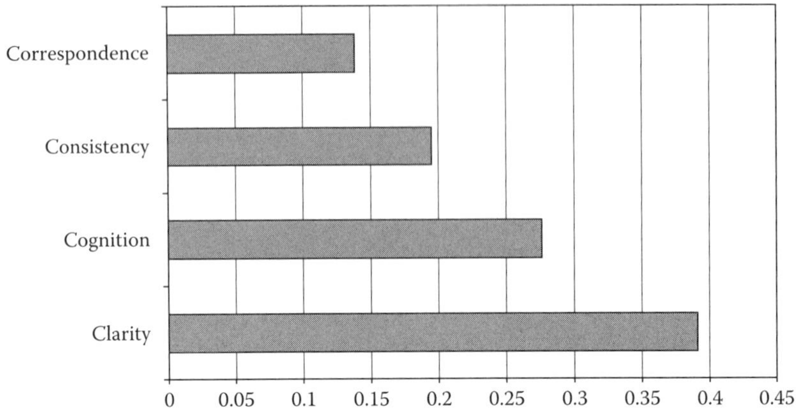

FIGURE 8.8 Weight factors for effectiveness subcriteria.

its meaning clearly and intuitively over many different kinds of displays and in many different viewing environments and conditions (e.g., high stress). This is analogous to displaying a standard digit (e.g., 5) on all kinds of displays and monitors and being able to recognize it without confusing it with an *S* or other symbol. The correspondence subcriterion, with a score of 0.14, describes whether the icon's design embeds the ground truth that it represents in its design (as a currently abstract *A* might resemble an ox head in an ancient alphabet). This specific way of imparting meaning to the user is not consequential if other attributes of the icon give it the ability to impart its meaning clearly to the user.

The efficiency subcriteria are evaluated in Table 8.4 and Figure 8.9 with the 4P subcriteria: perception, presentation, programming, and perturbation. The perception subcriterion, judged most important of the four, with a value of 0.33, is defined as the user's rate (or speed) in perceiving the intended meaning of an icon, and the probability that the user perceives the icon erroneously. The presentation subcriterion, deemed to be of equal importance with the perception subcriterion (with a value of 0.33), represents the simplicity (as opposed to complexity) of the icon's design (e.g., few components, linearity, simple colors, etc.). The program simplicity

TABLE 8.4
Pair-Wise Comparison of Efficiency Subcriteria according to the Analytic Hierarchy Process

	Clarity	Cognition	Consistency	Correspondence
Perception rate and error probability	1	1	2	2
Presentation simplicity	1	1	2	2
Programming simplicity	1/2	1/2	1	2
Perturbation probability	1/2	1/2	1	1

Note: 1 = equal; 3 = moderate; 5 = strong; 7 = very strong;, and 9 = extremely strong.

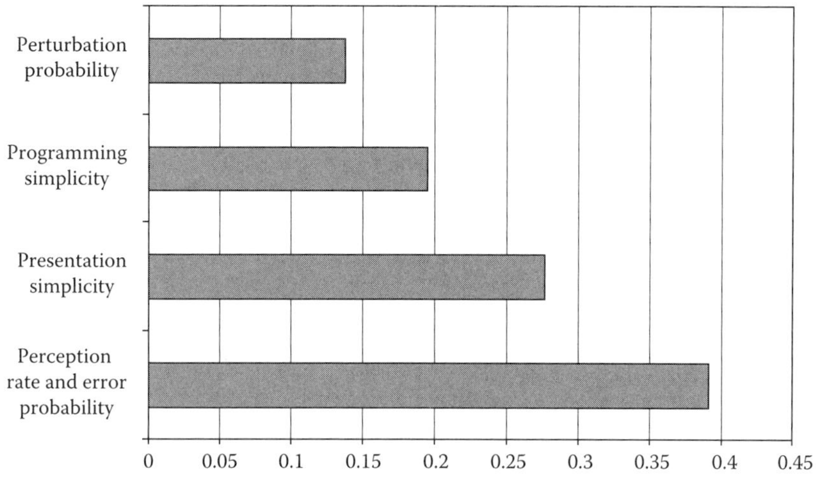

FIGURE 8.9 Weight factors for efficiency subcriteria.

subcriterion (with a value of 0.20) means that the icon can be generated on many different displays with a relatively simple software code. The perturbation probability subcriterion is the likelihood that the icon may be disrupted or causes an error or confusion because of its design in relation to its display or environment (e.g., a dropped pixel or insufficient resolution, or a monochrome display instead of color could alter the icon's meaning or the perception of its meaning).

8.3.3 Summary of Criteria, Subcriteria, and Weight Factors

Figure 8.10 shows a summary of the criteria and their respective subcriteria and computed weight factors. It should be noted that the weight factors at the lowest level, i.e., at the subcriteria level, add up to 1. These weight factors are also shown in Figure 8.11.

8.3.4 Design of Visualization Method

After the (initially filtered) prospective icons are selected, they will be evaluated against the weighted criteria and subcriteria. And in performing the pair-wise comparisons against the weighted criteria and subcriteria, relevant iconic attributes will be considered, including:

- Geometric form
- Orientation
- Color
- Texture
- Shading
- Size
- Position
- Intensity

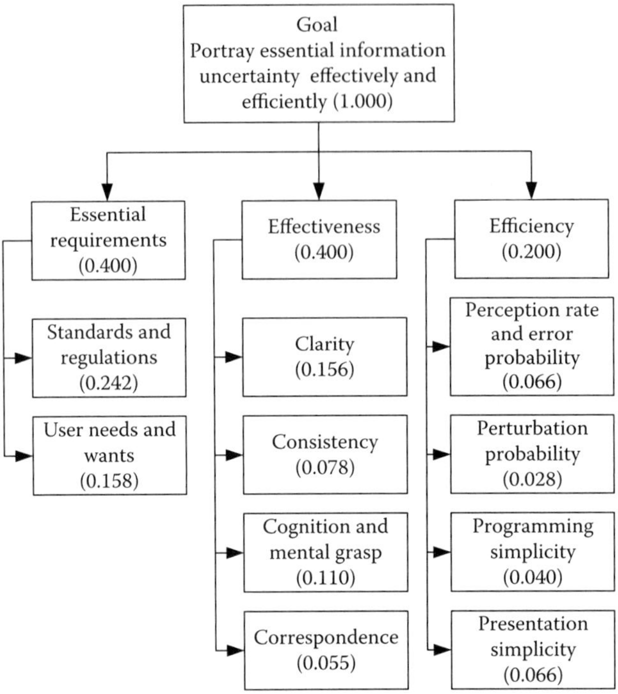

FIGURE 8.10 Summary of criteria and weight factors.

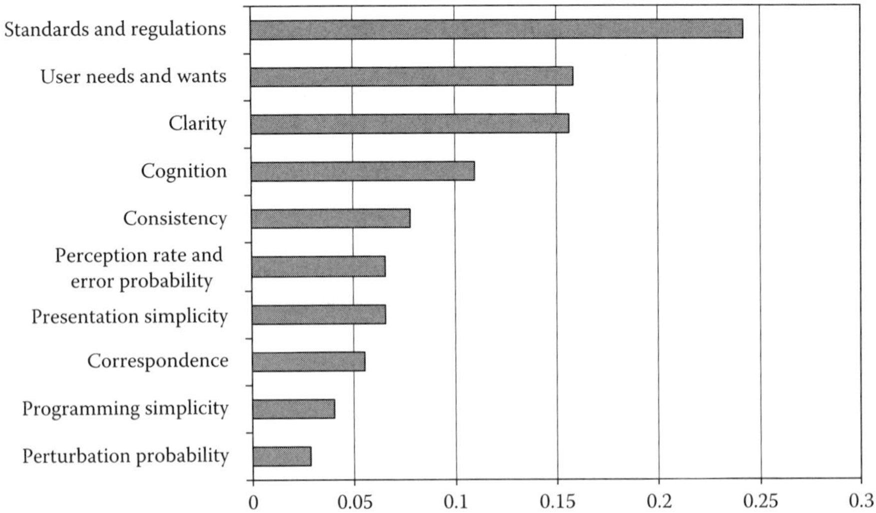

FIGURE 8.11 Synthesis of the subcriteria weight factors.

- Motion
- Special effects (e.g., blinking, animation)

Additional considerations include:

- Shapes that permit scaling
- Shapes that permit reduction
- Shapes that are common or have been used
- Color, but the gray scale version should be meaningful
- Shapes suitable for in-print and electronic media use

8.3.5 Assessment Methods and Experimental Protocol

Three assessments methods will be explored as part of the experimental protocol in Phase II of the project, as follows:

- Expert opinions (e.g., see Ayyub, 2001)
- Focus groups (Krueger and Casey, 2000)
- The analytic hierarchy process (AHP) (Saaty, 1996)

Some fundamental concepts of expert opinion elicitation and focus group use are provided by Ayyub (2001).

As shown in Figure 8.12, an icon selection protocol (A) can be used by employing the AHP (B) and generating a large list of prospective icons (C) to symbolically visualize uncertainty in data and information. As part of the AHP, we defined and weighted metrics and submetrics for evaluating the alternative icons. We then filtered down the initial large list of icons to a smaller list (D). If there were insufficient useful icons remaining (E), we could have generated more icons, but this was not necessary.

The experimental protocol (F) can be refined. We can start by selecting and assembling a focus group (G), where the subjects are prospective users and developers of tactical and strategic displays, as well as two groups of prospective users (for two different key applications) and one control group populated by undergraduate and graduate students (H). Key applications for the experimental subjects include the Command Post of the Future and the Future Combat System (especially for the robotic vehicle control system displays). The focus group subjects will evaluate the filtered, shorter list of icons, and their comments and suggestions will be recorded on videotape and then transcribed (the usual practice for focus groups). The focus group exercise will result in a short list of icons suitable for use with the experimental subjects (I); if not, the process will be iterated. The selected icons will be integrated with selected data streams (J) for the experimental subjects. Domain partners can assist us in defining the objectives and scope of the experiments, the evaluation criteria and metrics, and how the icons should be displayed in operational contexts (K).

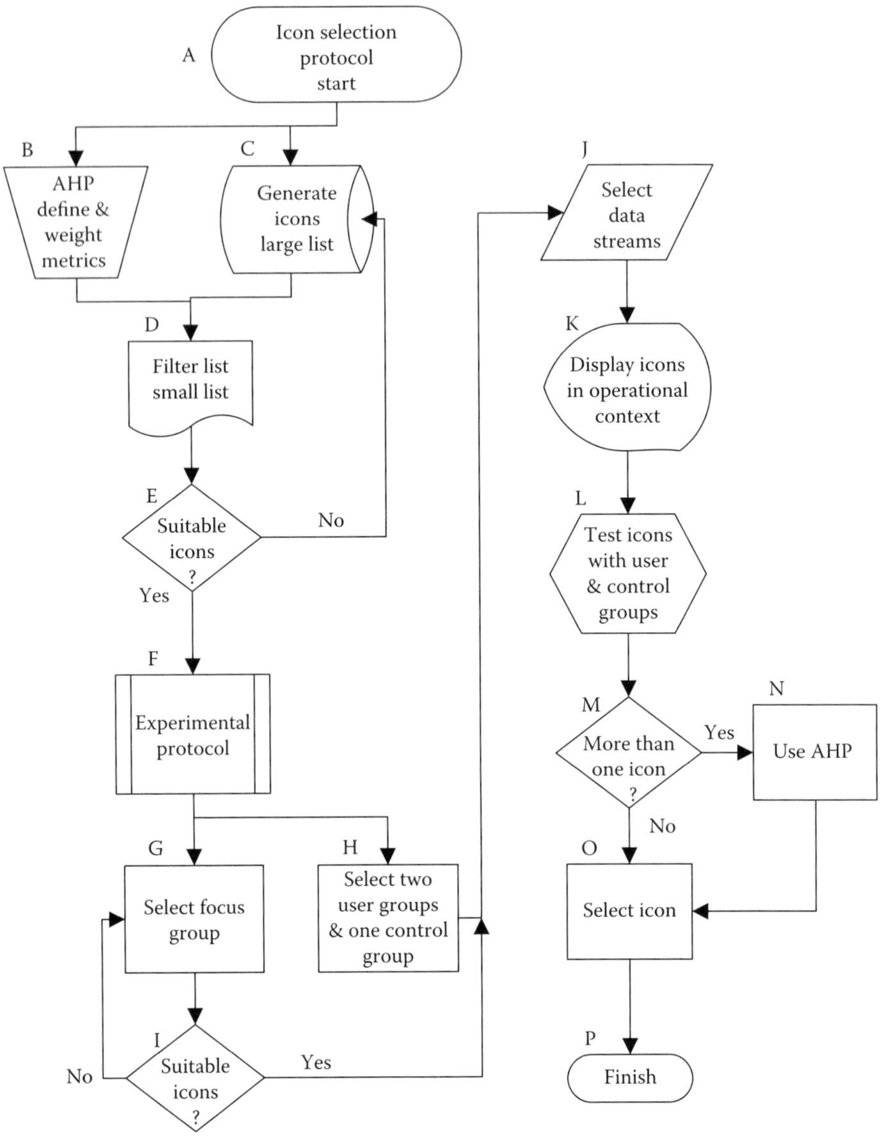

FIGURE 8.12 Icon selection and experimental protocol.

We can then design a series of experiments to select the ultimate icon (and its variants, such as color quadrants) for visualizing uncertainty. In testing the icons with the user and control groups (L), for example, the subjects might be shown a digital map display, with the usual terrain and military symbology, and with various friendly and enemy units displayed for a selected order of battle and tactical situation. A pair of different prospective ignoricons will be flashed on the display for a dwell time interval, after which the subject indicates, by pushing a button (for example), which icon is easier to perceive and how well the subject understands the information content of the symbol and its variants. Experimental variables include the subjects; display type, resolution, and size; icons and icon variants (e.g., color, size, position); and dwell time on the display. In addition to testing alternative icons, we will experiment with the use of textual blurbs in conjunction with mouse rolling over symbols (e.g., font type, size, and position). Such features offer the potential to communicate additional information on an as-needed basis without cluttering displays. The series of experiments should result in the selection of a suitable icon, but if the experiments result in more than one potentially suitable icon (M), we will use the AHP and our expert judgment for the final selection (O) and concluding the selection protocol (P).

Agreements will be negotiated and developed with one or two domain partners for the purpose of selecting and finalizing the ignorance and uncertainty symbology.

8.3.6 Candidate Shapes for Ignoricons and Uncerticons

This section provides candidate shapes of ignoricons and uncerticons that should be examined further using expert opinion or focus group studies. A comprehensive, long list of icons and symbols was filtered from a larger set of icons, consisting of military unit symbols; map symbols; written alphabets, pictograms, and other symbols from ancient languages; road and transportation symbols; consumer and warning symbols; religious symbols; heraldry; and modern symbols invented for various purposes (e.g., shorthand, Braille, science fiction alien languages, etc.). The long list contained several thousand icons. The sources used for this purpose include those listed at http://www.symbols.net/military.html, Muller (2003), Thomas (2000), Fishel and Gardner (2003), Maruyama (2003), and UN (2000). The long list contained several thousand icons. An intermediate list was subjectively culled from the long list based on initial criteria provided in Figure 8.11. The initial criteria include some of the metrics that can be used in the final evaluation of suitable icons. In particular, apprehension, clarity, consistency, efficiency, and perturbation potential influenced the initial filtering. Simple geometric shapes, able to display a number of different states (such as by color changes among, for example, red, yellow, and green) were preferred. For example, simple circles, ellipses, rectangles, squares, or diamonds with quadrants, where the quadrants might be filled with symbolic colors, perhaps with the addition pennants, also filled with color, would be visible, intuitive (with minimal instruction and training for operators), and robust. The short list was obtained by further filtering the intermediate list. The short list included the interrobang punctuation mark, which can be used for representation of symbology and creating the ignoricons and uncerticons by using parts of the mark, varying the sizes of its components, and filling or color-coding the shapes. The interrobang, a rarely used punctuation mark that has been in existence for several decades, can be modified to accommodate colors in three areas: the question mark, the exclamation mark, and the point that is a part of the exclamation mark. Depending on the number of colors and how they are distributed (e.g., red in the question mark would represent a different condition than red in the exclamation mark or point), many different states of information could be represented.

The interrobang is shown in Figure 8.13. It is combined exclamation mark and question mark. The interrobang has been described as an obscure punctuation

Or

FIGURE 8.13 The interrobang punctuation mark.

mark. The purpose of this page is to move the interrobang from the obscure to the ubiquitous.

As an advocate of precision in communication, the concept of the interrobang was introduced by Martin K. Speckter in 1962 in an article written for *TYPEtalks Magazine*. The interrobang was created to fill a gap in our punctuation system where writers often used typographically cumbersome and unattractive combinations of the question mark and exclamation mark to punctuate rhetorical statements where neither the question nor an exclamation alone exactly served the writer (e.g., How about that?!). Mr. Speckter called his mark interrobang from the Latin for query and the proofreader's term for exclamation. Most dictionaries have spelled the word correctly, although several other spellings with no logical genesis have appeared. At the time the interrobang was introduced in 1962, a number of graphic designs were sent to the magazine from many sources. Many newspapers, magazines, and talk shows reported on the new mark. In an April 1962 editorial, the *Wall Street Journal* deemed this punctuation exactly right for "'Who forgot to put gas in the car?' where the question mark alone just isn't adequate." The interrobang can convey in print an attitude, curiosity, and wonder. American Type Founders issued a metal typeface in 1966 called Americana, which included the interrobang. Remington Rand included the key as an option on its 1968 typewriters, commenting that the interrobang "expresses Modern Life's Incredibility." In 1996, a New York art studio designed variations of the mark for each of the fonts in its computer library. You can find an interrobang in Microsoft Word's Fonts: go to Format, choose Fonts, then Wingdings 2. You will find four different versions of the interrobang. Hit the ` ~ key, the] } key, the 6 ^ key, or the - _ key.

8.4 INTELLIGENT AGENTS FOR ICON SELECTION, DISPLAY, AND UPDATING

8.4.1 Intelligent Agents

An *intelligent agent* is a computer program or system situated in some environment that is capable of autonomous actions in this environment in order to meet its design objectives. An intelligent agent could have several of the following characteristics: autonomous behavior, information collection, information processing, decision making, adaptive internal state of logic, and communicative with users or agents. Autonomy herein means simply to indicate the agent's ability to act without the direct intervention of users (or other agents), and should have control over its own actions and internal state.

The notion of autonomy for an agent is different from the notion of encapsulation with respect to object-oriented computer program or system. An object encapsulates some states and has some control over these states in that they can only be accessed or modified via the methods that the object provides. Agents encapsulate states in just the same way; however, the agents also encapsulate behaviors, whereas an object does not encapsulate behaviors. Encapsulating behavior means that an agent has control over the execution; i.e., it has control over what actions it performs. Because of this distinction, incoming information does not invoke actions on agents, but rather results in requesting actions to be performed. The decision about whether to act upon the request lies with the recipient, i.e., the agent.

Agents could contain some level of intelligence, from fixed rules to learning engines that allow them to adapt to changes in the environment. Agents not only act reactively, but also sometimes proactively. Agents have communicative or social ability; that is, they communicate with the user, the system, and other agents as required (e.g., they may cooperate with other agents to carry out more complex tasks than they themselves can handle). In addition, agents may move from one system to another to access remote resources or even to meet other agents. Intelligent agents are expected to be flexible by meeting the following requirements:

- *Responsiveness*, by their ability to perceive their environment (which may be the physical world, a user, a collection of agents, the Internet, etc.) and respond in a timely fashion to changes that occur in it
- *Proactiveness*, by their ability to not only act in response to their environment, but also exhibit opportunistic, goal-directed behavior and take the initiative where appropriate
- *Communicativeness*, by their ability to interact, when they deem appropriate, with other artificial agents and humans in order to complete their own problem solving and to help others with their activities

Figure 8.14 shows a classification of agents (Jennings and Wooldridge, 1998).

8.4.2 Information Uncertainty Agent

Information uncertainty agents should be designed to process information from multiple sources, and manipulate or collate information from many distributed sources to assess inconsistency and incompleteness in the attributes of a symbol or unit, as provided in Table 8.1. The attributes could include the unit's:

- Type
- Size
- Intentions
- Location

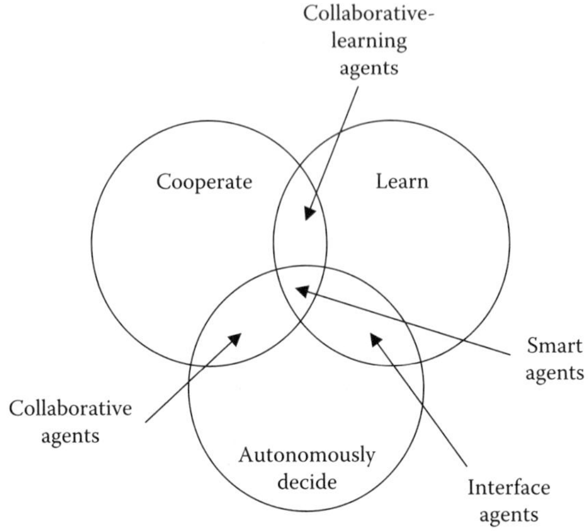

FIGURE 8.14 A classification of agents.

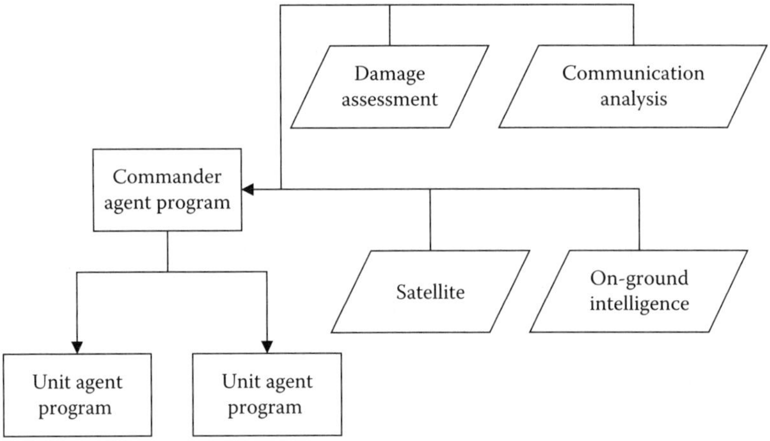

FIGURE 8.15 A classification of agents.

A schematic representation of the interfaces is shown in Figure 8.15.

8.4.3 Processing Information for Symbology Selection

Computational algorithms are needed to map incoming information into a selection among available symbology options. Uncertainty in information methods, including the uncertainty measures of Chapter 4, can be used for this purpose (Klir and Folger, 1988; Klir and Wierman, 1999). Uncertainty measures are available for three uncertainty classes: (1) imprecision or nonspecificity associated with sizes or cardinalities,

(2) fuzziness or vagueness associated with imprecision in boundaries, and (3) conflict or strife and discord among various sets.

8.5 IGNORANCE MARKUP LANGUAGE

Markup languages are used to display information in a way that is accessible to humans and machines. The most common markup language is the *Hypertext Markup Language* (HTML), which is designed to display information in a way that is accessible to humans for viewing via web browsers. Although HTML enables the visualization of information on the web, it does not provide much capability to describe the information in ways that facilitate the machine access. The *Extensible Markup Language* (XML) was therefore recently developed by the *World Wide Web Consortium* (W3C) to allow information to be more accurately described using tags. XML is formatted for machine processing but has a very limited semantics capability (for describing relationships or *ontologies*). The *Deference Advanced Research Projects Agency* (DARPA) *Agent Markup Language* (DAML) was recently developed to allow the use of ontologies, which provides very powerful means of describing objects and their relationships to other objects. The DAML, developed as an extension to XML, is formatted for machine processing with very rich semantics to support agents for intelligence analysis and production; military planning and operations; software agents; and sensor fusion (Anken, 2003). These markup languages discussed above do not consider ignorance or uncertainties associated with the domain knowledge; therefore, the development of a markup language based on uncertainty tagging of information will facilitate the communication of uncertainty and its visualization.

EXERCISE PROBLEMS

8.1. Develop alternate ignoricons to the ones provided in Figure 8.4.

8.2. Develop ignoricons and uncerticons similar to those in Figure 8.2 for others types of ignorance shown in Figure 1.19.

8.3. Provide examples to demonstrate the use of ignoricons in making decisions.

8.4. Develop a methodology to using uncertainty measures for making appropriate selection of ignoricons and uncerticons for the purpose of aiding decision makers.

8.5. Develop a research plan for proposing a markup language based on uncertainty tagging of information to facilitate the communication of uncertainty and its visualization.

Appendix A: Historical Perspectives on Knowledge

Philosophical views on knowledge evolved over time. This section summarizes these views on knowledge and describes the evolution of these views to contemporary schools. The presentation in this section is drawn on the works of selected philosophers who were either great influences or representatives of their respective periods, and it is not intended to provide a comprehensive coverage of all views. Solomon and Higgins (1996), Russell (1975), Popkin (2000), Durant (1991), and Honderich (1995) are recommended sources for additional details on any of the views presented in these sections.

The *pre-Socratics* period includes Gorgias (483 to 378 B.C.), Heraclitus (535 to 475 B.C.), and Empedocles (c. 450 B.C.), as summarized in Table A.1. The *Socrates* period includes Socrates (469 to 399 B.C.), Antisthenes (440 to 370 B.C.), and Euclides (430 to 360 B.C.). The works of these philosophers are summarized in Table A.2. The *Plato and Aristotle* period included Protagoras (485 to 415 B.C.), Plato (427 to 347 B.C.), and Aristotle (384 to 322 B.C.). Views on knowledge during this period are captured in Table A.3. The word *Platonism* refers both to the doctrines of Plato and to the manner or tradition of philosophizing that he founded. Often, in philosophy, *Platonism* is virtually equivalent to idealism or intrinsicism, since Plato was the first Western philosopher to claim that reality is fundamentally something ideal or abstract, and that knowledge largely consists of insight into or perception of the ideal. In common usage, the adjective *Platonic* refers to the ideal; for example, Platonic love is the highest form of love that is nonsexual or nonphysical. The works of Plato formed the basis for *Neoplatonism*, founded by Plotinus (205 to 270), which greatly influenced medieval philosophers. Aristotle followed Plato as his student; however, he maintained that knowledge can be derived from sense experiences — a departure from Plato's thoughts. Knowledge can be gained either directly or indirectly by deduction using logic. For Aristotle, form and matter were inherent in all things and inseparable. Aristotle rejected the Platonic doctrine that knowledge is innate and insisted that it can be acquired only by generalization from experiences, emphasizing empiricism by stating that "there is nothing in the intellect that was not first in the senses." Table A.3 provides a summary of the views during this period.

The *Hellenistic* period includes Epicurus (341 to 271 B.C.), Epictetus (55 to 135 B.C.), and Pyrrho (360 to 270 B.C.), with a summary of views provided in Table A.4. The *Medieval* period can be characterized as an Islamic-Arabic period that resulted in translating, preserving, commenting on, and providing Europe with the works of Greek philosophers. Also, the philosophers of this period maintained and strength-

TABLE A.1
Knowledge Views during the Pre-Socratics Period

Philosophers (Year)	Nature of Knowledge
Gorgias (483–378 B.C.)	Stated that knowledge does not exist, nor can it be communicated if it does exists
Heraclitus (535–475 B.C.)	Maintained that wisdom is not the knowledge of many things; it is the clear knowledge of one thing only; perfect knowledge is only given to the Gods, but a progress in knowledge is possible for humans
Empedocles (c. 450 B.C.)	Distinguished between the world as presented to our senses (*kosmos aisthetos*) and the intellectual world (*kosmos noetos*)

TABLE A.2
Knowledge Views during the Socrates Period

Philosophers (Year)	Nature of Knowledge
Antisthenes (440–370 B.C.)	Maintained that happiness is a branch of knowledge that could be taught, and that once acquired could not be lost
Euclides (430–360 B.C.)	Maintained that virtue is knowledge; if virtue is knowledge, therefore, it can only be the knowledge of the ultimate being

TABLE A.3
Knowledge Views during the Plato and Aristotle Period

Philosophers (Year)	Nature of Knowledge
Protagoras (485–415 B.C.)	Maintained that knowledge is relative since it is based on individual experiences
Plato (427–347 B.C.)	Maintained that knowledge can exist based on unchanging and invisible forms or ideas; objects that are sensed are imperfect copies of the pure forms; genuine knowledge about these forms can only be achieved by abstract reasoning through philosophy and mathematics
Aristotle (384–322 B.C.)	Followed Plato, however maintained that knowledge is derived from sense experiences; knowledge can be gained either directly or indirectly by deduction using logic

TABLE A.4
Knowledge Views during the Hellenistic Period

Philosophers (Year)	Nature of Knowledge
Epicurus (341–271 B.C.) and Epictetus (55–135 C.E.)	Said philosophy is a means not an end
Pyrrho (360–270 B.C.)	Argued for skepticism in logic and philosophy

TABLE A.5
Knowledge Views during the Medieval Period

Philosophers (Year)	Nature of Knowledge
Plotinus (205–270)	Created Plotinus' principle with assumptions stated crudely as follows: (1) truth exists and is the way the world exists in the mind or the intellect; (2) the awareness of the world as it exists in the intellect is knowledge; and (3) two kinds of truth exist, the contingent and the necessary truth, for example, the contingent truth that 10 coins are in my pocket, and the necessary truth that 4 + 6 equals 10
Al-Kindi (800–873)	Translated, preserved, and commented on Greek works
Al-Farabi (870–950)	Carried the thoughts of Aristotle and was named the second teacher, with Aristotle being the first; according to him, logic was divided into idea and proof
Ibn Sina (980–1037)	Synthesized Aristotelian, Neoplatonic, and Islamic thoughts
Ibn Rushd (1128–1198)	Wrote a primary work (*Tuhafut al-Tuhafut*) that was critical of the works of medieval philosophers on limiting rationalism and moving to faith, and prepared Europe for modern philosophy
Aquinas (1224–1274)	Followed the schools of Plato and Aristotle, and added religious belief and faith

TABLE A.6
Knowledge Views during the Renaissance Period

Philosophers (Year)	Nature of Knowledge
Bacon (1561–1626)	Criticized Aristotelian logic as useless for the discovery of new laws, and formulated rules of inductive inference
Galileo (1564–1642)	Explained and defended the foundations of a thoroughly empirical view of the world by creating the science of mechanics, which applied the principles of geometry to the motions of bodies
Newton (1642–1727)	Applied mathematics to the study of nature
Montaigne (1533–1592)	Belonged to the skepticism school with his motto "What do I know?"

ened the school of rationalism and laid down the foundation of *empiricism*. The philosophers of this period were influenced by Plato, Aristotle, and Plotinus, who founded *Neoplatonism*. This period included leading philosophers such as Al-Kindi (800 to 873), Al-Farabi (870 to 950), Ibn Sina (named Avicenna by the West, 980 to 1037), Ibn Rushd (named Averroes by the West, 1128 to 1198), and Aquinas (1224 to 1274), as summarized in Table A.5.

The *Renaissance* period included Bacon (1561 to 1626), Galileo (1564 to 1642), Newton (1642 to 1727), and Montaigne (1533 to 1592), as summarized in Table A.6, and was followed by the *17th-century* period of Descartes (1596 to 1650), Spinoza (1632 to 1677), and Locke (1632 to 1704), as summarized in Table A.7. The *18th-century* period includes leading philosophers such as Berkeley (1685 to

TABLE A.7
Knowledge Views during the 17th-Century Period

Philosophers (Year)	Nature of Knowledge
Descartes (1596–1650)	As the father of modern philosophy, identified rationalism as a system of thought that emphasized the role of reason *and a priori* principles in obtaining knowledge; he also believed in the dualism of mind (thinking substance) and body (extended substance)
Spinoza (1632–1677)	Termed metaphysical (i.e., cosmological) concepts such as substance and mode, thought and extension, causation and parallelism, and essence and existence
Locke (1632–1704)	Identified empiricism as a doctrine that affirms all knowledge is based on experience, especially sense perceptions, and on *a posteriori* principles; Locke believed that human knowledge of external objects is always subject to the errors of the senses, and concluded that one cannot have absolutely certain knowledge of the physical world

TABLE A.8
Knowledge Views during the 18th-Century Period

Philosophers (Year)	Nature of Knowledge
Berkeley (1685–1753)	Agreed with Locke that knowledge comes through ideas (i.e., sensation of the mind), but denied Locke's belief that a distinction can be made between ideas and objects
Hume (1711–1776)	Asserted that all metaphysical things that cannot be directly perceived are meaningless; divided all knowledge into two kinds: relations of ideas (i.e., the knowledge found in mathematics and logic, which is exact and certain but provides no information about the world) and matters of fact (i.e., the knowledge derived from sense perceptions); furthermore, he held that even the most reliable laws of science might not always remain true
Kant (1724–1804)	Provided a compromise between empiricism and rationalism by combining both types, and distinguished three knowledge types: (1) an analytical *priori*, (2) a synthetic *posteriori*, and (3) a synthetic *priori*

1753), Hume (1711 to 1776), and Kant (1724 to 1804), with views relating to knowledge summarized in Table A.8. The *19th-century* period includes leading philosophers such as Hegel (1770 to 1831), Comte (1798 to 1857), Marx (1818 to 1883), Engels (1820 to 1895), and Nietzsche (1844 to 1900), with views summarized in Table A.9.

The *20th-century* period includes leading philosophers such as Bradley (1846 to 1924), Royce (1855 to 1916), Peirce (1839 to 1914), Dewey (1859 to 1952), Husserl (1859 to 1938), Russell (1872 to 1970), Wittgenstein (1889 to 1951), and Austin (1911 to 1960), with views relating to knowledge summarized in Table A.10.

Bradley maintained that reality was a product of the mind rather than an object perceived by the senses. Like Hegel, he also maintained that nothing is altogether

TABLE A.9
Knowledge Views during the 19th-Century Period

Philosophers (Year)	Nature of Knowledge
Hegel (1770–1831)	Claimed as a rationalist that absolutely certain knowledge of reality can be obtained by equating the processes of thought, nature, and history; his absolute idealism was based on a dialectical process of thesis, antithesis, and synthesis as cyclical and ongoing process
Comte (1798–1857)	Brought attention to the importance of sociology as a branch of knowledge, and extended the principles of positivism, the notion that empirical sciences are the only adequate source of knowledge
Marx (1818–1883) and Engels (1820–1895)	Developed the philosophy of dialectical materialism, based on the logic of Hegel
Nietzsche (1844–1900)	Concluded that traditional philosophy and religion are both erroneous and harmful, and traditional values (represented primarily by Christianity) had lost their power in the lives of individuals; therefore, there are no rules for human life, no absolute values, no certainties on which to rely

real except the *absolute*, the totality of everything that transcends contradiction. Everything else, such as religion, science, moral precept, and even common sense, is contradictory. Royce believed in an *absolute truth* and held that human thought and the external world were unified. Peirce developed *pragmatism as a theory of meaning*, in particular, the meaning of concepts used in science. The only rational way to increase knowledge is to form mental habits that would test ideas through *observation and experimentation*, leading to an *evolutionary process* of knowledge for *humanity* and *society*, i.e., a *perpetual state of progress*. He believed that the truth of an idea or object could only be measured by empirical investigation of its usefulness. *Pragmatists* regarded all theories and institutions as tentative hypotheses and solutions, and that efforts to improve society must be geared toward problem solving in an ongoing process of progress. Pragmatism sought a middle ground between traditional metaphysical ideas about the nature of reality, and the radical theories of *nihilism* and *irrationalism*, which had become popular in Europe at that time. Pragmatists did not believe that a single absolute idea of goodness or justice existed, but rather that these concepts were relative and depended on the context in which they were being discussed. Peirce influenced a group of philosophers, called *logical positivists*, who emphasized the importance of *scientific verification* and rejected personal experience as the basis of true knowledge. Dewey further developed *pragmatism* into a comprehensive system of thought that he called *experimental naturalism*, or *instrumentalism*. *Naturalism* regards human experience, intelligence, and social communities as ever-evolving mechanisms; therefore, human beings could solve social problems using their experience and intelligence, and through *inquiry*. He considered traditional ideas about knowledge and absolute reality or absolute truth to be incompatible with a *Darwinian* worldview of progress, and therefore, they must be discarded or revised. Husserl developed *phenomenology* as an elaborate procedure by which

TABLE A.10
Knowledge Views during the 20th-Century Period

Philosophers (Year)	Nature of Knowledge
Bradley (1846–1924)	Maintained that reality was a product of the mind rather than an object perceived by the senses; like Hegel, nothing is altogether real except the absolute, the totality of everything that transcends contradiction; everything else, such as religion, science, moral precept, and even common sense, is contradictory
Royce (1855–1916)	Believed in an absolute truth and held that human thought and the external world were unified
Peirce (1839–1914)	Developed pragmatism as a theory of meaning, in particular the meaning of concepts used in science; the only rational way to increase knowledge was to form mental habits that would test ideas through observation and experimentation, leading to an evolutionary process for humanity and society, i.e., a perpetual state of progress. He believed that the truth of an idea or object could only be measured by empirical investigation of its usefulness
Dewey (1859–1952)	Further developed pragmatism into a comprehensive system of thought that he called experimental naturalism, or instrumentalism; naturalism regards human experience, intelligence, and social communities as ever-evolving mechanisms; therefore, human beings could solve social problems using their experience and intelligence and through inquiry
Husserl (1859–1938)	Developed phenomenology as an elaborate procedure by which one is said to be able to distinguish between the way things appear to be and the way one thinks they really are
Russell (1872–1970)	Revived empiricism and expanded to epistemology as a field
Wittgenstein (1889–1951)	Developed logical positivism that maintained that only scientific knowledge exists verifiable by experience; he viewed philosophy as a linguistic analysis and "language games," leading to his work *Tractatus Logico-Philosophicus* (1921), which asserted language and the world are composed of complex propositions or facts that can be analyzed into less complex propositions arriving at elementary propositions, and into less complex facts arriving at simple "picture atomic facts or states of affairs," respectively
Austin (1911–1960)	Developed the speech-act theory, where language utterances might not describe reality and can have an effect on reality

one is said to be able to distinguish between the way things appear to be and the way one thinks they really are.

Russell revived *empiricism* and expanded to *epistemology* as a field. He attempted to explain all factual knowledge as constructed out of immediate experiences. Wittgenstein developed *logical positivism* that maintained (1) only scientific knowledge exists, (2) any valid knowledge must be verifiable in experience, and (3) a lot of previous philosophy was neither true nor false but literally meaningless, expressed by him as "philosophy is a battle against the bewitchment of our intelligence by means of language." He viewed philosophy as a *linguistic analysis* and "language games," leading to his work *Tractatus Logico-Philosophicus* (1921), which asserted language is composed of *complex propositions* that can be analyzed into less complex propositions until one arrives at simple or elementary propositions. This view of decomposing complex language propositions has a parallel in our view of the world to be composed of *complex facts* that can be analyzed into less complex facts until one arrives at simple "picture atomic facts or states of affairs." His *picture theory of meaning* required and built on atomic facts pictured by the *elementary propositions*. Therefore, only propositions that picture facts are the propositions of science that can be considered cognitively meaningful. Metaphysical, ethical, and theological statements, on the other hand, are not meaningful assertions. Wittgenstein's work influenced that of Russell in developing the *theory of logical atomism*. Russell, Wittgenstein, and others formed the core of the *Vienna Circle*, which *developed logical positivism*, with philosophy being defined by its role in clarification of meaning, not the discovery of new facts or the construction of traditional metaphysics. They introduced strict principles of *verifiability* to reject as meaningless the nonempirical statements of metaphysics, theology, and ethics, and regarded as meaningful only statements reporting empirical observations, taken together with the *tautologies* of *logic* and *mathematics*.

Austin developed the *speech-act theory*, where he considered that many utterances do not merely describe reality, but they also have an effect on reality, insofar as they too are the performance of some act. In addition, the period includes the foundational work of Keynes, Borel, Von Mises, Ramsey, Von Neumann and Morgenstern, Popper, De Finetti, and Savage. The modern philosophy of science, including semantic analysis, formulated by Mach and applied by Einstein and Bohr to enable the revolutions of relativity and quantum mechanics, is a central component in the modern philosophy of science. It could supply conceptual tools for a critical evaluation of the many putative representations of uncertainty.

The question regarding the meaning of systems is one of the most fundamental epistemological issues of science, particularly of systems science. Two opposing positions on this issue have been advanced and debated since the emergence of systems science and are based on two very different views about the nature of knowledge: realism and constructivism. According to realism, each system that is obtained by applying correctly the principles and methods of science represents some aspect of the real world. According to constructivism, all systems are artificial abstractions and are not made by nature, but we construct them by our perceptual and mental capabilities within the domain of our experiences.

The philosopher Karl Popper (Popper, 1963) introduced the concept of three worlds: World 1 consists of physical objects and phenomena; World 2 consists of subjective experiences and mental phenomena; and World 3 consists of mathematical structures ("truths in themselves"). These three worlds form a base for developing his philosophical views. According to him, scientific hypotheses cannot be validated; they can only be falsified.

Bibliography

Abraham, D.M., Bernold, L.E., and Livingston, E.E. 1989. Emulation for control system analysis in automated construction. *J. Comput. Civ. Eng.*, 3, 320–332.

Al-Fadhala, M.M. and Ayyub, B.M. 2002. Decision Analysis Using Belief Functions for Assessing Existing Structures. Paper presented at the International Conference on Structural Safety and Reliability (ICOSSAR), 2001. Newport Beach, CA.

Alpert, M. and Raiffa, H. 1982. A progress report on the training of probability assessors. In *Judgement Under Uncertainty, Heuristics and Biases*, Kahneman, D. et al., Eds. Cambridge University Press, Cambridge, pp. 294–306.

Alvi, I.A. and Ayyub, B.M. 1990. Analysis of System Performance under Stochastic and Fuzzy Uncertainty. Paper presented at the North American Fuzzy Information Processing Society (NAFIPS), University of Toronto, June 1990.

Amendola, A. 1986. *System Reliability Benchmark Exercise Parts I and II*, EUR-10696. EN/I, Joint Research Center of Ispra, Italy.

American Institute of Aeronautics and Astronautics. 1998. *Guide for Verification and Validation of Computational Fluid Dynamics Simulation*, AIAA G-077-1998.

American Psychological Association. 1985. *Standards for Educational and Psychological Testing*. Washington, D.C.

American Society of Mechanical Engineers (ASME). 1993. *The Use of Decision-Analytic Reliability Methods in Codes and Standards Work*, CRTD-Vol. 23. New York.

Ammerman, R.R., Ed. 1970. *Belief, Knowledge, and Truth*. Charles Scribners Sons, New York.

Ang, A. and Tang, W. 1975. *Probability Concepts in Engineering Planning and Design*, Vol. I. John Wiley, New York.

Ang, A. and Tang, W. 1984. *Probability Concepts in Engineering Planning and Design*, Vol. II. John Wiley, New York.

Anken, C.S. 2003. Information Understanding. Workshop on Strategic Research Directions in AI: Knowledge Representation, Discovery, and Integration, Cornell University.

Armstrong, D. 1973. *Belief Truth and Knowledge*. Cambridge University Press, Cambridge.

Austin, D.F., Ed. 1998. *Philosophical Analysis: A Defense by Example*, Philosophical Studies Series, Vol. 39. D. Reidel Dordrecht, Holland.

Ayyub, B.M. 1991. Systems framework for fuzzy sets in civil engineering. *Int. J. Fuzzy Sets Syst.*, 40, 491–508.

Ayyub, B.M. 1992a. Generalized Treatment of Uncertainties in Structural Engineering. In *Analysis and Management of Uncertainty: Theory and Applications*, Ayyub, B.M. and Gupta, M.M., Eds. Elsevier Science Publisher, New York, pp. 235–246.

Ayyub, B.M. 1992b. *Fault Tree Analysis of Cargo Elevators Onboard Ships*, BMA Engineering Report. Prepared for Naval Sea System Command, U.S. Navy, Crystal City, VA, 1992.

Ayyub, B.M. 1993. *Handbook for Risk-Based Plant Integrity*, BMA Engineering Report. Prepared for Chevron Research and Technology Corporation, Richmond, CA, 1993.

Ayyub, B.M. 1994. The Nature of uncertainty in structural engineering. In *Uncertainty Modelling and Analysis: Theory and Applications*, Ayyub, B.M. and Gupta, M.M., Eds. North-Holland-Elsevier Scientific Publishers, Amsterdam, pp. 195–210.

Ayyub, B.M., Ed. 1998. *Uncertainty Modeling and Analysis in Civil Engineering*. CRC Press, Boca Raton, FL.

Ayyub, B.M. 2001. *Elicitation of Expert Opinions for Uncertainty and Risks: Theory, Applications, and Guidance*, CRC Press, Boca Raton, FL.

Ayyub, B.M. 2003. *Risk Analysis in Engineering and Economics*. Chapman & Hall, CRC Press, Boca Raton, FL.

Ayyub, B.M. and Chao, R.-J. 1998. Uncertainty modeling in civil engineering with structural and reliability applications. In *Uncertainty Modeling and Analysis in Civil Engineering*, Ayyub, B.M., Ed. CRC Press, Boca Raton, FL, chap. 1, pp. 1–32.

Ayyub, B.M. and Gupta, M.M., Eds. 1997. *Uncertainty Analysis in Engineering and the Sciences: Fuzzy Logic, Statistics, and Neural Network Approach*. Kluwer Academic Publisher, Dordrecht, The Netherlands.

Ayyub, B.M. and Haldar, A. 1984. Practical structural reliability techniques. *J. Structural Eng.*, 110, 1707–1724.

Ayyub, B.M. and Hassan, M.H.M. 1992a. Control of construction activities. I. Systems identification. *Civ. Eng. Syst.*, 9, 123–146.

Ayyub, B.M. and Hassan, M.H.M. 1992b. Control of construction activities. II. Condition assessment of attributes. *Civ. Eng. Syst.*, 9, 147–204.

Ayyub, B.M. and Hassan, M.H.M. 1992c. Control of construction activities. III. A fuzzy-based controller. *Civ. Eng. Syst.*, 9, 275–297.

Ayyub, B.M. and Lai, K.-L. 1992. Structural reliability assessment with ambiguity and vagueness in failure. *Nav. Eng. J.*, 104, 21–35.

Ayyub, B.M. and McCuen, R. 1996. *Numerical Methods for Engineers*. Prentice Hall, Upper Saddle River, NJ.

Ayyub, B.M. and McCuen, R. 2003. *Probability, Statistics, and Reliability for Engineers and Scientists*. Chapman & Hall/CRC Press, Boca Raton, FL.

Ayyub, B.M., Gupta, M.M., and Kanal, L.N. 1992. *Analysis and Management of Uncertainty: Theory and Applications*. Elsevier Science Publishing Company, New York.

Ayyub, B.M., Guran, A., and Haldar, A., Eds. 1997. *Uncertainty Modeling in Vibration, Control, and Fuzzy Analysis of Structural Systems*. World Scientific, Singapore.

Ayyub, B.M., Moser, D.A., and Davis, S.A. 2000. *Economic Consequence Assessment of Floods in the Feather River Basin of California Using Expert-Opinion Elicitation*, Technical Report. Institute for Water Resources, USACE.

Ayyub, B.M., Riley, B.C., and Hoge, M.T. 1996. *Expert Elicitation of Unsatisfactory-Performance Probabilities and Consequences for Civil Works Facilities*, Technical Report. USACE, Pittsburgh District, PA.

Bae, H.-R., Grandhi, R.V., and Canfield, R.A. 2004. *Structural Uncertainty Quantification and Optimization Using Evidence Theory*, Technical Report. Wright State University and Air Force Institute of Technology at Wright Patterson Air Force Base, Dayton, OH.

Bailey, K.D. 1994. *Methods of Social Research*. The Free Press, Maxwell Macmillan, New York.

Beacher, G. 2000. *Expert Elicitation in Geotechnical Risk Assessment*, USACE Draft Report. University of Maryland, College Park.

Beck, L.J. 1965. *The Metaphysics of Descartes*. Oxford University Press, Oxford.

Bell, T.E. and Esch, K. 1989. The Space Shuttle: A Case of Subjective Engineering. *IEEE Spectrum*, June, pp. 42–46.

Bennett, J. 1990. Why Is Belief Involuntary? *Analysis*, Vol. 50, March.

Benton, J.R., Iyengar, S.S., Deng, W., Brenner, N., and Subrahmanian, V.S. 1995. Tactical route planning: new algorithms for decomposing the map. In *IEEE International Conference on Tools for AI*, Herndon, VA, November 6–8, 1995, pp. 268–277.

Bertalanffy, L. 1968. *General Systems Theory: Foundations, Development, Applications*. George Braziller, New York.

Blair, A.N. and Ayyub, B.M. 1999. Fuzzy stochastic cost and schedule risk analysis: MOB case study. In *Proceedings of the Symposium on Very Large Floating Structures*. Elsevier, North Holland, Amsterdam.

Blanchard, B.S. 1998. *System Engineering Management*, 2nd ed. John Wiley & Sons, New York.

Blinn, J.F. 1982. A generalization of algebraic surface drawing. *ACM Trans. Graphics*, 1, 235–256.

Blockley, D.I. 1975. Predicting the likelihood of structural accidents. *Proc. Inst. Civ. Eng.*, 59, 659–668.

Blockley, D.I. 1979a. The calculations of uncertainty in civil engineering. *Proc. Inst. Civ. Eng.*, 67, 313–326.

Blockley, D.I. 1979b. The role of fuzzy sets in civil engineering. *Fuzzy Sets Syst.*, 2, 267–278.

Blockley, D.I. 1980. *The Nature of Structural Design and Safety*. Ellis Horwood, Chichester, U.K.

Blockley, D.I., Pilsworth, B.W., and Baldwin, J.F. 1983. Measures of uncertainty. *Civ. Eng. Syst.*, 1, 3–9.

Bogler, G. 1987. Shafer-Dempster reasoning with applications to multi-sensor target identification systems. *IEEE Trans. Syst. Man Cybern.*, SMC-17, 968–977.

Boltzmann, L. 1894. Zur Integration der Diffusionsgleichung bei variablen Diffusion-skoeffizienten. *Ann. Phys.*, 53, 959–964.

Bouissac, P. 1992. The Construction of Ignorance and the Evolution of Knowledge. *University of Toronto Quarterly*, Vol. 61, Summer.

Bowles, D. 1990. Risk assessment in dam safety decision making. In *Proceedings of the Fourth Conference of Risk-Based Decision Making in Water Resources*, Haimes, Y.Y. and Stakhiv, E.Z., Eds., ASCE, Reston, VA, pp. 254–283.

Bradie, M. 1986. Assessing evolutionary epistemology. *Biol. Philos.*, 1, 401–450.

Bremermann, H.J. 1962. Optimization through evolution and recombination. In *Self-Organizing Systems*, Yovits, M.C., Jacobi, G.T., and Goldstein, G.D., Eds. Spartan Books, Washington, D.C., pp. 93–106.

Brown, C.B. 1979. A fuzzy safety measure. *J. Eng. Mech. Div.*, 105, 855–872.

Brown, C.B. 1980. The merging of fuzzy and crisp information. *J. Eng. Mech. Div.*, 106, 123–133.

Brown, C.B. and Yao, J.T.P. 1983. Fuzzy sets and structural engineering. *J. Struct. Eng.*, 109, 1211–1225.

Brune, R., Weinstein, M., and Fitzwater, M. 1983. *Peer Review Study of the Draft Handbook for Human Reliability Analysis with Emphasis on Nuclear Power Plant Applications*, Nuclear Regulatory Commission, Washington, D.C.

Budnitz, R.J., Apostolakis, G., Boore, D.M., Cluff, L.S., Coppersmith, K.J., Cornell, C.A, and Morris, P.A. 1997. Recommendations for Probabilistic Seismic Hazard Analysis: Guidance on Uncertainty and Use of Expert. Vol.1. U.S. Nuclear Regulatory Commission, Washington, D.C.

Campbell, D.T. 1974. Evolutionary epistemology. In *The Philosophy of Karl Popper*, Vol. 1, Schilpp, P.A., Ed. Open Court Publishing, LaSalle, IL, pp. 413–463.

Carey, M.R. and Johnson, D.S. 1979. *Computers and Intractability: A Guide to the Theory of NP-Completeness*. W.H. Freeman, San Francisco.

Carter, E.C. and Carter, R.C. 1981. Color and conspicuousness. *J. Optical Soc. Am.*, 71, 723–729.

Chen, S. and Goodman, J. 1996. An empirical study of smoothing techniques for language modeling. In *Proceedings of the Thirty-Fourth Annual Meeting of the Association for Computational Linguistics*. Morgan Kaufmann Publishers, San Francisco, pp. 310–318.

Chestnut, H. 1965. *Systems Engineering Tools*. John Wiley & Sons, New York.

Chomsky, N. 1968. *Language and Mind*. Harcourt Brace Jovanovich, New York.

Christensen, R. 1985. Entropy manimax multivariate statistical modeling. *Int. J. Gen. Syst.*, 11, 231–277.

Clemen, R.T. 1989. Combining forecasts: a review and annotated bibliography. *Int. J. Forecasting*, 5, 559–583.

Colglazier, E.W. and Weatherwax, R.K. 1986. Failure estimates for the space shuttle. In *Abstracts from the Society of Risk Analysis*, annual meeting, Boston, November 9–12, 1986, p. 80.

Committee on Safety Criteria for Dams. 1985. *Safety of Dams: Flood and Earthquake Criteria*. National Academy Press, Washington, D.C.

Committee on the Safety of Existing Dams. 1983. *Safety of Existing Dams, Evaluation and Improvement*. National Research Council, National Academy Press, Washington, D.C.

Cooke, R.M. 1986. Problems with empirical Bayes. *Risk Anal.*, 6, 269–272.

Cooke, R.M. 1991. *Experts in Uncertainty*. Oxford University Press, Oxford.

Cui, W.C. and Blockley, D.I. 1990. Interral probability theory for evidential support. *Int. J. Intelligent Syst.*, 5(2), June, 183–192.

Dancy, J. 1985. *Contemporary Epistemology*. Basil Blackwell, Oxford.

Davis, J.P. and Hall, J.W. 2003. A software-supported process for assembling evidence and handling uncertainty in decision-making. *Decision Support Syst.*, 35, 415–433.

De Cooman, G. 1997. Possibility theory. *Int. J. Gen. Syst.*, 25, 291–371.

De Corte, W.D. 1986. Optimal colors, phosphors, and illuminant characteristics for CRD displays: the algorithmic approach. *Hum. Factors*, 28, 39–47.

Defense Acquisition University. 1998. *Risk Management Guide*. Defense Systems Management College Press, Fort Belvoir, VA.

De Finetti, B. 1937. *Studies in Subjective Probabilities*. Wiley, New York. (English translation in 1964 by Kyburg, H. and Smokler, H., Eds.)

DeKay, M.L. and McClelland, G.H. 1993. Predicting loss of life in cases of dam failure and flash flood. *Risk Anal.*, 13, 193–205.

Dempster, A.P. 1976a. Upper and lower probabilities induced by multivalued mapping. *Ann. Math. Stat.*, 38, 325–339.

Dempster, A.P. 1976b. Upper and lower probabilities inferences based on a sample from a finite univariate population. *Biometrika*, 54, 515–528.

Deng, J. 1993. *Grey Control System*. HUST Press, Wuhan, China.

Department of the Army. 1995. *Military Operations Battlefield Visualization Concept*, TRADOC Pamphlet 525-70. Headquarters, U.S. Army, Training and Doctrine Command, Fort Monroe, VA.

Devillers, O. 1992. *Robust and Efficient Implementation of the Delaunay Tree*. Rapport de recherche 1619, INRIA.

Dewey, J. 1929. *The Quest for Certainty*. Minton, Balch and Co., New York.

di Carlo, C.W. 1998. Evolutionary Epistemology and the Concept of Ignorance. Ph.D. thesis, University of Waterloo, Ontario, Canada.

Di Nola, A., Sessa, S., Pedrycz, W., and Sauchez, E. 1989. Fuzzy Relation Equations and Their Applications to Knowledge Engineering. Kluwer, Boston.

Dombi, J. 1982. A general class of fuzzy operators, the De Morgan class of fuzzy operators and fuzziness measures induced in fuzzy operators. *Fuzzy Sets Syst.*, 8, 149–163.

Dong, W. and Shah, H.C. 1987. Vertex methods for computing functions of fuzzy variables. *Fuzzy Sets Syst.*, 24, 79–86.

Dong, W.-M. and Wong, F.S. 1986a. From uncertainty to approximate reasoning. Part 1. Conceptual models and engineering interpretations. *Civ. Eng. Syst.*, 3, 143–154.

Dong, W.-M. and Wong, F.S. 1986b. From uncertainty to approximate reasoning. Part 2. Reasoning with algorithmic rules. *Civ. Eng. Syst.*, 3, 192–202.

Dong, W.-M. and Wong, F.S. 1986c. From uncertainty to approximate reasoning. Part 3. Reasoning with conditional rules. *Civ. Eng. Syst.*, 3, 112–121.

Doore, G.S. et al. 1993. Guidelines for using color to depict meteorological information. *Bull. Am. Meteor. Soc.*, 74, 1709–1713.

Dretske, F. 1981. *Knowledge and the Flow of Information.* MIT Press, Cambridge, MA.

Driver, J. 1989. The virtues of ignorance. *J. Philos.*, 86, 373–384.

Dubois, D. 1981. A new definition of the cardinality of finity fuzzy sets preserving the classic addition property. *Busefal*, 8, 65–67.

Dubois, D. and Prade, H. 1980. On several representations of an uncertain body of evidence. In *Fuzzy Information and Decision Processes*, Gupta, M.M. and Sanchez, E., Eds. North-Holland, New York, pp. 167–181.

Dubois, D. and Prade, H. 1985. Fuzzy cardinality and the modeling of imprecise quantification. *Fuzzy Sets Syst.*, 16, 199–230.

Dubois, D. and Prade, H. 1986a. On the unicity of Dempster rule of combination. *Int. J. Intell. Syst.*, 1, 133–142.

Dubois, D. and Prade, H. 1986b. A set-theoretic view of belief function. *Int. J. Gen. Syst.*, 12, 193–226.

Dubois, D. and Prade, H. 1988. *Possibility Theory.* Plenum Press, New York.

Dubois, D. and Prade, H. 1991. Random sets and fuzzy interval analysis. *Fuzzy Sets Syst.*, 42, 87–101.

Dubois, D. and Prade, H. 1997. Focusing vs. belief revision: a fundamental distinction when dealing with generic knowledge. In *Proceedings of the First International Joint Conference on Qualitative and Quantitative Practical Reasoning*, ECSQARU-FAPR '97, Bad Honnef, Germany, pp. 96–107.

Duncan, R. and Weston-Smith, M., Eds. 1977. *The Encyclopedia of Ignorance*. Pocket Books, New York.

Durant, W. 1991. *Story of Philosophy: The Lives and Opinions of the World's Greatest Philosophers*. Pocket Books, New York.

Durrett, H.J. 1987. *Color and the Computer*, Academic Press, New York.

Eldukair, Z.A. and Ayyub, B.M. 1991. Analysis of recent U.S. structural and construction failures. *J. Performance Constructed Facilities*, 5, 57–73.

Feller, W. 1968. *An Introduction to Probability Theory and Its Applications*, Vol. I. John Wiley & Sons, New York.

Ferrell, W.R. 1985. Combining individuals' judgments. In *Behavioral Decision Making*, Wright, G., Ed. Plenum, New York.

Ferrell, W.R. 1994. Discrete subjective probabilities and decision analysis: elicitation, calibration, and combination. In *Subjective Probability*, Wright, G. and Ayton, P., Eds. John Wiley & Sons, New York.

Ferson, S. and Hajagos, J.G. 2004. Arithmetic with uncertain numbers: rigorous and (often) best possible answers. J. *Reliability Eng. Syst. Saf.*, 85, 135–152.

Ferson, S., Nelson, R.B., Hajagos, J., Berleant, D.J., Zhang, J., Tucker, W.T., Ginzburg, L.R., and Oberkampf, W.L. 2004. *Dependence in Probabilistic Modeling, Dempster-Shafer Theory, and Probability Bounds Analysis*, Sandia Report SAND2004-3072. Sandia National Laboratories, Department of Energy, Oak Ridge, TN.

Ferson, S., Root, W., and Kuhn, R. 1999. *RAMAS Risk Calc: Risk Assessment with Uncertain Numbers*. Applied Biomathematics, Setauket, NY.

Fishel, C. and Gardner, B. 2003. *Logo Lounge: 2000 International Idenities by Leading Designers*. Rockport Publishers, Cambridge, MA, www.rockpub.com.

Fisher, R.A. 1930. *The Genetical Theory of Natural Selection*. Oxford University Press, Oxford.

Flake, G.W. 1998. *The Computational Beauty of Nature: Computer Explorations of Fractals, Chaos, Complex Systems, and Adaptation*. A Bradford Book, MIT Press, Cambridge, MA.

Forrest, J. 1997. Some Samples of Application of a Quantitative Systemic Analysis, http://www.geocities.com/athens/delphi/4695/quansa.html#grey.

Frank, M.J. 1979. On the simultaneous associativity of F(x,y) and x+y-F(x,y). *Aequationes Mathematicae*, 19, 194–226.

Frank, M.J., Nelson, R.B., and Schweizer, B. 1987. Best-possible bounds for the distribution of a sum: a problem of Kolmogorov. *Probab. Theor. Related Fields*, 74, 199–211.

Fréchet, M. 1935. "Généralisations du Théoréme des Probabilitiés Totales. *Fundamenta Mathematica*, 25, 379–387.

Fréchet, M. 1951. Sur les Tableaux de Corrélation dont les Marges sont Données. *Ann. Univ. Lyon Sect. A*, 9, 53–77.

French, S. 1985. Group consensus probability distributions: a critical survey. In *Bayesian Statistics*, Bernardo, J.M. et al., Eds. Elsevier, North Holland, Amsterdam, pp. 183–201.

Furuta, H., Fu, K.S., and Yao, J.T.P. 1985. Structural engineering applications of expert systems. *Comput. Aided Design*, 17, 410–419.

Furuta, H., Shiraishi, N., and Yao, J.T.P. 1986. An expert system for evaluation of structural durability. *Proc. Fifth OMAE Symp.*, 1, 11–15.

Galanter, E. 1962. The direct measurement of utility and subjective probability. *Am. J. Psychol.*, 75, 208–220.

Gale, W. and Church, K. 1994. What is wrong with adding one? In *Corpus Based Research into Language*, Oostdijk, N. and de Haan, P., Eds. Rodopi, Amsterdam.

Garey, M.R. and Johnson, D.S. 1979. *Computers and Intractability: A Guide to the Theory of NP-Completeness*. W.H. Freeman, San Francisco.

Geer, J.F. and Klir, G.J. 1992. A mathematical analysis of information preserving transformations between probabilistic and possibilistic formulations of uncertainty. *Int. J. Gen. Syst.*, 20, 143–176.

Genest, C. and Zidek, J. 1986. Combining probability distributions: critique and an annotated bibliography. *Stat. Sci.*, 1, 114–148.

Goldman, A. 1993. *Philosophical Applications of Cognitive Science*. Westview Press, Boulder, CO.

Goldman, S. 1968. *Information Theory*. Dover, New York.

Goodman, J. and O'Rourke, J. 1997. *Handbook on Discrete and Computational Geometry*. CRC Press, Boca Raton, FL.

Grossman, J.D. 1992. Color conventions and application standards. In *Color in Electronic Displays*, Widdel, H. and Post, D.L., Eds. Plenum Press, New York, pp. 209–218.

Gustafson, D.H., Shukla, R.K., Delbecq, A., and Walster, G.W. 1973. A comparative study of differences in subjective likelihood estimates made by individuals, interacting groups, Delphi groups, and nominal groups. *Org. Behav. Hum. Performance*, 9, 200–291.

Hackett, C. 1969. The origin of speech. *Sci. Am.*, 203, 88–111.

Haldar, A., Guran, A., and Ayyub, B.M., Eds. 1997. *Uncertainty Modeling in Finite Element, Fatigue, and Stability of Systems*. World Scientific, Singapore.

Hall, A.D. 1962. *A Method for Systems Engineering*. D. Van Nostrand Company, Princeton, NJ.

Hall, A.D. 1989. *Metasystems Methodology, a New Synthesis and Unification*. Pergamon Press, New York.

Hall, J.W., Blockley, D.I., and Davis, J.P. 1998. Uncertain inference using interval probability theory. *Int. J. Approximate Reasoning*, 19, 247–264.

Hallden, S. 1986. *The Strategy of Ignorance: From Decision Logic to Evolutionary Epistemology*. Library of Theoria, Stockholm.

Halmos, P. 1950. *Measure Theory*, Van Nostrand, Princeton, NJ.

Hamacher, P.R. 1978. Uber Logische Verknupfungen Unscharfer Aussagen und Deren Zugehorige Bewertungs-Funktionen. In *Progress in Cybernetics and Systems Research*, Trappl, R., Klir, G.J., and Ricciardi, L., Eds. Hemisphere, Washington, D.C., pp. 276–287.

Hansen, R. 1990. Breaking the color barrier. *Comput. Graphics World*, 13, 39–48.

Harmanec, D., Resconi, G., Klir, G., and Pan, Y. 1996. On the computation of uncertainty measure in Dempster-Shafer theory. *Int. J. Gen. Syst.*, 25, 153–163.

Harr, M.E. 1987. *Reliability-Based Design in Civil Engineering*. McGraw-Hill Book Company, New York.

Hartford, D.N.D. 1995. *How Safe Is Your Dam? Is It Safe Enough?* B.C. Hydro, Maintenance, Engineering, and Projects, Burnaby, BC.

Hartley, R.V.L. 1928. Transmission of information. *Bell Syst. Tech. J.*, 7, 535–563.

Hassan, M.H.M. and Ayyub, B.M. 1993a. Multi-attribute control of construction activities. *Civ. Eng. Syst.*, 10, 37–53.

Hassan, M.H.M. and Ayyub, B.M. 1993b. A fuzzy controller for construction activities. *Fuzzy Sets Syst.*, 56, 253–271.

Hassan, M.H.M. and Ayyub, B.M. 1994. Multi-attribute fuzzy control of construction activities. In *Uncertainty Modelling and Analysis: Theory and Applications*, Ayyub, B.M. and Gupta, M.M., Eds. North-Holland-Elsevier Scientific Publishers, Amsterdam, pp. 271–286.

Hassan, M.H.M. and Ayyub, B.M. 1997. Structural fuzzy control. In *Uncertainty Modeling in Vibration, Control and Fuzzy Analysis of Structural Systems*, Ayyub, B.M., Guran, A., and Haldar, A., Eds. World Scientific, Singapore, chap. 7, pp. 179–232.

Hassan, M.H.M., Ayyub, B.M., and Bernold, L. 1992. Fuzzy-based real-time control of construction activities. In *Analysis and Management of Uncertainty: Theory and Applications*, Ayyub, B.M., Gupta, M.M., and Kanal, L.N., Eds. North-Holland-Elsevier Scientific Publishers, Amsterdam, pp. 331–350.

Helmer, O. 1968. Analysis of the future: the Delphi method; The Delphi method: an illustration. In *Technological Forecasting for Industry and Government*, Bright, J., Ed. Prentice Hall, Englewood Cliffs, NJ.

Henley, E.J. and Kumamoto, H. 1981. *Reliability Engineering and Risk Assessment*. Prentice Hall, Englewood Cliffs, NJ.

Higashi, M. and Klir, G.J. 1983. Measures of uncertainty and information based on possibility distributions. *Int. J. Gen. Syst.*, 8, 43–58.

Hoffman, C.M., Kim, Y.-J., Winkler, R.P., Walrath, J.D., and Emmerman, P.J. 1998. Visualization for situation awareness. In *Workshop on New Paradigms in Information Visualization and Manipulation*, Bethesda, MD, pp. 36–40.

Hoffman, J.E. 1980. Interaction between global and local levels of a form. *J. Exp. Psychol. Hum. Percept. Performance*, 6, 222–234.

Hoffman, R.R. 1991. Human factors psychology in the support of forecasting: the design of advanced meteorological workstations. *Weather Forecasting*, 6, 98–110.

Hoffman, R.R. and Conway, J.A. 1989. Psychological factors in remote sensing: a review of some recent research. *GEOCARTO Int.*, 4, 3–21.

Hoffman, R.R. and Lipton, K.S. 1992. Color Displays for Meteorological Information Processing Systems. Essay, Department of Psychology, Adelphi University, Garden City, NY.

Honderich, H., Ed. 1995. *The Oxford Companion to Philosophy.* Oxford University Press, Oxford.

Horgan, J. 1996. *The End of Science*. Addison-Wesley Publishing Co., Reading, MA.

Horgan, T., Ed. 1995. Vagueness. The Spindel Conference 1994. *South. J. Philos.*, 33 (Suppl.).

Horwich, P. 1987. *Asymmetries in Time: Problems in the Philosophy of Science*. MIT Press, Cambridge, MA.

Inagaki, T. 1991. Interdependence between safety-control policy and multiple-sensor scheme via Dempster-Shafer theory. *IEEE Trans. Reliability*, 40, 182–188.

Ishizuka, M., Fu, K.S., and Yao, J.T.P. 1981. A Rule-Inference Method for Damage Assessment, ASCE Preprint 81-502. ASCE, St. Louis, MO.

Ishizuka, M., Fu, K.S., and Yao, J.T.P. 1983. Rule-based damage assessment system for existing structures. *Solid Mech. Arch.*, 8, 99–118.

Itoh, S. and Itagaki, H. 1989. Application of fuzzy-Bayesian analysis to structural reliability. In *Proceedings of the Fifth International Conference on Structural Safety and Reliability*, ICOSSAR, Vol. 3, Ang, A.H.-S., Shinozuka, M., and Schuëller, G.I., Eds. ASCE, San Francisco, pp. 1771–1774.

Jennings, N.R. and Wooldridge, M.J., Eds. 1998. *Agent Technology: Foundations, Applications, and Markets*. Springer, Berlin.

Johnson, O.A. 1979. Ignorance and irrationality: a study in contemporary scepticism. *Philos. Res. Arch.*, 5, 10–13.

Johnson-Laird, P. 1988. *The Computer and the Mind: An Introduction to Cognitive Science*. Harvard University Press, Cambridge, MA.

Kahn, H. 1960. *On Thermonuclear War.* Free Press, New York.

Kahn, H. and Wiener, A.J. 1967. *The Year 2000: A Framework for Speculation*. Macmillan, New York.

Kahneman, D., Slovic, P., and Tversky, A., Eds. 1982. *Judgment under Uncertainty: Heuristics and Biases*. Cambridge University Press, Cambridge, U.K.

Kaiser, M.K. and Proffitt, D.R. 1989. *Perceptual Issues in Scientific Visualization. Three Dimensional Visualization and Display Technologies*. SPIE, Bellingham, WA.

Kaneyoshi, M., Tanaka, H., Kamei, M., and Furuta, H. 1990. Optimum cable tension adjustment using fuzzy regression analysis. In *The Third WG 7.5 Working Conference on Reliability and Optimization of Structural Systems*. International Federation for Information Processing, University of California, Berkeley, 11 pp.

Kapur, J.N. 1994. *Measures of Information and Their Applications*. John Wiley, New York.

Kaufmann, A. 1975. *Introduction to the Theory of Fuzzy Subsets*, Swanson, D.L., Trans. Academic Press, New York.

Kaufmann, A. and Gupta, M.M. 1985. *Introduction to Fuzzy Arithmetic, Theory and Applications*. Van Nostrand Reinhold Co., New York.

Keynes, J.M. 1921. *A Treatise on Probability.* MacMillan, New York.

Kinney, J.S. and Huey, B.M., Eds. 1990. *Application Principles for Multicolored Displays,* Workshop Report. National Research Council, National Academic Press, Washington, D.C., pp. 10–27.

Klir, G.J. 1969. *An Approach to General Systems Theory.* Van Nostrand Reinhold Company, New York.

Klir, G.J. 1972. *Trends in General Systems Theory.* Wiley Interscience, New York.

Klir, G.J. 1985. *Architecture of Systems Problem Solving.* Plenum Press, New York (second edition, Kluwer Academic/Plenum Publishers, New York, 2003).

Klir, G.J. 1990. A principle of uncertainty and information invariance. *Int. J. Gen. Syst.*, 17, 249–275.

Klir, G.J. 1991a. Generalized information theory. *Fuzzy Sets Syst.*, 40, 127–142.

Klir, G.J. 1991b. *Facets of Systems Science.* Plenum Press, New York (second edition, 2001).

Klir, G.J. 1997a. Fuzzy arithmetic with requisite constraints. *Fuzzy Sets Syst.*, 91, 165–175.

Klir, G.J. 1997b. The role of constrained fuzzy arithmetic in engineering. In *Uncertainty Modeling and Analysis in Engineering and the Sciences*, Ayyub, B.M. and Gupta, M.M., Eds. Kluwer, New York, pp. 1–19.

Klir, G.J. 1999a. On fuzzy-set interpretation of possibility theory. *Fuzzy Sets Syst.*, 108, 263–273.

Klir, G.J. 1999b. On the complementarity of systems sciences and classical science. In *Toward New Paradigm of Systems Science*, Rhee, W.P., Ed. Seoul National University Press, Korea.

Klir, G.J. 2003. Lecture Notes on Generalized Uncertainty-Based Information Theory: Aims, Results, and Open Problems. Short course at the International Symposium on Uncertainty Modeling and Analysis, September.

Klir, G.J. 2004. Generalized information theory: aims, results, and open problems. *J. Reliability Eng. Syst. Saf.*, 85, 21–38.

Klir, G.J. 2005. Measuring Uncertainty Associated with Convex Sets of Probability Distributions: A New Approach. Paper presented at the Proceedings of the North American Fuzzy Information Processing Society, Ann Arbor, MI.

Klir, G.J. 2006. *Uncertainty and Information: Generalized Information Theory.* John Wiley, Hoboken, NJ.

Klir, G.J. and Cooper, J.A. 1996. On constraint fuzzy arithmetic. In *Proceedings of the Fifth IEEE International Conference on Fuzzy Systems*, New Orleans, pp. 1693–1699.

Klir, G.J. and Folger, T.A. 1988. *Fuzzy Sets, Uncertainty, and Information.* Prentice Hall, Englewood Cliffs, NJ.

Klir, G.J. and Pan, Y. 1998. Constrained fuzzy arithmetic: basic questions and some answers. *Soft Comput.*, 2, 100–108.

Klir, G.J. and Parviz, B. 1992. Probability-possibility transformations: a comparison. *Int. J. Gen. Syst.*, 21, 291–310.

Klir, G.J. and Smith, R.M. 1999. Recent developments in generalized information theory. *Int. J. Fuzzy Syst.*, 1, 1–13.

Klir, G.J. and Wierman, M.J. 1999. *Uncertainty-Based Information: Elements of Generalized Information Theory. Studies in Fuzziness and Soft Computing.* Physica-Verlag, a Springer-Verlag Company, New York.

Klir, G.J. and Yuan, B. 1995a. *Fuzzy Sets and Fuzzy Logic: Theory and Applications.* Prentice Hall, Upper Saddle River, NJ.

Klir, G.J. and Yuan, B. 1995b. On nonspecificity of fuzzy sets with continuous membership functions. *Proc. IEEE Conf. on Systems, Man, and Cybernetics,* Vancouver.

Klir, G.J. and Yuan, B. 1996. *Fuzzy Sets, Fuzzy Logic, and Fuzzy Sets: Selected Papers by Lotfi Zadeh*. World Scientific, Singapore.

Korta, J.P., Lee, M.P., Eisenberg, N.A., DeWispelare, A.R. 1996. Branch Technical Position on the Use of Expert Elicitation in the High-Level Radioactive Waste Program. U.S. Nuclear Regulatory Commission, Washington, D.C.

Krebs, M.J. and Wolf, J.D. 1979. Design principles for the use of color in displays. *Proc. Soc. Inf. Display*, 20, 10–15.

Krebs, M.J., Wolf, J.D., and Sandvig, J.H. 1978. *Color Display Design Guide*, Report N00014-77-C-0349. Office of Naval Research, Arlington, VA.

Krueger, R.A. and Casey, M.A. 2000. *Focus Groups: A Practical Guide for Applied Research*, 3rd ed. Sage Publication, Thousands Oaks, CA.

Kumamoto, H. and Henley, E.J. 1996. *Probabilistic Risk Assessment and Management for Engineers and Scientists*, 2nd ed. IEEE Press, New York.

Lai, K.-L. 1992. Generalized Uncertainty in Structural Reliability Assessment. Ph.D. dissertation, University of Maryland, College Park.

Lai, K.-L. and Ayyub, B.M. 1994. Generalized uncertainty in structural reliability assessment. *Civ. Eng. Syst.*, 11, 81–110.

Langer, E. 1975. The illusion of control. *J. Personality Soc. Psychol.*, 32, 311–328.

Laplace, P. 1825. *Philosophical Essays on Probabilistic Analysis*, Springer-Verlag, New York. (Translated by A.I. Dale from 5th edition in 1995.)

Levi, I. 1977. Four Types of Ignorance. *Social Research*, 44, Winter.

Levkowitz, H. 1988. Color in Computer Graphic Representation of Two-Dimensional Parameter Distributions. Ph.D. thesis, University of Pennsylvania, Philadelphia, pp. 62–78.

Levkowitz, H. and Herman, G.T. 1992. Color scales for image data. *IEEE Comput. Graphics Appl.*, 17, 72–80.

Lichtenstein, S. and Newman, J.R. 1967. Empirical scaling of common verbal phrases associated with numerical probabilities. *Psychonometric Sci.*, 9, 563–564.

Lin, Y. 1999. *General Systems Theory: A Mathematical Approach*. Kluwer/Plenum, New York.

Lindley, D. 1970. *Introduction to Probability and Statistics from a Bayesian Viewpoint*. Cambridge University Press, Cambridge, U.K.

Linstone, H.A. and Turoff, M. 1975. *The Delphi Method, Techniques, and Applications*. Addison Wesley, Reading, MA.

Lisy, I.J. 2003. Profiles of World Economists: Ludwig Von Mises. *BIATEC*, Vol. XI, July.

Lucas, D.J. 1995. Default correlation and credit analysis. *J. Fixed Income*, 4, 76–87.

Maruyama, M. 2003. New Technologies for Brain Study: A Meeting Ground for Social, Biological, and Psychological Sciences. *Futures Research Quarterly*, Summer, pp. 13–22.

Mehlhorn, K. and Naher, S. 1998. *Leda: A Platform for Combinatorial and Geometric Computing*. Cambridge University Press, Cambridge, U.K.

Merton, R.K., Fiske, M., and Kendall, P.L. 1956. *The Focused Interview*. Free Press, Glencoe, IL.

Mesarovic, M.D. and Takahara, Y. 1975. *General Systems Theory: Mathematical Foundations*. Academic Press, New York.

Mesarovic, M.D. and Takahara, Y. 1988. *Abstract Systems Theory*. Springer-Verlag, New York.

Miller, J.G. 1978. *Living Systems*. McGraw Hill, New York.

Miller-Jacobs, H.H. 1984. Color displays and the color deficient operator. In *Advances in Display Technology*, Schlar, E., Ed. International Society of Optical Engineering, Bellingham, WA, pp. 14–15.

Mitchell, H.B. and Estrakh, D.D. 1997. On the best scalar approximation of cardinality of a fuzzy set. *Int. J. Uncertainty Fuzziness Knowledge-Based Syst.*, 5, 681–688.

Montagu, A. and Darling, E. 1970. *The Ignorance of Certainty.* Harper & Row Publishers, New York.

Moore, R.E. 1966. *Interval Analysis.* Prentice Hall, Englewood Cliffs, NJ.

Moore, R.E. 1979. *Methods and Applications of Interval Analysis.* SIAM, Philadelphia.

Morgan, M.G. and Henrion, M. 1992. *Uncertainty: A Guide to Dealing with Uncertainty in Quantitative Risk and Policy Analysis.* Cambridge University Press, New York.

Morris, J.M. and D'Amore, R.J. 1980. *Aggregating and Communicating Uncertainty.* Pattern Analysis and Recognition Corp., Rome, New York.

Morris, S. 1997. Risk, uncertainty and hidden information. *Theory and Decision*, 42, 235–269.

Motta, R.J. 1991a. Color encoding computer images. *Inf. Display*, 7, 4–7.

Motta, R.J. 1991b. Color encoding computer images: part two. *Inf. Display*, 7, 16–19.

Muller, R. 2003. *The LosLogos: A Collection of Selected Logos.* Die Gestalten Verlag, Berlin, www.loslogos.org.

Murphy, A. and Daan, H. 1984. Impact of feedback and experience on the quality of subjective probability forecasts: comparison of the results from the first and second years of the Zierikzee experiment. *Mon. Weather Rev.*, 112, 413–423.

National Research Council. 1995. *Flood Risk Management and the American River Basin: An Evaluation.* National Academy Press, Washington, D.C.

Nau, R.F. 2001. De Finetti was right: probability does not exist. *Theory Decision*, 51, 89–124.

Navon, D. 1977. Forest before trees: the precedence of global features in visual perception. *Cognit. Psychol.*, 9, 353–383.

Nelson, R.B. 1999. *An Introduction to Couplas*, Lecture Notes in Statistics, Vol. 139. Springer-Verlag, New York.

New School University. 2003a. The von Neuman-Morgenstern Expected Utility Theory. New School University (The History of Economic Thought). Retrieved on October 1, 2003, from cepa.newschool.edu/het/essays/uncert/vnmaxioms.htm.

New School University. 2003b. Theory of Risk Aversion. New School University (The History of Economic Thought). Retrieved on October 1, 2003, from cepa.newschool.edu/het/essays/uncert/aversion.htm.

New School University. 2003c. Profile of Emile Borel. New School University (The History of Economic Thought). Retrieved on October 1, 2003, from http://cepa.newschool.edu/het/.

Newman, J.R. 1961. Thermonuclear War. *Scientific America*, March.

Nguyen, H.T. 1978. On random sets and belief functions. *J. Math. Anal. Appl.*, 65, 531–542.

Nofi, A.A. 2000. *Defining and Measuring Shared Situational Awareness*, CRM D0002895.A1/Final. Center for Naval Analyses, Alexandria, VA.

Nuclear Regulatory Commission. 1975. *Reactor Safety Study*, WASH-1400, NUREG 751014.

Nuclear Regulatory Commission. 1997. *Recommendations for Probabilistic Seismic Hazard Analysis: Guidance on Uncertainty and Expert Use*, NUREG/CR-6372, UCRL-ID-122160, Vols. 1 and 2. Prepared by the Senior Seismic Hazard Analysis Committee, Washington, D.C.

Oberkampf, W.L., Helton, J.C., Joslyn, C.A., Wojtkiewicz, S.F., and Ferson, S. 2003. Challenge problems: uncertainty in system response given uncertain parameters. *Reliability Eng. Syst. Saf.*, 85(1), 11–19.

Orlitsky, A., Santhanam, N.P., and Zhang, J. 2003. Always good Turing: asymptotically optimal probability estimation. *Science*, 302, 427–431.

Pal, S.K. and Skowron, A., Eds. 1999. *Rough Fuzzy Hybridization.* Springer-Verlag, New York.

Paté-Cornell, E. 1996. Uncertainties in Risk Analysis and Comparability of Risk Estimates. Paper presented at the Society for Risk Analysis 1996 Annual Meeting, McLean, VA.

Pawlak, Z. 1991. *Rough Sets: Theoretical Aspects of Reasoning about Data*. Kluwer, Boston.
Pawlak, Z. 1999. Rough sets, rough functions, and rough calculus. In *Rough Fuzzy Hybridization*, Pal, S.K. and Skowron, A., Eds. Springer-Verlag, New York.
Pearcey, N.R. and Thaxton, C.B. 1994. *The Soul of Science: Christian Faith and Natural Philosophy (Turning Point Christian Worldview)*. Crossway Books, Wheaton, IL.
Pearl, J. 1990. Reasoning with belief functions: an analysis of compatibility. *Int. J. Approximate Reasoning*, 4, 363–390.
Pearl, J. 1992. Rejoinders to comments on "Reasoning with belief functions: an analysis of compatibility." *Int. J. Approximate Reasoning*, 6, 425–443.
Piattelli-Palmarini, M. 1994. *Inevitable Illusions: How Mistakes of Reason Rule Our Minds*. John Wiley & Sons, New York. (Translated from Italian.)
Pléh, C. 2003. Ernst Mach and Daniel Dennett: Two Evolutionary Models of Cognition. Center for Advanced Studies in the Behavioral Sciences, Stanford. Retrieved on October 1, 2003, from http://www.staff.u-szeged.hu/~pleh/english/articles/mach.html.
Ponce, V.M. 1989. *Engineering Hydrology Principles and Practices*. Prentice Hall, Englewood Cliffs, NJ, 640 pp.
Popkin, R.H., Ed. 2000. *The Columbia History of Western Philosophy*. Mjf Books, New York.
Popper, K.R. 1963. *Conjectures and Refutations*. Rutledge & Kegan Paul, London.
Popper, K. *The Logic of Scientific Discovery*. Review of the book by Routledge Classics. Released on March, 29, 2002.
Preyssl, C. and Cooke, R. 1989. Expert Judgment: Subjective and Objective Data for Risk Analysis for Space-Flight Systems. Paper presented at the Proceedings of PSA 1989, Pittsburgh, PA, April 2–7.
Ramsey, F. 1931. Truth and probability. In *The Foundation of Mathematics and Other Logical Essays*, Braithwaite, R.B., Ed. Kegan Paul, London, pp. 156–198.
Regan, H.M., Ferson, S., and Berleant, D. 2000. Equivalence of Five Methods for Reliable Uncertainty Propagation, draft paper. Applied Biomathematics, Setauket, NY.
Reichenbach, H. 1968. *The Rise of Scientific Philosophy*. University of California Press, Berkeley.
Reza, F.M. 1961. *An Introduction to Information Theory*. McGraw-Hill, New York.
Rice, J.F. 1991. Ten rules for color coding. *Inf. Display*, 7, 12–14.
Rice, S.A., Ed. 1931. *Methods in Social Research*. University of Chicago Press, Chicago, p. 561.
Robertson, A.R. 1992. Color perception. *Phys. Today*, 45, 24–29.
Rosen, R. 1985. *Anticipatory Systems: Philosophical, Mathematical, and Methodological Foundations*. Pergamon Press, Oxford, MA.
Rowe, G. 1992. Perspectives on expertise in aggregation of judgments. In *Expertise and Decision Support*, Wright, G. and Bolger, F., Eds. Plenum Press, New York, pp. 155–180.
Royden, H. 1988. *Real Analysis*. Prentice Hall, Upper Saddle River, NJ.
Russell, B. 1975. *A History of Western Philosophy*. Simon & Schuster, New York.
Saaty, T.L. 1996. *The Analytic Hierarchy Process*. RWS Publications, Pittsburgh, PA.
Sackman, H. 1975. *Delphi Critique: Expert Opinion, Forecasting, and Group Process*. Lexington Books, Lexington, MA.
Samet, M.G. 1975. Quantitative interpretation of two qualitative scales used to rate military intelligence. *Hum. Factors*, 17, 192–202.
Schaefer, H.F., III. 2003. Quantum Mechanics and Postmodernism, lecture on this topic. Retrieved on October 1, 2003, from http://www.westminsterhall.us/hfs3/qm_postmod.doc.
Schweizer, B. and Sklar, A. 1983. *Probability Metric Spaces*. North-Holland, New York. *Science*. 1983. 222, 1293.

Scientific American. 1969. *Science, Conflict and Society: Readings from* Scientific American. W.H. Freeman and Co., San Francisco.

Shafer, G. 1976. *A Mathematical Theory of Evidence*. Princeton University Press, Princeton, NJ.

Shannon, C.E. 1948. The mathematical theory of communication. *Bell Syst. Tech. J.*, 27, 379–423, 623–656.

Shewchuk, J.R. 1996. Triangle: engineering a 2D quality mesh generator and Delaunay triangulator. In *Applied Computational Geometry: Towards Geometric Engineering*, Vol. 1148 of Lecture Notes in Computer Science, Ming, C., Manocha, L., and Manocha, D., Eds. Springer-Verlag, Berline, pp. 203–222. (From the First ACM Workshop on Applied Computational Geometry.)

Shiraishi, N. and Furuta, H. 1983. Reliability analysis based on fuzzy probability. *J. Eng. Mech.*, 109, 1445–1459.

Shiraishi, N., Furuta, H., and Sugimoto, M. 1985. Integrity Assessment of Structures Based on Extended Multi-Criteria Analysis. Paper presented at the Proceedings of the Fourth ICOSSAR, Kobe, Japan.

Silberschatz, A., Stonebreaker, M., and Ullman, J.D. 1991. Database systems: achievements and opportunities. *Commun. ACM*, 34, 110–119.

Smets, P. 1988. Belief functions. In *Non-Standard Logics for Automated Reasoning*, Smets, P., Mamdani, A., Dubois, D., and Prade, H., Eds. Academic Press, New York, pp. 253–286.

Smets, P. 1990. The combination of evidence in the transferable belief model. *IEEE Trans. Pattern Anal. Mach. Intell.*, 12, 447–458.

Smets, P. 1992. Resolving misunderstandings about belief functions. *Int. J. Approximate Reasoning*, 6, 321–343.

Smets, P. 1993. Belief functions: the disjunctive rule of combination and the generalized Bayesian theorem. *Int. J. Approximate Reasoning*, 9, 1–35.

Smets, P. 1998. Application of the transferable belief model to diagnostic problems. *Int. J. Intell. Syst.*, 13, 127–157.

Smithson, M. 1985. Towards a social theory of ignorance. *J. Theory Soc. Behav.*, 15, 151–172.

Smithson, M. 1988. *Ignorance and Uncertainty*, Springer-Verlag, New York.

Smuts, J.C. 1926. *Holism and Evolution*. Macmillan, London.

Sober, E. 1991. *Core Questions in Philosophy*. Macmillan Publishing Company, New York.

Solomon, R. and Higgins, K. 1996. *A Short History of Philosophy*. Oxford University Press, Oxford.

Spetzler, C.S. and Stael von Holstein, C.-A.S. 1975. Probability encoding in decision analysis. *Manage. Sci.*, 22, 340–358.

Starkweather, G.K. 1992. Digital color printing. *Phys. Today*, 45, 60–65.

Stillings, N. et al. 1995. *Cognitive Science*, 2nd ed. MIT Press, Cambridge, MA.

Stone, M.C., Cowan, W.B., and Beatty, J.C. 1988. Color gamut mapping and the printing of digital color images. *ACM Trans. Graphics*, 7, 249–292.

Sugeno, M. 1974. Theory of Fuzzy Intervals and Its Applications. Ph.D. dissertation, Tokyo Institute of Technology, Tokyo.

Sugeno, M. 1977. Fuzzy measures and fuzzy integrals: a survey. In *Fuzzy Automata and Decision Processes*, Gupta, M.M., Saridis, G.N., and Gaines, B.R., Eds. North-Holland, Amsterdam, pp. 89–102.

Tannas, L.E., Jr. 1992. Color in electronic displays. *Phys. Today*, 45, 52–57.

Thagard, P. 1996. *Mind: Introduction to Cognitive Science*. MIT Press, Cambridge, MA.

Thomas, G. 2000. *How to Design Logos, Symbols, and Icons*. How Design Books, Cincinnati, www.HowDesign.com.

Thorell, L.G. and Smith, W.J. 1990. *Using Computer Color Effectively: An Illustrated Reference*. Prentice Hall, Englewood Cliffs, NJ.

Thys, W. 1987. Fault Management. Ph.D. dissertation, Delft University of Technology, Delft, The Netherlands.

Toulmin, S.E. 2003. Philosophy of science. *In Encyclopedia Britannica 2003, Ultimate Reference Suite*. Retrieved on October 1, 2003, from http://www.compilerpress.atfreeweb.com/Anno%20Toulmin%20EB%20Philosophy%20of%20Science.htm.

Trauth, K.M., Hora, S.C., and Guzowski, R.V. 1993. *Expert Judgement on Markers to Deter Inadvertent Human Intrusion into the Waste Isolation Pilot Plant*, Report SAND92-1382. Sandia National Laboratories, Albuquerque, NM.

Travis, D. 1991. *Effective Color Displays: Theory and Practice*. Academic Press, New York, pp. 113–151.

Tribus, M. 1969. *Thermostatics and Thermodynamics*. Van Nostrand, Princeton, NJ.

Tufte, E. 1990. *Envisioning Information*. Graphics Press, Cheshire, CT.

Tufte, E. 1997. *Visual Explanation: Images and Quantities, Evidence and Narrative*. Graphics Press, Cheshire, CT.

Unger, P. 1975. *Ignorance*. Clarendon Press, Oxford.

United Nations. 2000. *United Nations Military Symbols Handbook*. Department of Peacekeeping Operations, New York.

U.S. Army Corps of Engineers. 1965. *Standard Project Flood Determinations*, Civil Engineer Bulletin 52-8, Engineering Manual EM 1110-2-1411.

U.S. Army Corps of Engineers. 1982. *National Program of Inspection of Nonfederal Dams*, final report to Congress, Engineering Report ER 1110-2-106.

U.S. Army Corps of Engineers. 1997. *Guidelines for Risk-Based Assessment of Dam Safety*, draft Engineering Pamphlet EP 1110-1-XX, CECW-ED.

U.S. Bureau of Reclamation. 1989. *Policy and Procedures for Dam Safety Modifications*. USBR, Denver.

U.S. Interagency Advisory Committee on Water Data, Hydrology Subcommittee. 1982. *Guidelines for Determining Flood Flow Frequency*, Bulletin 17B. USGS, Reston, VA.

Von Eckardt, B. 1993. *What Is Cognitive Science*? MIT Press, Cambridge, MA.

von Glasersfeld, E. 1995. *Radical Constructivism: A Way of Knowing and Learning*. The Farmer Press, London.

Waldrop, M.M. 1992. *Complexity: The Emerging Science at the Edge of Order and Chaos*. Simon & Schuster, New York.

Walley, P. 1991. *Statistical Reasoning with Imprecise Probabilities*. Chapman & Hall, London.

Wang, Z. and Klir, G.J. 1992. *Fuzzy Measure Theory*. Plenum Press, New York.

Ware, C. and Beatty, J.C. 1988. Using color dimensions to display data dimensions. *Hum. Factors*, 30, 127–142.

Weaver, W. 1948. Science and complexity. *Am. Sci.*, 36, 536–544.

Webster's New World College Dictionary. 1995. Zane Publishing, Atlanta.

Weicheselberger, K. and Pohlemann, S. 1990. *A Methodology for Uncertainty in Knowledge-Based Systems*. Springer Verlag, New York.

Weinberg, G.M. 1975. *An Introduction to General Systems Thinking*. John Wiley, New York.

Weinstein, D. and Weinstein, M.A. 1978. The sociology of non-knowledge: a paradigm. In *Research in the Sociology of Knowledge, Science, and Art*, Vol. 1, Jones, R.A., Ed. JAI Press, New York.

Weitzman, D. 1985. Color coding re-viewed. In *Proceedings of the Human Factors Society 29th Annual Meeting*, Santa Monica, CA, pp. 1079–1083.

Wertheimer, M. 1958. Principles of perceptual organization. In *Readings in Perception*, Beardslee, D.S. and Wertheimer, M., Eds. Van Nostrand-Reinhold, Princeton, NJ, pp. 15–137. (Abridged translation by M. Wertheimer; original work published 1923.)

White, G.J. and Ayyub, B.M. 1985. Reliability methods for ship structures. *Nav. Eng. J.*, 97, 86–96.

Wiener, N. 1948. *Cybernetics*. John Wiley, New York.

Wiggins, J. 1985. *ESA Safety Optimization Study*, HEI-685/1026. Hernandez Engineering, Houston.

Williams, B. 1995. Philosophy and the understanding of ignorance. *Diogenes*, Vol. 4311, Spring.

Williams, J. 1992. The making of the weather page. *Weatherwise*, 45, 12–18.

Williamson, R. and Downs, T. 1990. Probabilistic arithmetic I: numerical methods for calculating convolutions and dependency bounds. *Int. J. Approximate Reasoning*, 4, 89–158.

Williamson, R.C. 1989. Probabilistic Arithmetic. Ph.D. dissertation, University of Queensland, Australia.

Williamson, T. 1994. *Vagueness*. Routledge, London.

Wilson, B. 1984. *Systems: Concepts, Methodologies, and Applications*. John Wiley & Sons, New York.

Winkler, R.N. and Murphy, A. 1968. Good probability assessors. *J. Appl. Meteorol.*, 7, 751–758.

Wise, B.P. and Henrion, M. 1986. A framework for comparing uncertain inference systems to probability. In *Uncertainty in Artificial Intelligence*, Kanal, L.N. and Lemmer, J.F., Eds. Elsevier Science Publishers, Amsterdam.

Wuketits, F.M. 1990. *Evolutionary Epistemology and Its Implications for Humankind*. State University of New York Press, Albany.

Wymore, A.W. 1976. *Systems Engineering Methodology for Interdisciplinary Teams*. John Wiley, New York.

Yager, R.R. 1979. On the measure of fuzziness and negation. Part I. Membership in the unit interval. *Int. J. Gen. Syst.*, 5, 189–200.

Yager, R.R. 1980a. On the general class of fuzzy connectives. *Fuzzy Sets Syst.*, 4, 235–242.

Yager, R.R. 1980b. On the measure of fuzziness and negation. Part II. Lattices. *Inf. Control*, 44, 236–260.

Yager, R.R. 1983. Entropy and specificity in a mathematical theory of evidence. *Int. J. Gen. Syst.*, 9, 249–260.

Yager, R.R. 1986. Arithmetic and other operations on Dempster-Shafer structures. *Int. J. Man-Mach. Stud.*, 25, 357–366.

Yager, R.R. 1987a. Cardinality of fuzzy sets via bags. *Math. Modeling*, 9, 441–446.

Yager, R.R. 1987b. Toward a theory of conjunctive variables. *Int. J. Gen. Syst.*, 13, 203–227.

Yager, R.R. 1987c. On the Dempster-Shafer framework and new combination rules. *Inf. Sci.*, 41, 93–137.

Yao, J.T.P. 1979. Damage assessment and reliability evaluation of existing structures. *Eng. Struct.*, 1, 245–251.

Yao, J.T.P. 1980. Damage assessment of existing structures. *J. Eng. Mech. Div.*, 106, 785–799.

Yao, J.T.P. and Furuta, H. 1986. Probabilistic treatment of fuzzy events in civil engineering. *Probab. Eng. Mech.*, 1, 58–64.

Zadeh, L.A. 1965. Fuzzy sets. *Inf. Control*, 8, 338–353.

Zadeh, L.A. 1968. Probability measures of fuzzy events. *J. Math. Anal.*, 23, 421–427.

Zadeh, L.A. 1973. Outline of a new approach to the analysis of complex systems and decision processes. *IEEE Trans. Syst. Man Cybern.*, SMC-3, 28–44.

Zadeh, L.A. 1975. The concept of linguistic variable and its application to approximate reasoning, Parts I, II, and III. *Inf. Control*, 8, 199–249, 301–357; 9, 43–80.

Zadeh, L.A. 1978. Fuzzy sets as a basis for a theory of possibility. *Fuzzy Sets Syst.*, 1(1), 3–28.

Zadeh, L.A. 1979a. A theory of approximate reasoning. In *Machine Intelligence*, Vol. 9, Hayes, J., Michie, D., and Mikulich, L.I., Eds. Halstead Press, New York, pp. 149–194.

Zadeh, L.A. 1979b. On the Validity of Dempster's Rule of Combination of Evidence, Memo UCB/ERL 97/24. University of California, Berkeley.

Zadeh, L.A. 1979c. Theory of fuzzy sets. In *Operations Research Support Methodology*, Holzman, A.G., Ed. Marcel Dekker, New York, pp. 569–606.

Zadeh, L.A. 1981. Test score semantics for natural languages and meaning representation via PRUF. In *Empirical Semantics*, Reiger, B., Ed. Brockmeyer, Bochum, Germany, pp. 281–349.

Zadeh, L.A. 1986. A simple view of the Dempster rule. *Artif. Intell. Mag.*, VII, 85–90.

Zadeh, L.A. 1987. Fuzzy sets as a basis for a theory of possibility. *Fuzzy Sets Syst.*, 1, 3–28.

Zadeh, L.A., Fu, K.S., Tanaka, K., and Shimara, J. 1975. *Fuzzy Sets and Their Application to Cognitive and Decision Processes*. Academic Press, New York.

Zeigler, B.P. 1976. *Theory of Modelling and Simulation*. John Wiley, New York.

Zimmerman, H.J. 1985. *Fuzzy Set Theory — And Its Applications*. Kluwer-Nijhoff Publishing, Boston.

Index

A

B

C

D

E

F

G

H

I

J

K

L

Q

R

S

T

U

V

W

X

Y